AiGX6
RESEARCH LAB

Transformation Blueprint Series

Published by EDGE Publish, www.EdgePublish.com

Generative AI tools were used during the research and creative process to support topic exploration, case study identification, content structuring, narrative refinement, and image generation. All AI-assisted outputs were reviewed, edited, validated, and finalized by the author in collaboration with the editorial, design, and operations teams of the Ai6x6 Research Lab across New York, London, and Kolkata. All interpretations, conclusions, structural decisions, and final narrative determinations represent the collective institutional judgment of the Ai6x6 Research Lab. No AI system holds rights to any portion of this work. All clipart and vector art in this book are AI-generated and included solely to enhance clarity and engagement. They are not intended to depict real individuals or groups.

Operations Transformation for the AI Era, First Edition, 2026
ISBN-13: 978-1-971148-05-2, Printed in the United States of America.

Ai6X6 RESEARCH LAB

Transformation Blueprint Series

OPERATIONS TRANSFORMATION FOR THE AI ERA

Engineering the Engine Room of Transformation.
The 6x6 Blueprint for Autonomous Efficiency and Resilient Scaling.

SUBRATA MUKHERJEE

EDGE PUBLISH
www.EdgePublish.com

"For the grace of God, who lights the path.
And, in loving memory of my mother Rupali and my father Dilip, who taught me the importance of kindness and perseverance. This work is born from the love you gave, the sacrifices you made, and the quiet wisdom that still echoes in my life. This one is for you."

Personal message from our Founder & Executive Director

SUBRATA MUKHERJEE

For over twenty-five years, I have sat at the heart of global digital transformations—from shifting a 180-year-old institution like *The Economist* into a digital powerhouse, to architecting hyper-automation for Fortune 500 leaders. I have seen the rise of the internet, the mobile revolution, and the cloud. But none of these shifts compare to the era we have just entered.

The Ai6x6 Research Lab was born out of a singular, urgent observation: Most organizations are trying to solve tomorrow's problems with yesterday's mental models. Traditional transformation is often treated as a "project"—a linear path with a start and an end. But in an era of Agentic AI and self-directing systems, transformation is a living organism. It is a system of behaviors, decisions, and capabilities that must evolve in unison. I founded this Lab to move beyond the "hype" and the "tools." My motivation is to provide leaders with the Cohesion, Schema, and Custody required to navigate a world where the boundary between human intuition and machine intelligence is disappearing. People often ask why "6×6." Throughout history, the number six has been the sacred geometry of our existence—from the sexagesimal systems of ancient Sumerian astronomy to the hexagonal patterns found in the farthest reaches of our cosmos. Mathematically, six is the first "perfect number" because its divisors (1, 2, and 3) add up to itself, making it uniquely balanced. In my journey across the U.S., Europe, and Asia, I realized that this ancient symmetry holds the key to modern enterprise. By distilling organizational complexity into six fundamental pillars and viewing them through six transformative lenses, we activate a collective "sixth sense" for the organization—the ability to perceive and act on what others miss. This architecture forms the foundation of our three flagship methodologies—the "*guided navigation system every transformation needs*".

My journey to this blueprint was forged in the engine rooms of the world's most complex organizations. I have led wholesale business transformations for *The Economist*, reinvented Bloomberg's digital presence following the *Businessweek* acquisition, revitalized the iconic *Dummies* brand within Wiley, and optimized the global ratings processes within S&P Global and McGraw-Hill. This strategic work was balanced by deep technical execution: implementing RPA, AI, IoT, and Digital Twins within Raytheon, and architecting CRM, BPM, and digital workplace automations for the global clients of Cognizant and Unisys.

This perspective was further refined through rigorous academic pursuit, drawing on my MBA in Marketing from Jadavpur University, my Master's in Technology Management from Columbia University, and executive certifications from Harvard University and the Indian Institute of Management (IIM-C) to bridge the gap between high-level theory and industrial-scale execution.

Today, through the Ai6x6 Research Lab, I lead the development of the frameworks that are now being adopted by executive teams globally. At the center of this work is three methodologies: 6X6 Blueprint, IGNITE Methodology and REVEAL Methodology our multidimensional navigation system that unifies architecture, execution, and evaluation across the enterprise.

However, my background as a Tech Journalist for *The Telegraph* for over a decade and a columnist for *CIO Magazine* has taught me that true insight requires total independence. I have built this Lab as a space where my team and I can express our opinions freely, without the interference of corporate sponsors. Our goal is not just to report the news, but to "fold" it into the Ai6x6 Blueprints—giving you the critical "so-what" and the "what-next" to make global shifts actionable for your organization.

We are democratizing this knowledge because I believe that the ability to thrive in the AI era should not be reserved only for those who can afford expensive global consulting firms. Our mission is to empower businesses of all sizes with the targeted, practical, and battle-tested information that I often wish I had at my disposal during my own leadership assignments. Through the Ai6x6 Transformation Blueprints and our Executive Mentor Network, we are scaling AI literacy to ensure that as systems become more autonomous, leaders become more intentional. We are not just adopting new tools; we are reimagining the nature of the enterprise itself.

AI6X6 RESEARCH LAB

NEW YORK | LONDON | KOLKATA

To our esteemed Industry Thought Leaders: Thank you!

Your generous time and profound insights have been the bedrock of this book. Each interview has been converted into sections and one-page cards that have been sprinkled across six books. You have provided a vital perspective, transforming our work from theory into a deeply practical guide rooted in real-world leadership in the AI-Era. Your willingness to share your hard-won wisdom, navigating complex challenges and revealing the future of your industries, enriches every page. You all have successfully navigated the AI transition in your careers, your companies and your industries. This book is a testament to your expertise. We are immensely grateful for your unwavering support and belief in this project. Thanks a lot, from me and my team at Ai6x6 and Executive Mentor Network!

TABLE OF CONTENTS

guided navigation for transformation

BLUEPRINT SERIES

SECTION A

Introduction

Ai6x6™ Transformation Blueprint

Ai6x6 Research Lab is a global intellectual institution dedicated to advancing the science and practice of organizational transformation. We develop original frameworks, diagnostic systems, capability playbooks, and research models that help leaders navigate complexity with clarity, precision, and confidence. Our work spans **Leadership, Workforce, Data, Customer, Product, and Operations**—the six pillars that define modern enterprise maturity. We are based in New York, London & Kolkata.

OUR MISSION FOR ORGANIZATIONAL TRANSFORMATION

To build a globally recognized body of knowledge that helps organizations think clearly, act decisively, and transform responsibly in an age of AI acceleration. We essentially play **six roles** for the industry:

THE CATALYST BRING ALIGNMENT TO LEADERSHIP	**THE STRATEGIST** BRING STRUCTURE TO AMBIGUITY	**THE ARCHITECT** BRING CLARITY TO COMPLEXITY
THE ENGINEER BRING DISCIPLINE TO TRANSFORMATION	**THE ETHICIST** BRING INTEGRITY TO OUTCOMES	**THE GUARDIAN** BRING OVERSIGHT TO AUTONOMY

WHAT WE DO?

1. Build Transformation Frameworks: We architect the Ai6x6 Transformation Blueprints— a structured, multi-lens transformation model that helps orgs decode patterns, risks, behaviors, and decision pathways.
2. Develop Diagnostic Systems: Our maturity diagnostics and radar models translate organizational signals into actionable insights, identifying Growth, Risk, Kill, and Shift zones.
3. Publish Research & Methodologies: We codify best practices in enterprise design through multi-volume series and research playbooks, available in bookstores or on our website ai6x6.com.
4. Create Executive Tools: From RACI matrices to transformation roadmaps, we build high-utility models for immediate executive application. We have many accelerated off-the-shelf offerings on our website.
5. Support Global Organizations: Our frameworks empower enterprises, consulting teams, and innovation labs. Our advisory team worldwide including our Executive Mentor Network is here to help.

ACCOMPLISHMENTS

Subrata Mukherjee has delivered enterprise-scale transformation across the Americas, Europe, and Asia, shaping modernization programs for some of the world's most influential organizations. His work spans AI/ML, automation, digital product strategy, enterprise architecture, and large-scale organizational transformation. He has led digital reinvention in global publishing, driven automation and AI adoption in aerospace and defense, modernized ratings and compliance processes in financial services, and guided product and data transformation across multiple industries. In 2016, he was recognized as one of Folio's Top 100 People in Publishing for his impact on digital innovation.

EDUCATION

Subrata holds a Master's in Technology Management from Columbia University, an MBA in Marketing from Jadavpur University, and a Business Communications certification from Harvard University. He also holds a Management by Values certification from the Indian Institute of Management, Calcutta, deepening his grounding in leadership behavior, organizational culture, and values-driven transformation. He is a Certified Product Owner. His academic foundation spans technology, strategy, communication, and organizational leadership, informing his multidisciplinary approach to AI-era transformation.

INDUSTRIES

His career spans aerospace and defense, financial services, publishing, healthcare, consulting, education, manufacturing, and technology services. He has delivered transformation in highly regulated environments, mission-critical engineering contexts, global media organizations, and fast-moving digital businesses. This cross-industry exposure gives him a multidimensional perspective on how AI reshapes strategy, operations, and organizational identity.

COMPANIES

Across his career, Subrata has delivered transformation programs for organizations including The Economist, Raytheon Technologies, S&P Global, Bloomberg, Wiley, Cognizant, Unisys, and Gujarat Ambuja Cement. His work has ranged from leading The Economist's digital transformation and modernizing Bloomberg's subscription and data infrastructure, to launching Raytheon's enterprise-wide Citizen Development program for Automation and AI, transforming the Dummies brand at Wiley, modernizing S&P's Ratings Process after the sub-prime mortgage crisis, scaling CRM and Automation/AI consulting portfolios at Cognizant and Unisys, and establishing the technology backbone and smart-building systems for the Gujarat Ambuja conglomerate.

EXPERTISE

Subrata's expertise spans the full arc of enterprise transformation, beginning with early work in CAD/CAM for steel plants and smart-city infrastructure, which shaped his systems-thinking foundation. He then became a specialist in enterprise CRM and BPM transformation, leading large-scale modernization programs that re-architected customer operations, workflow orchestration, and enterprise process design. Over time, he transitioned into a strategist, product thinker, and transformation architect, guiding organizations on designing, adopting, and scaling next-generation digital capabilities. Today, he is recognized as a leading voice in AI and Emerging Technologies, with deep expertise across AI/ML, Generative AI, LLMs, RAG architectures, intelligent automation, hyper-automation, conversational AI, chatbots, agentic systems, and decision-automation frameworks. His work is rooted in strategy, product design, operating-model transformation, and enterprise-scale adoption—not coding or engineering execution—and focuses on translating emerging AI capabilities into practical, scalable, enterprise-grade systems that reshape how organizations operate and compete.

THOUGHT LEADERSHIP

Subrata is a prolific speaker who has presented at more than 100 global conferences and events. His writing has appeared in CIO Magazine, and he served as a weekly technology columnist for The Telegraph. Through the Ai6x6 Research Lab, he publishes frameworks, diagnostics, and transformation blueprints that help organizations navigate the AI economy. In 2026–27, he will launch the Executive Mentor Network to develop a global community of AI Strategy–certified coaches and expand AI literacy among senior leaders. Beyond his professional work, he is an avid traveler who has explored more than 40 countries and an accomplished portrait artist.

www.linkedin.com/in/sm360

Ai6x6 RESEARCH LAB – 10 GUIDING PRINCIPLES

1. GIVE **VALUE**, REPEATEDLY, BECOME A GOTO BOOK

   The 18-book series brings in a lot of content from a practitioner perspective. The advice we are providing have worked for us in our past projects across industries; so we are preaching what we practiced before

2. BE **BRIEF**, PROVIDE CONTENT IN SMALL NUGGETS

   I have never read a book cover to cover. I have short attention span, and more and more people are suffering from the same syndrome. Hence each micro-topic we have introduced is covered in two to three pages

3. BE **COMPREHENSIVE** FOR REAL WORLD IMPACT

   Transformation is not easy. In the real world there are a lot of inter-connected challenges. We have proposed detailed frameworks that will connect the dots and provide comprehensive solutions to your team/client

4. BE **PRACTICAL**, PROVIDE EXAMPLES TO DEMYSTIFY

   Most companies and leaders know they and their companies need to transform, but don't know where to start. I wanted to provide a step-by-step guide or a MOSCOW priority list on getting started

5. PUSH **BOUNDARIES** AND DARE LEADERS TO DREAM

   Real transformation is not tinkering around the edges and playing safe; its often about self disrupting and pushing boundaries and reinventing oneself. Applies to individuals, teams and organizations.

6. PUSH LEADERS TO **INTROSPECT** & ASK QUESTIONS

   Since the world around us is moving so fast, leaders often have a fear of been exposed about their lack of knowledge and ask "stupid" questions. These book educates you enough, so you can ask intelligent questions

7. BE **EXCITING**, DON'T BE BORING

   I have tried to keep the book interesting, small text, lot of structured tables, quite a few images, bold formatting, case studies, Tips and Tricks, Interviews, RACI matrix, buying guides, assignment questions etc.

8. BRING **OTHERS**, HEAR FROM MY COMMUNITY

   Tech sector worldwide is very collaborative. We learn from each other every day and we use the knowledge to impact lives positively. This book attempts to bring a few voices from my network and hear their perspective

9. PROMOTE **AI ECONOMY**, USE AI TOOLS YOURSELF

   Generative AI has been used extensively for formatting content in a repeatable format & page size control. Without GenAI such an extensive book series would not have been possible in a short time.

10. BE **DIFFERENT**, CHARTER YOUR OWN COURSE

    Most management self-help books are pages and pages of long text with guidelines provided by publishers that are a bin antiquated. Modern learning happens byte sized and is multi-modal and multi-channel..

The **Guided Navigation for Transformation** framework is a comprehensive, end-to-end methodology designed to de-risk and accelerate AI adoption within an organization. It moves away from "random acts of digital" by providing a disciplined, three-stage approach that balances human capability with technical execution. The core of the system is the **6x6 Lens Sequence**, which ensures that every strategic move is evaluated through six critical dimensions: **Leadership, Workforce, Data, Customer, Product, and Operations**.

By applying these lenses consistently across the Blueprint, Reveal, and Ignite phases, the framework creates a "Golden Thread" that links high-level strategy to on-the-ground deliverables. This is not just a consulting model, but an educational ecosystem with an 18-book series that ensures that AI transformation is both measurable and sustainable.

"Guided Navigation for Transformation" in the AI Era

We provide guided navigation through three distinct phases: First, we establish your **Transformation Blueprint** to align skills. Second, we run the **Reveal Diagnostic** to assess your current maturity. Finally, we execute the **Consulting Playbook** to turn the 'Art of the Possible' into a 'Science of Deliverables'.

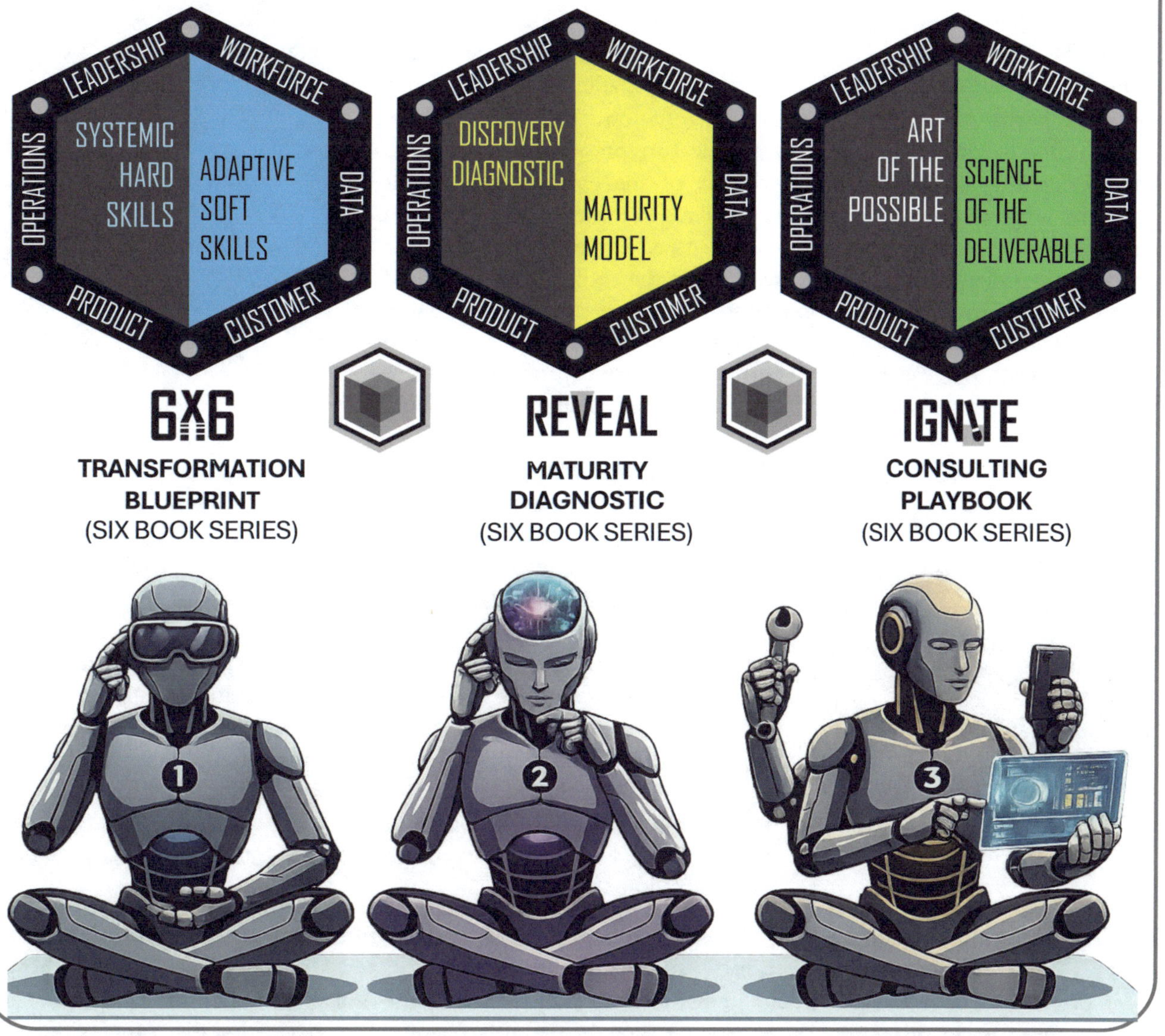

The **AI6x6 Transformation Engine™** is the flagship construct of the AI6x6 Research Lab—a precision-engineered ecosystem that serves as the definitive "Operating System" for the modern enterprise. Built on a proprietary 6×6 geometry, it unifies architecture, execution, and calibration into a single, high-performance unit. The Engine provides leaders with the high-torque machinery and **Guided Navigation** required to convert organizational inertia into scalable momentum in the era of autonomous systems.

The **Transformation Blueprint Series** defines the engine's architecture and establishes the structural logic of the entire system. By introducing the **Six Pillars** and **Six Lenses**, it forms the foundational geometry of enterprise modernization. Written for CEOs, Boards, and Strategy Leaders, this series acts as the **Navigation Deck**, providing the "Architectural North Star." It ensures that every move is strategically calibrated and ethically anchored before the gears are engaged, offering a shared mental model for aligning leadership teams. This intelligence layer removes the guesswork from transformation, providing a rigid yet adaptive framework that allows organizations to visualize their future state with absolute clarity. It is the vital starting point where complex, high-stakes ambition is codified into a disciplined, institutional reality that can withstand the pressures of rapid technological shift.

The **Maturity Diagnostic Series** provides the engine's accountability and calibration system, ensuring peak performance through proprietary **REVEAL ™** methodology. . Acting as the **Sensor-Driven Intake**, it utilizes 36 diagnostic cylinders to quantify organizational readiness and prioritize investments based on evidence rather than intuition. This series allows executive teams to identify hidden gaps and measure progress with scientific clarity, functioning as the **Measuring Layer** of the entire ecosystem. It ensures the engine remains tuned for resilience through a data-driven maturity model that tracks evolution over time. By providing rigorous assessment instruments and scoring logic, it offers the accountability required to sustain a long-term transformation journey. This layer acts as the engine's nervous system, constantly scanning the internal and external environment to calibrate the Blueprint's trajectory and ensure that every unit of energy spent drives the organization toward sustained excellence and institutional stability.

The **Consulting Playbook Series** translates architecture into action by operationalizing the engine through the proprietary **IGNITE™** methodology. It serves as the **Heavy-Duty Gearbox**, providing the repeatable consulting patterns, execution models, and directional force required to activate transformation across diverse domains. Designed for PMOs and delivery leads, these playbooks convert high-level strategy into tangible enterprise-scale outcomes, ensuring that the "torque" of the engine is effectively applied to the road. They function as the essential doing layer, reducing internal friction and shifting the organization from abstract concept to deep-seated conviction. By providing step-by-step pathways and intervention models, the series bridges the gap between executive intent and operational reality. This transmission layer ensures that momentum is not just created, but sustained and scaled, allowing the enterprise to accelerate through complex transitions with precision and measurable impact.

The **Transformation Engine** is built for scale across the global economy. Future **Industry Evolution** volumes will apply this 6×6 engine to specific sectors—such as Healthcare, Banking, and Manufacturing—revealing how the model drives specialized, high-velocity transformation. By adapting the Pillars and Lenses to sector-specific demands, these extensions ensure the Engine remains the definitive standard for navigating the complexities of the age of intelligent systems.

SERIES 1 6 BOOKS

SERIES 2 6 BOOKS

SERIES 3 6 BOOKS

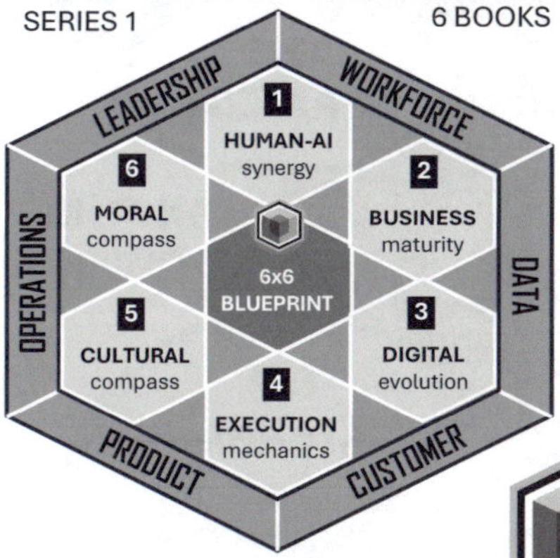

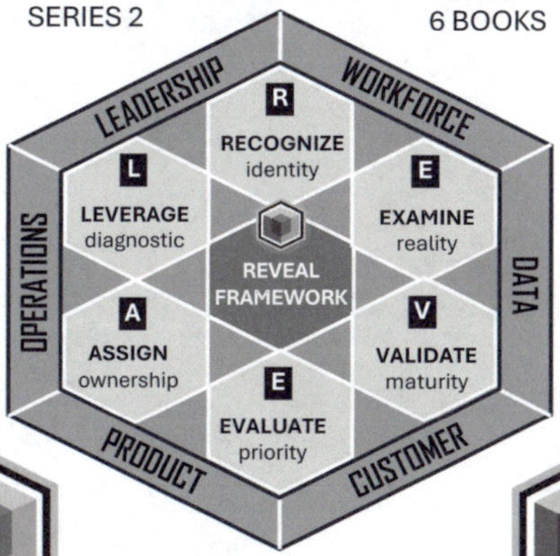

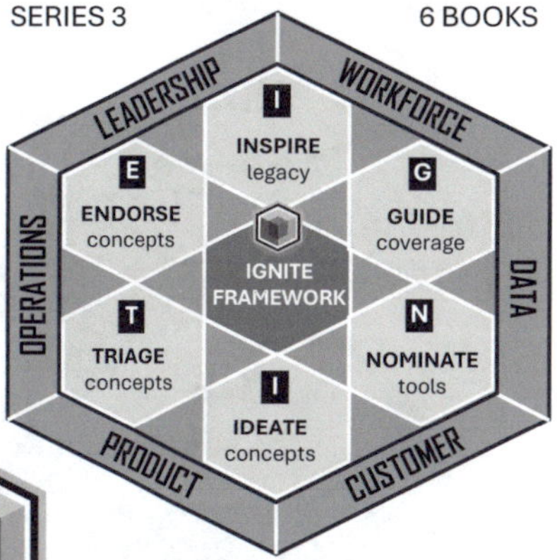

6X6 TRANSFORMATION BLUEPRINT

REVEAL MATURITY DIAGNOSTIC

IGNITE CONSULTING PLAYBOOK

see **think** **act**

The Ai6x6 palette is a semantic system, not a decorative choice. Each color encodes the emotional and operational signature of its lens: **pink** captures the human spark that ignites potential; **green** reflects the strategic discipline required for maturity; **purple** signals the depth and computational power of digital evolution; **orange** embodies the kinetic urgency of execution; **blue** anchors the cultural soul in trust and cohesion; and **yellow** illuminates the ethical clarity of the moral audit. Together, these colors form a coherent visual language that guides the reader through the transformation cycle with intuitive, emotionally aligned cues.

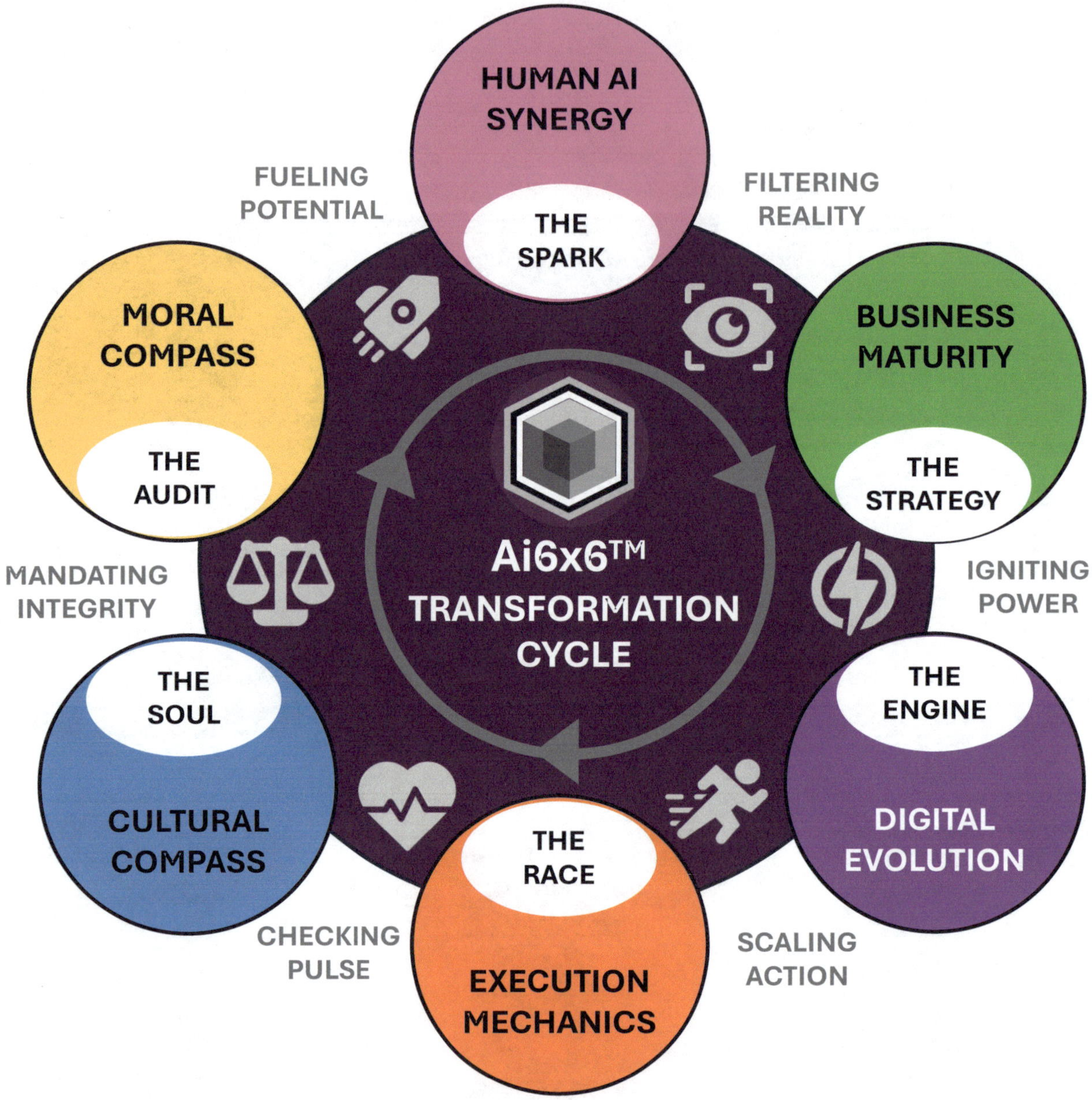

The **COLOR TABS ALONG THE PAGE EDGES IN THE MIDDLE** serve as the physical navigation system of the Ai6x6 series. Each tab mirrors the chromatic identity of its lens, allowing readers to locate themselves instantly within the transformation cycle. This creates a spatial reading experience where orientation is continuous and intuitive. The tabs turn our book into a modular interface—one where Human–AI Synergy, Business Maturity, Digital Evolution, Execution Mechanics, Cultural Compass, and Moral Compass become distinct territories with their own visual signatures. The result is a playbook that reads like a designed system, not a linear text.

6x6™ Transformation Blueprint

The Prologue to a Purpose

"The only way to go forward is to let go of the past."
The Myth of Orpheus and Eurydice

The tragic lesson of this myth is a cautionary tale for business. Orpheus's failure to trust his journey and looking back results in losing everything, underscoring the importance of fully committing to a forward-looking transformation.

The mission of this book series is simple: to transform leaders from passive observers of the AI Era into its active pioneers. We will ignite your thinking with exciting, highly practical content that pushes boundaries and provides the essential tools and frameworks you need to build real-world impact. We won't just tell you what to do; we'll show you how, using a diverse community's experiences and real-world examples that bring the abstract to life. This isn't a passive read; it's an active partnership and a dynamic dialogue designed to turn your natural curiosity into a definitive competitive advantage. This book is the first of a six-part series : Leadership Transformation; which governs your journey on how you should lead from the top and how you should inspire others to take on the AI Era and excel in it.

The **AI6x6 Transformation Engine** is a precision-engineered ecosystem designed to deliver **Guided Navigation** for enterprises facing high-stakes digital evolution. By integrating three proprietary systems—the **Transformation Blueprint**, the **Consulting Playbook**, and the **Maturity Diagnostic**—the Engine provides the robust "Operating System" required to dominate the AI era.

The architecture begins with the **Transformation Blueprint**, the Engine's intelligence layer. Here, the system utilizes **Six Lenses**—including the Cultural and Moral Compasses—to calibrate your trajectory and codify a master architectural plan. This strategic map then engages the **Consulting Playbook**, where the gears of the **IGNITE** methodology translate executive intent into immediate directional force through rapid-response execution cycles. Finally, the **Maturity Diagnostic** acts as a sensor-driven intake, utilizing six diagnostic cylinders to compress raw organizational data into high-pressure strategic insights that validate the transformation's impact. Bolted to a rigid **Chassis of Six Pillars**, the Engine ensures the enterprise remains resilient, balanced, and unshakeable even at peak acceleration. It is more than a strategy; it is a proprietary architecture for sustained institutional excellence.

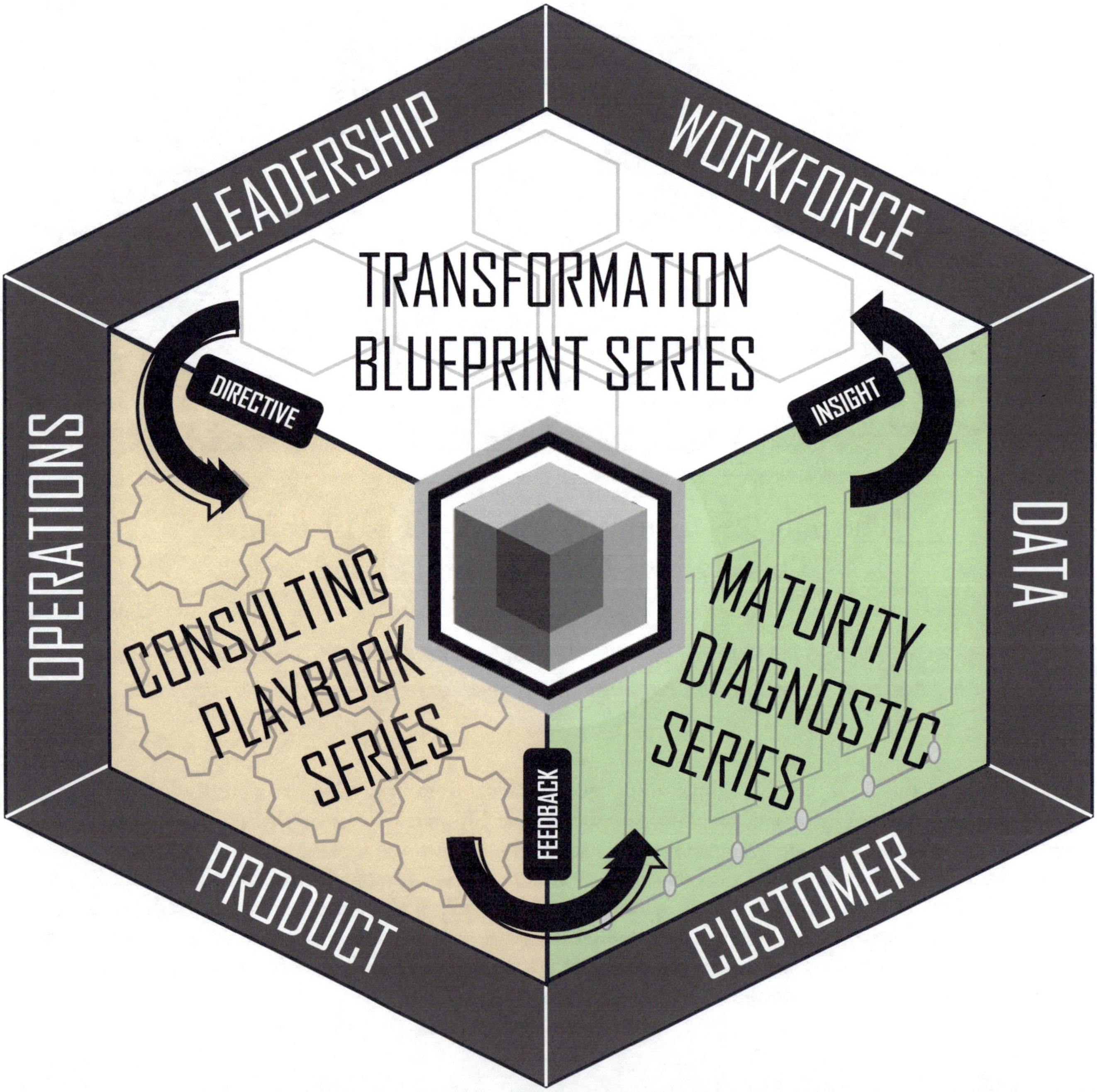

The AI6x6 Transformation Engine is a precision-engineered ecosystem that replaces guesswork with **Guided Navigation**. It is designed to convert organizational inertia into scalable momentum through a disciplined, mechanical approach to digital evolution.

CHASSIS (SIX PILLARS)
The entire system is anchored to a rigid structural frame. This chassis ensures that even as you reach peak acceleration, your enterprise remains resilient and unshakeable.

NAVIGATION DECK (SIX LENSES):
This is the intelligence layer. By applying our **Six Lenses**, we provide the **Guided Navigation** required to calibrate your trajectory, preventing costly drifts in strategy.

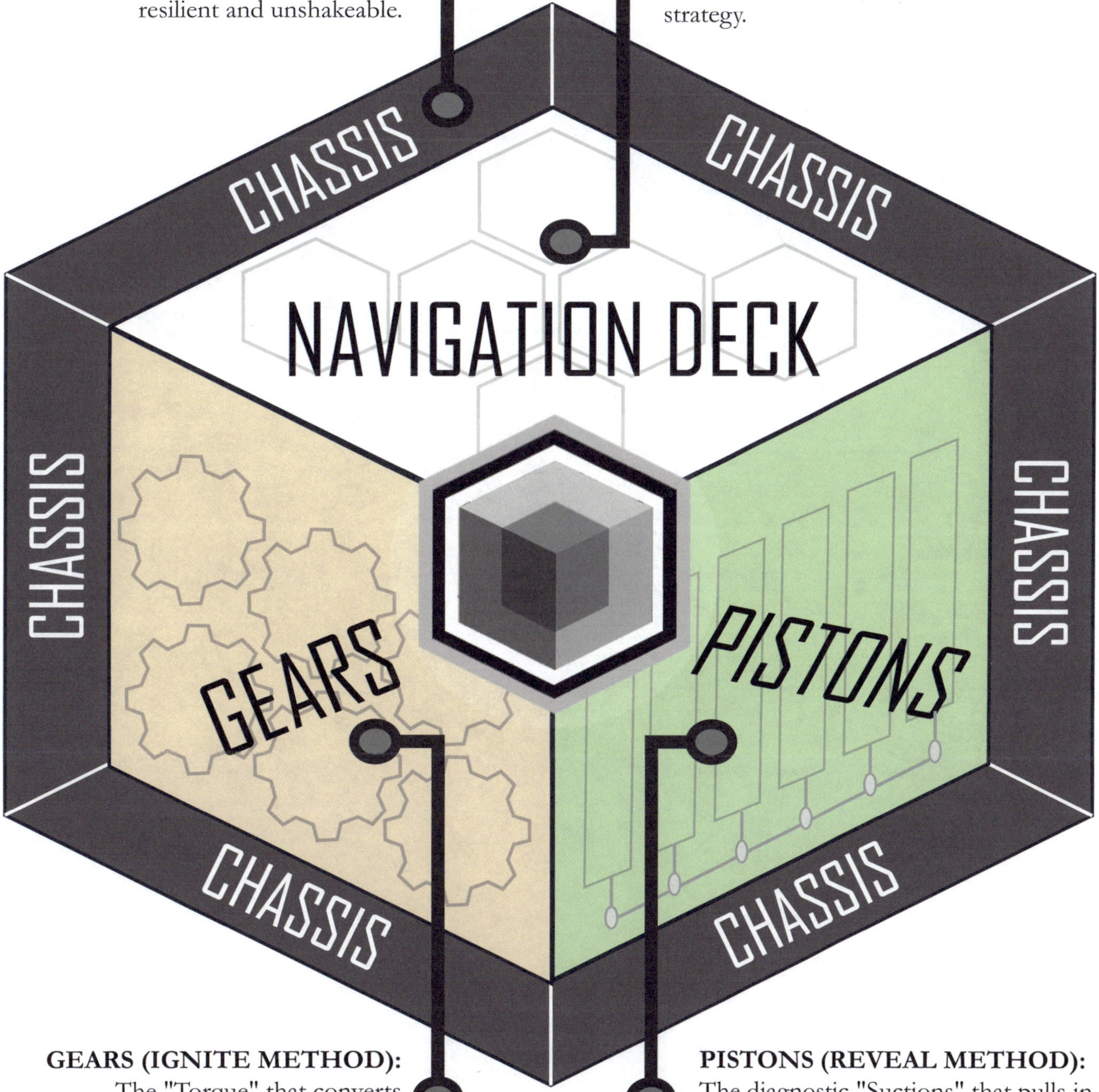

GEARS (IGNITE METHOD):
The "Torque" that converts diagnostic energy into immediate directional force through pilot programs and playbooks.

PISTONS (REVEAL METHOD):
The diagnostic "Suctions" that pulls in raw organizational data and compresses it into high-pressure actionable insights.

A car works best when its actions are coordinated through systems that generate power, transmit motion, and guide direction. Each component plays a precise mechanical role that keeps the vehicle stable, responsive, and moving with purpose.

CHASSIS
Supports vehicle structure, absorbs road stress, anchors mechanical systems, maintains stability, and ensures safe, balanced handling across varying speeds, terrains, and driving conditions.

NAVIGATION DECK
Displays real-time driving information, maps, routes, and alerts, helping the driver maintain direction, anticipate conditions, and make confident, informed decisions.

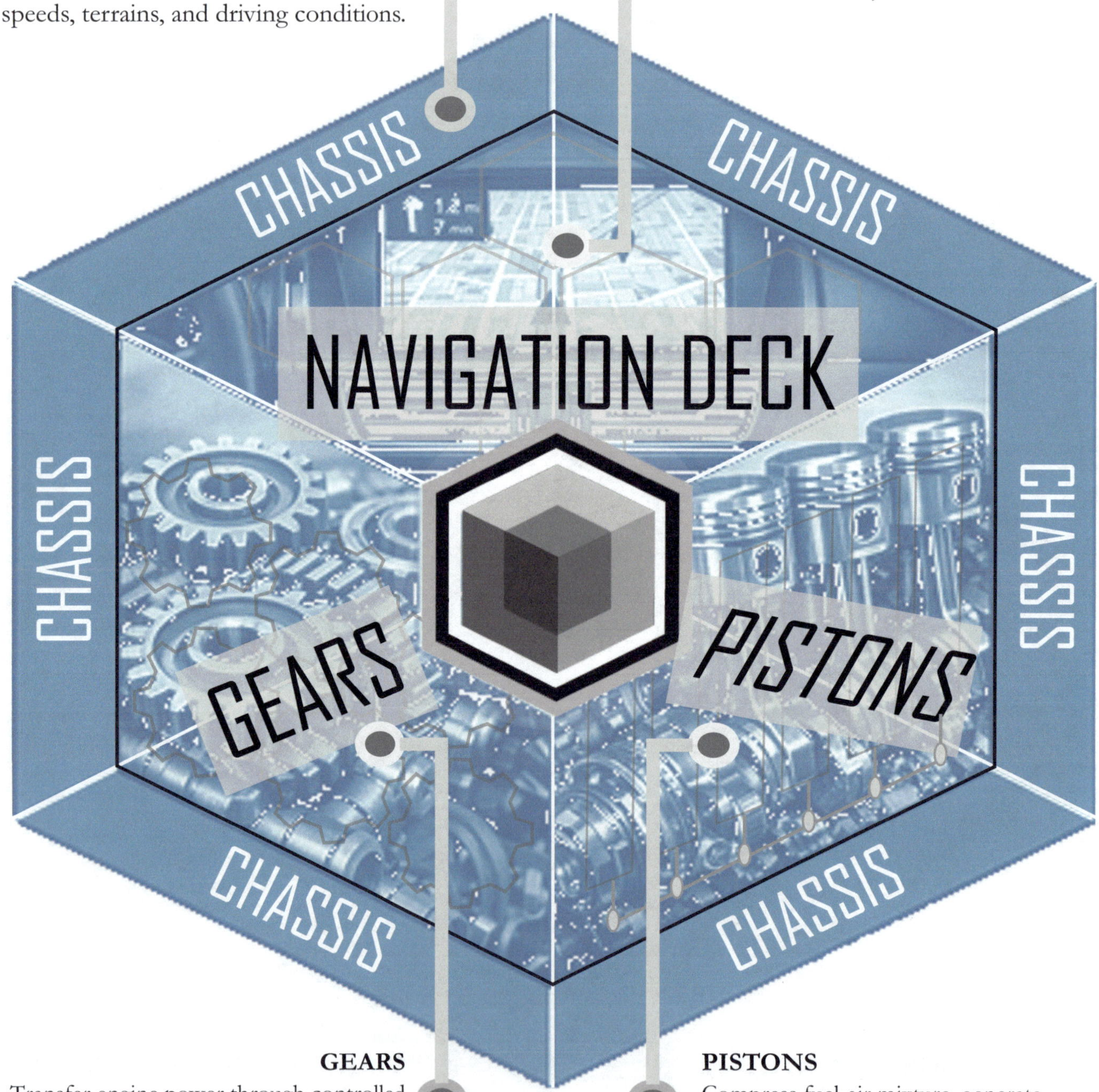

GEARS
Transfer engine power through controlled ratios, regulating speed and torque, ensuring smooth acceleration, efficient performance, and seamless coordination between mechanical components.

PISTONS
Compress fuel-air mixture, generate combustion force, drive the crankshaft, & convert explosive energy into continuous motion powering the vehicle's entire drivetrain system.

The **AI6x6 Transformation Engine** replaces the ambiguity of traditional consulting with mechanical certainty, converting organizational inertia into scalable momentum. By integrating three proprietary systems—the **REVEAL** diagnostic intake, the **IGNITE** consulting transmission, and the **6x6 Blueprint** navigation deck—the Engine provides the robust "Operating System" required to dominate the AI era. The **REVEAL** methodology acts as a sensor-driven intake, utilizing 6x6 diagnostic cylinders to compress raw organizational data into high-pressure strategic insights. This "data fuel" then powers the **IGNITE** phase, where consulting playbooks serve as a heavy-duty gearbox, translating executive intent into immediate directional force through rapid-response execution cycles. This entire power plant is governed by the **6x6 Blueprint**, the intelligence layer that uses **Six Lenses**—including the Cultural and Moral Compasses—to calibrate your trajectory and ensure ethical, operational alignment. Bolted to a rigid **Chassis** of **Six Pillars**, the Engine ensures the enterprise remains resilient, balanced, and unshakeable even at peak acceleration. It is more than a strategy; it is a proprietary architecture for sustained institutional excellence.

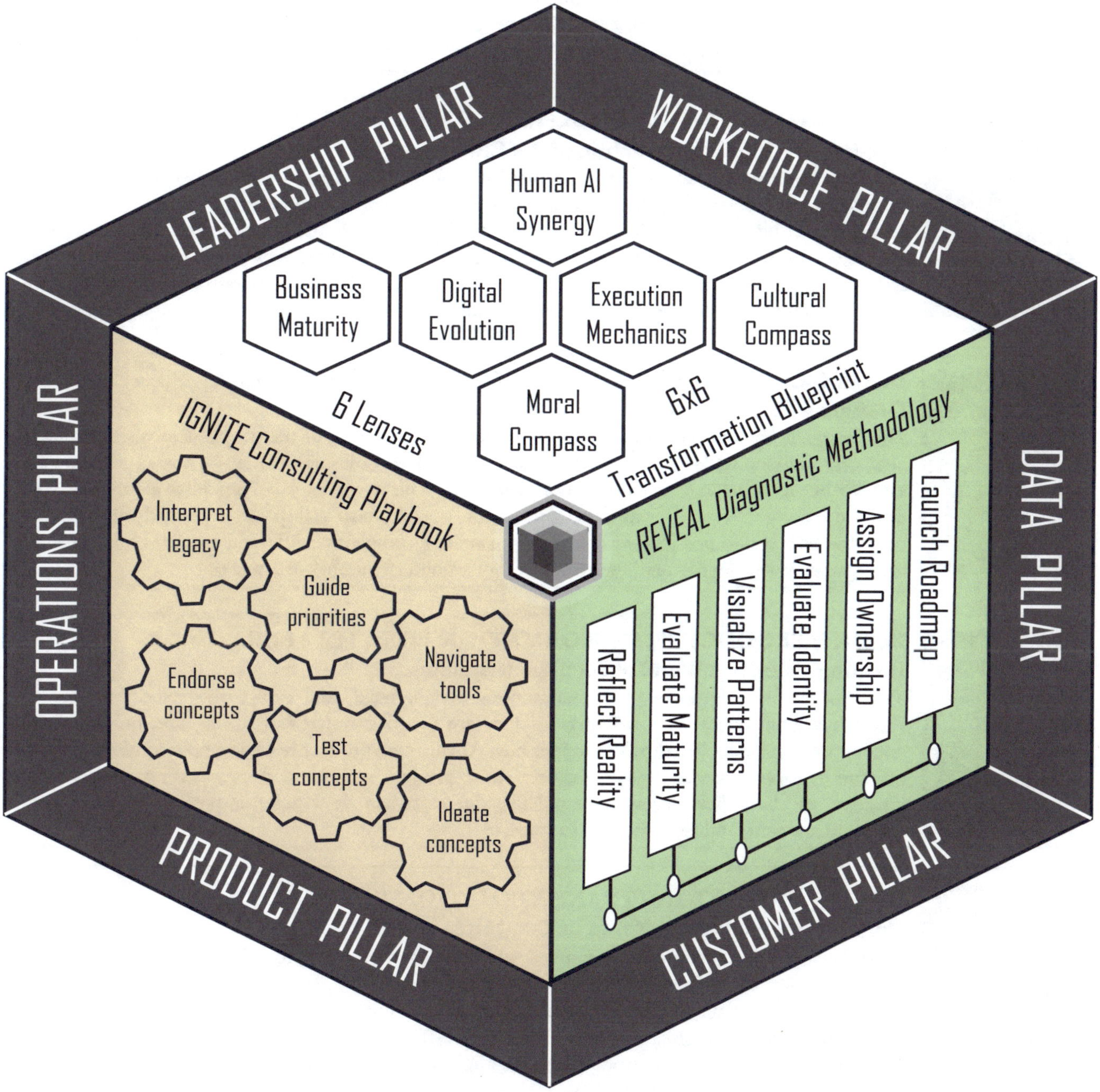

BOOK 1: LEADERSHIP TRANSFORMATION FOR THE AI ERA

Designing the Vision, Not Just Directing It

The Leadership book reframes what it means to lead in an era defined by intelligence, acceleration, and ambiguity. It explores how leaders think, decide, communicate, and shape meaning across the enterprise, emphasizing that leadership is not a role but an architecture. The narrative examines how leaders make decisions systemically, exercise judgment in an AI-augmented world, create alignment through narrative clarity, and uphold ethical stewardship as the foundation of trust. This book sets the tone for the entire series by positioning leadership as the integrator of all transformation forces.

BOOK 2: WORKFORCE TRANSFORMATION FOR THE AI ERA

Elevating Human Capability in a Technology-Driven Era

The Workforce book examines the human system of the enterprise — the skills, roles, mindsets, structures, and experiences that determine how people show up to the work of transformation. It explores how organizations support people as AI reshapes every function, and how leaders build environments where individuals feel empowered rather than displaced. The narrative dives into the evolution of adaptive talent ecosystems, the emergence of AI-augmented roles, the importance of team cohesion and psychological safety, and the recognition of human capability as a strategic asset.

BOOK 3: DATA TRANSFORMATION FOR THE AI ERA

Building the Intelligence Fabric of the Enterprise

The Data book establishes the informational backbone of the organization. It explores how data is collected, governed, interpreted, and transformed into intelligence, emphasizing that data is no longer a technical asset but the raw material of decision-making, personalization, automation, and innovation. The narrative examines how trustworthy data governance is built, how intelligence flows across systems, how data architectures are designed, and how data becomes a source of enterprise-level advantage. This book gives leaders the clarity to build an intelligent, insight-driven organization.

BOOK 4: CUSTOMER TRANSFORMATION FOR THE AI ERA

Designing Experiences That Earn Trust and Loyalty

The Customer book anchors the enterprise in the lived experience of the people it serves. It explores how organizations understand, anticipate, and respond to customer needs — not as a marketing exercise but as a cultural and operational discipline. The narrative examines how emotion shapes experience design, how AI enables personalization at scale, how transparency strengthens customer trust, and how end-to-end journey coherence becomes a competitive differentiator. This book ensures the organization evolves around the customer, not around its internal structures.

BOOK 5: PRODUCT TRANSFORMATION FOR THE AI ERA

Creating Value in a World of Intelligent Systems

The Product book explores how organizations design, build, and evolve the value they deliver. Products are no longer static offerings — they are dynamic systems of intelligence, behavior, and continuous adaptation. This book examines how AI enhances product behavior, how continuous value creation replaces episodic feature delivery, how product strategy and product operations must integrate, and how digital ecosystems redefine differentiation. It transforms product thinking from output to outcome, and from features to value.

BOOK 6: OPERATIONS TRANSFORMATION FOR THE AI ERA

Engineering the Engine Room of Transformation

The Operations book addresses the systems, workflows, and processes that determine whether strategy becomes reality. It explores how organizations orchestrate complexity, manage risk, and maintain resilience in environments defined by acceleration and uncertainty. The narrative examines how intelligent workflows are designed, how operational resilience is built, how AI optimizes processes, and how execution scales with precision. This is the book where transformation becomes tangible.

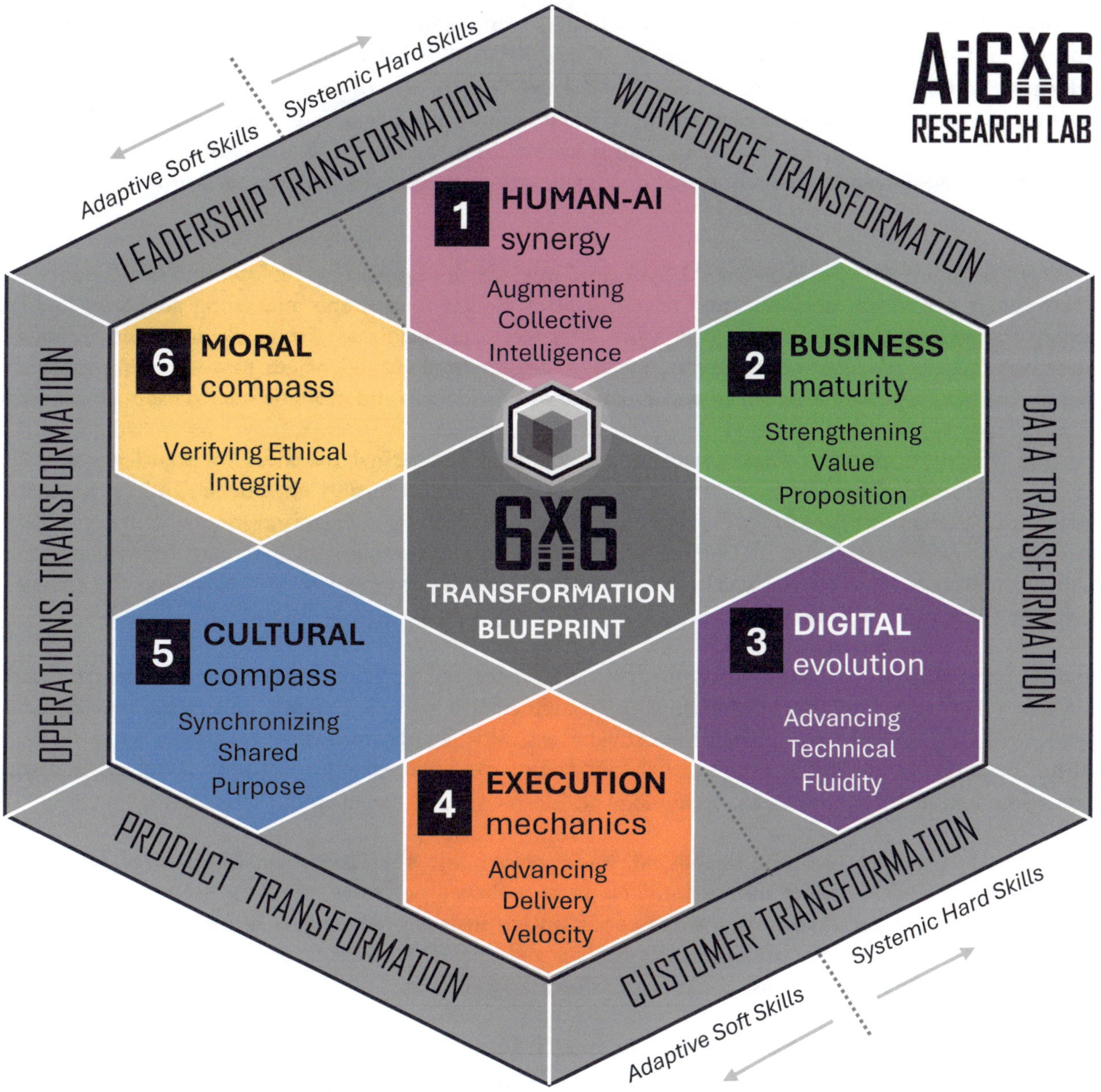

The **Ai6x6 TRANFORMATION BLUEPRINT™** provides the multidimensional structural integrity required for sustainable, high-stakes evolution. Serving as the **Navigation Deck** and **Chassis** of the Transformation Engine, the BLUEPRINT™ moves beyond episodic change to establish a permanent, scalable "Operating System" for the enterprise. The architecture is defined by the intersection of **Six Pillars** and **Six Lenses**. The Pillars—Leadership, Workforce, Data, Operations, Product, and Customer—provide the horizontal breadth and load-bearing strength of the organization. These are bisected vertically by the Six Lenses to reveal the true depth of the enterprise. Leaders apply **Human–AI Synergy** to augment collective intelligence and **Business Maturity** to instill systemic discipline. The **Digital Evolution** lens ensures technical fluidity, while **Execution Mechanics** masters the art of precision delivery. This is anchored by the **Cultural Compass**, which synchronizes shared purpose, and the **Moral Compass**, which verifies ethical integrity.

By aligning these twelve dimensions, the BLUEPRINT™ eliminates strategic drift and structural fragmentation. It ensures that every initiative is not just an isolated project, but a calibrated component of a unified, resilient, and unshakeable corporate architecture. Also remember to follow the Tab colors on the edges to navigate quickly.

The **Ai6x6 Transformation Blueprint™** gives leaders a multidimensional, architecture-grade view of how transformation behaves inside their enterprise. Where REVEAL™ diagnoses the present and IGNITE™ accelerates the future, the Blueprint defines the *structural truth* of the organization. Built on six proprietary Lenses, it converts fragmented observations, cultural noise, and operational ambiguity into a coherent, systems-level understanding. The Blueprint equips leaders with the clarity, depth, and structural intelligence required before designing any transformation roadmap, ensuring decisions are grounded in reality rather than aspiration.

The Blueprint begins with the **Human–AI Synergy Lens**, which examines how people and intelligent systems interact across workflows, decisions, and responsibilities. Rather than simply measuring adoption, this Lens analyzes augmentation, displacement risk, trust dynamics, and readiness for human-machine teaming. It distinguishes where AI enhances capability, where it creates friction, and where new operating models are required. Leaders begin with a grounded understanding of how human and machine roles must evolve together.

Next, the **Business Maturity Lens** evaluates the enterprise's structural readiness for transformation. This includes governance, financial discipline, decision velocity, and leadership alignment. The Lens separates foundational strength from structural fragility, revealing whether the organization can support sustained transformation without collapse. It clarifies which capabilities are stable, which require reinforcement, and which must be redesigned before acceleration. Business Maturity ensures leaders understand the chassis on which transformation will operate.

The third Lens, **Digital Evolution**, maps the depth, coherence, and scalability of the digital ecosystem. Leaders often face fragmented platforms, technical debt, and integration gaps that silently undermine transformation. Digital Evolution cuts through this complexity by assessing architecture, interoperability, data flow, and platform extensibility. It clarifies which digital foundations can scale, which must be modernized, and which create systemic risk. This Lens ensures technology decisions are grounded in architectural truth.

With structural and digital realities established, leaders move into the **Execution Mechanics Lens**, which examines how strategy becomes movement. This Lens evaluates processes, rhythms, accountability systems, and operational precision. It identifies where execution is consistent, where it is brittle, and where it breaks under pressure. Execution Mechanics ensures leaders understand the organization's true capacity to deliver, adapt, and sustain momentum across complex transformation cycles.

The fifth Lens, **Cultural Compass**, explores the beliefs, behaviors, incentives, and informal power structures that shape how people respond to change. Culture is often the silent accelerator or the invisible brake. This Lens reveals alignment, resistance patterns, psychological safety, and leadership influence networks. It distinguishes cultural strengths from cultural liabilities, giving leaders a clear view of the human forces that determine whether transformation will be embraced or rejected.

The final Lens, **Moral Compass**, evaluates ethics, trust, transparency, and responsible AI principles. As organizations adopt intelligent systems, the stakes of ethical decision-making rise dramatically. This Lens ensures transformation is not only effective but also equitable, accountable, and aligned with long-term societal and organizational values. It clarifies where safeguards are strong, where risks exist, and where governance must evolve.

Together, these Six Lenses form a complete, repeatable, consulting-grade architectural framework. The Ai6x6 Transformation Blueprint™ is not a checklist or maturity model; it is a structural intelligence system that moves leaders from assumption to understanding, from understanding to design, and from design to confident transformation planning. It gives leaders the conviction to move forward — not because the future is predictable, but because their decisions are grounded in multidimensional truth, structural clarity, and a deep understanding of how transformation behaves across the enterprise.

Ai6x6™ Transformation Blueprint

Understand Pillars & Lenses

"We don't see things as they are, we see them as we are." — Anaïs Nin

The **Ai6x6 TRANSFORMATION BLUEPRINT** is fundamentally structured around two core concepts: **Pillars and Lenses**. Together, they create a dual-axis architecture that brings both breadth and depth to enterprise transformation. The Pillars provide the WIDTH of the framework, defining the major domains where transformation must take root for the organization to evolve coherently. These domains—Customer, Product, Operations, Data, Workforce, and Leadership—represent the full span of enterprise behavior. Each Pillar receives its own dedicated book, allowing leaders to explore the complexities, tensions, and opportunities of that domain without dilution or compromise.

This ensures that transformation is not treated as a single initiative but as a system of interconnected disciplines that must advance in parallel. Complementing this horizontal structure is the DEPTH created by the Six Lenses, the universal imperatives that cut vertically through every Pillar. These Lenses represent the non-negotiable dimensions of modern transformation—forces that must be activated in every domain, regardless of industry, maturity, or strategic ambition. The six Lenses are: Human–AI Synergy (L1), Business Maturity (L2), Digital Evolution (L3), Execution Mechanics (L4), Cultural Compass (L5), and Moral Compass (L6). Each Lens exposes a different layer of organizational capability, ensuring that no domain is approached superficially, and no critical dimension is overlooked. The goal of the blueprint is to be holistic and very comprehensive so that executives are not blind-sighted by a key aspects being overlooked. Since a lot of ground has to be covered, the books may look very dense, but I have tried my best to keep it brief & engaging.

Transformation has become one of the most OVERUSED words in modern leadership, yet one of the least understood. Organizations declare it, teams attempt it, consultants package it—but few enterprises ever truly achieve it. Not because they lack ambition or intelligence, but because they approach transformation as a sequence of disconnected initiatives rather than a shift in how the organization thinks, behaves, and evolves. I created the **Ai6x6 Transformation Blueprint** because I have seen the same pattern repeat across industries, functions, and eras of technology: transformation fails not at the point of strategy, but at the point of structure. It fails when organizations modernize technology but ignore culture. It fails when they redesign processes but overlook trust. It fails when they invest in AI but fail to prepare people. It fails when they build strategy but cannot execute it. It fails when leaders underestimate the complexity of aligning human capability with machine intelligence.

The Ai6x6 Transformation Blueprint exists to solve this fragmentation. It brings together **six foundational lenses** and **six transformation pillars** into a single, coherent architecture. Each lens represents a dimension of capability that every organization must master. Each pillar represents a domain— Leadership, Workforce, Data, Customer, Product, Operations —where transformation must take root. Together, they form a 6×6 matrix that reveals not only what must change, but how change becomes sustainable, repeatable, and scalable.

This blueprint is intentionally multidimensional. Transformation is not linear. It is not sequential. It is not a checklist. It is a living system—human, technical, cultural, operational, ethical—moving together or not at all. The Ai6x6 model acknowledges this complexity and gives leaders a structure strong enough to hold it.

- It is called a blueprint because it is both **architectural** and **actionable**.
- This blueprint is not a methodology per se. It is not a maturity model in the traditional sense. It is not a checklist as you commonly see it. It is a **HOLISTIC** way of seeing things that are often overlooked.
- It is called Ai6x6 because it integrates **AI-driven intelligence** with **six human-centered disciplines**, applied across **six transformation pillars**.
- It is a system designed for organizations that want to evolve with intention rather than react with urgency.

The Six Lenses emerged from a simple but unavoidable truth: "**no single discipline can carry transformation on its own**". Not AI. Not culture. Not governance. Not execution. Not digital modernization. Not ethics. Each is necessary, but none is sufficient. Real transformation requires all six lenses working together—reinforcing one another, correcting one another, and elevating one another.

It reveals the forces that shape organizational behavior, the tensions that undermine progress, and the capabilities that determine whether transformation becomes sustainable or collapses under its own ambition. It provides leaders with a structural map—one that makes complexity navigable and makes evolution repeatable.

Each lens represents a dimension of organizational life:

1. Human–AI Synergy: how intelligence is created, shared, and amplified.
2. Business Maturity: how discipline, readiness, and structure turn ambition into capability.
3. Digital Evolution: how technology becomes fluid, integrated, and trustworthy.
4. Execution Mechanics: how strategy becomes action and action becomes advantage.
5. Cultural Compass: how beliefs, behaviors, and meaning shape the human system.
6. Moral Compass: how ethics, transparency, and responsibility anchor everything else.

Together, they form a complete operating philosophy for modern transformation—one that is as human as it is technical, as strategic as it is operational, as aspirational as it is grounded. The Ai6x6 Transformation Blueprint is not about predicting the future. It is about preparing organizations to shape it. It is about giving leaders a language for what they instinctively know but rarely articulate. It is about building enterprises that can evolve continuously, coherently, and confidently. Most of all, it is about honoring the truth that **transformation is not something organizations *do*—it is something they *become*.**

The Pillars of the **Ai6x6 Transformation Blueprint** define the WIDTH of enterprise evolution. They represent the six domains where transformation must become real, measurable, and operationally grounded. While the Lenses provide the depth of capability, the Pillars provide the breadth of application—ensuring that transformation is not isolated to a single function but becomes a structural shift across the entire enterprise. Each Pillar receives its own dedicated book, allowing leaders to confront the full complexity of that domain with precision and without compromise.

The Leadership Pillar sits at the top of the architecture because leadership is the force that shapes every other domain. It defines how decisions are made, how priorities are set, how accountability is distributed, and how meaning is communicated. This Pillar examines the evolution of leadership in an era defined by AI, acceleration, and ambiguity. It challenges leaders to move from managing activities to designing systems, from directing work to enabling capability, and from protecting the past to shaping the future. Leadership is not a separate function—it is the integrator of the entire Blueprint.

The Workforce Pillar focuses on the human system that powers the enterprise. It explores how roles evolve, how teams collaborate, how skills are developed, and how people experience their work. In a world where AI augments every function, the Workforce Pillar examines how organizations support human capability, build adaptive talent ecosystems, and create environments where people feel empowered rather than displaced. This Pillar forces leaders to confront the truth that transformation succeeds or fails based on whether people feel equipped, supported, and connected to the mission.

The Data Pillar establishes the informational foundation of the enterprise. It examines how data is collected, governed, interpreted, and transformed into intelligence. In an AI-driven world, data is no longer a byproduct of operations—it is the raw material of decision-making, personalization, automation, and innovation. This Pillar explores how organizations build trustworthy data ecosystems, eliminate silos, and ensure that intelligence flows across the enterprise. It challenges leaders to treat data not as a technical asset but as a strategic capability that shapes every other Pillar.

The Customer Pillar anchors the organization in the lived experience of the people it serves. It forces leaders to confront the reality that customer expectations evolve faster than internal processes, and that loyalty is earned through coherence, trust, and emotional resonance. This Pillar examines how organizations understand, anticipate, and respond to customer needs—not as a marketing exercise, but as a cultural and operational discipline. It challenges leaders to build enterprises where customer experience is not a department, but a shared responsibility embedded in every decision.

The Product Pillar focuses on how organizations design, build, and evolve the value they deliver. Products are no longer static offerings—they are dynamic systems of intelligence, behavior, and continuous adaptation. This Pillar explores how product strategy, product design, and product operations must work together to create experiences that feel alive and responsive. It examines how AI reshapes product behavior, how digital ecosystems redefine differentiation, and how organizations must shift from feature delivery to continuous value creation.

The Operations Pillar addresses the engine room of the enterprise—the systems, workflows, and processes that determine whether strategy becomes reality. It examines how organizations orchestrate complexity, manage risk, and maintain resilience in environments defined by acceleration and uncertainty. This Pillar forces leaders to recognize that operational excellence is not about efficiency alone; it is about adaptability, intelligence, and the ability to respond to change without losing coherence. Operations is where transformation becomes tangible.

Together, these six Pillars form the structural WIDTH of the Ai6x6 Transformation Blueprint. They define the terrain on which the Six Lenses operate, ensuring that transformation is comprehensive, coherent, and enterprise-wide.

4
Earned relationships
CUSTOMER
"Who we serve?"
5
Brand Experience
PRODUCT
"What we serve?"
6
Engine room of service
OPERATIONS
"How we serve?"
2
Human Capability
WORKFORCE
"Human Assets"
3
Information foundation
DATA
"Insights"
1
Vision, mission & agenda
LEADERSHIP
"C-Suite"
Each Pillar = "Width" of the Transformation

4
Discerning Diners
CUSTOMER
"The Patrons"
5
Food, Drinks, Ambiance
PRODUCT
"The Feast"
6
Ingredients, Purchase
OPERATIONS
"The Logistics"
2
Cooks, Stewards, Servers
WORKFORCE
"The other staff"
3
Menu, Recipe & Reviews
DATA
"The Menu"
1
Lead Culinary aspect
LEADERSHIP
"The Exec Chef"

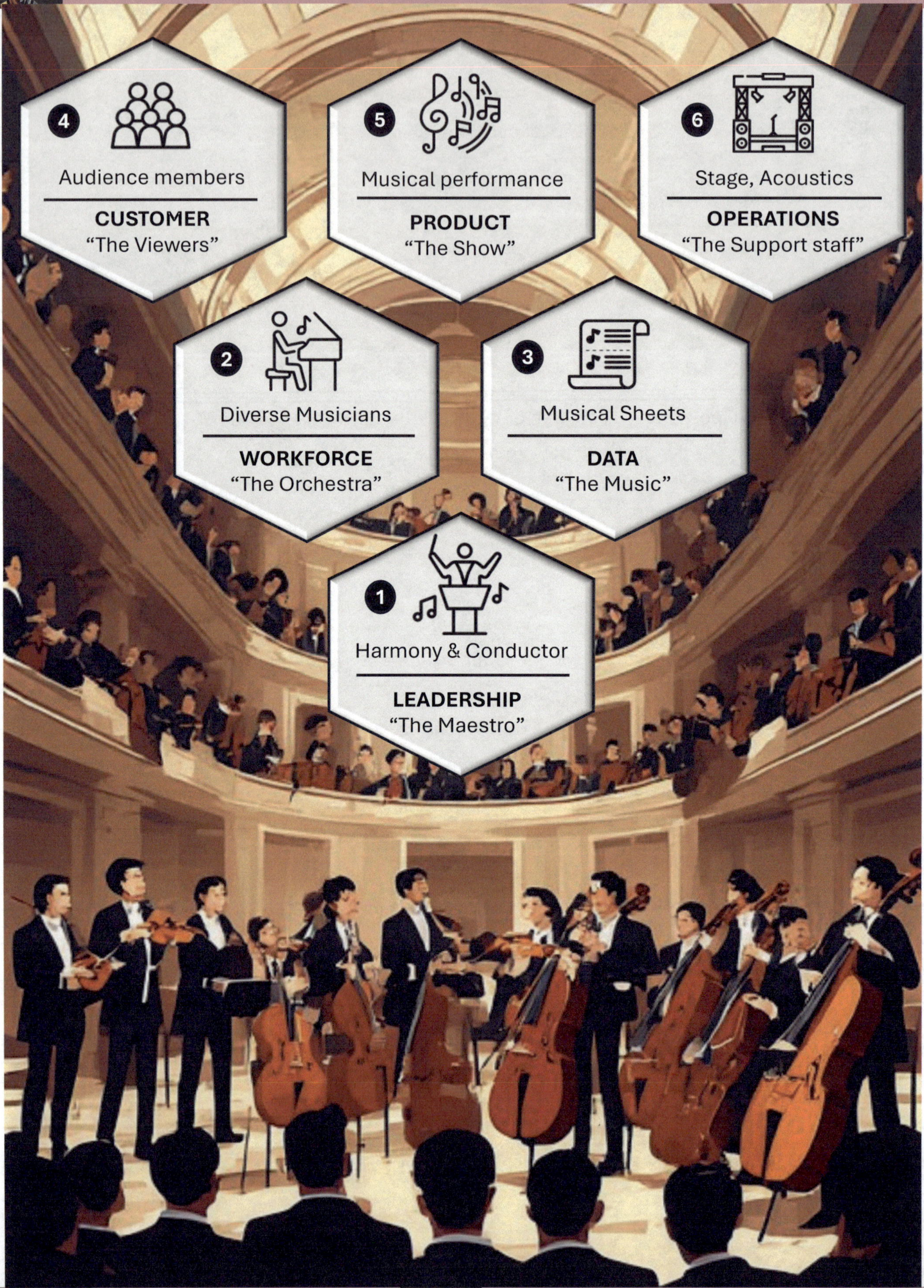
4
Audience members
CUSTOMER
"The Viewers"
5
Musical performance
PRODUCT
"The Show"
6
Stage, Acoustics
OPERATIONS
"The Support staff"
2
Diverse Musicians
WORKFORCE
"The Orchestra"
3
Musical Sheets
DATA
"The Music"
1
Harmony & Conductor
LEADERSHIP
"The Maestro"

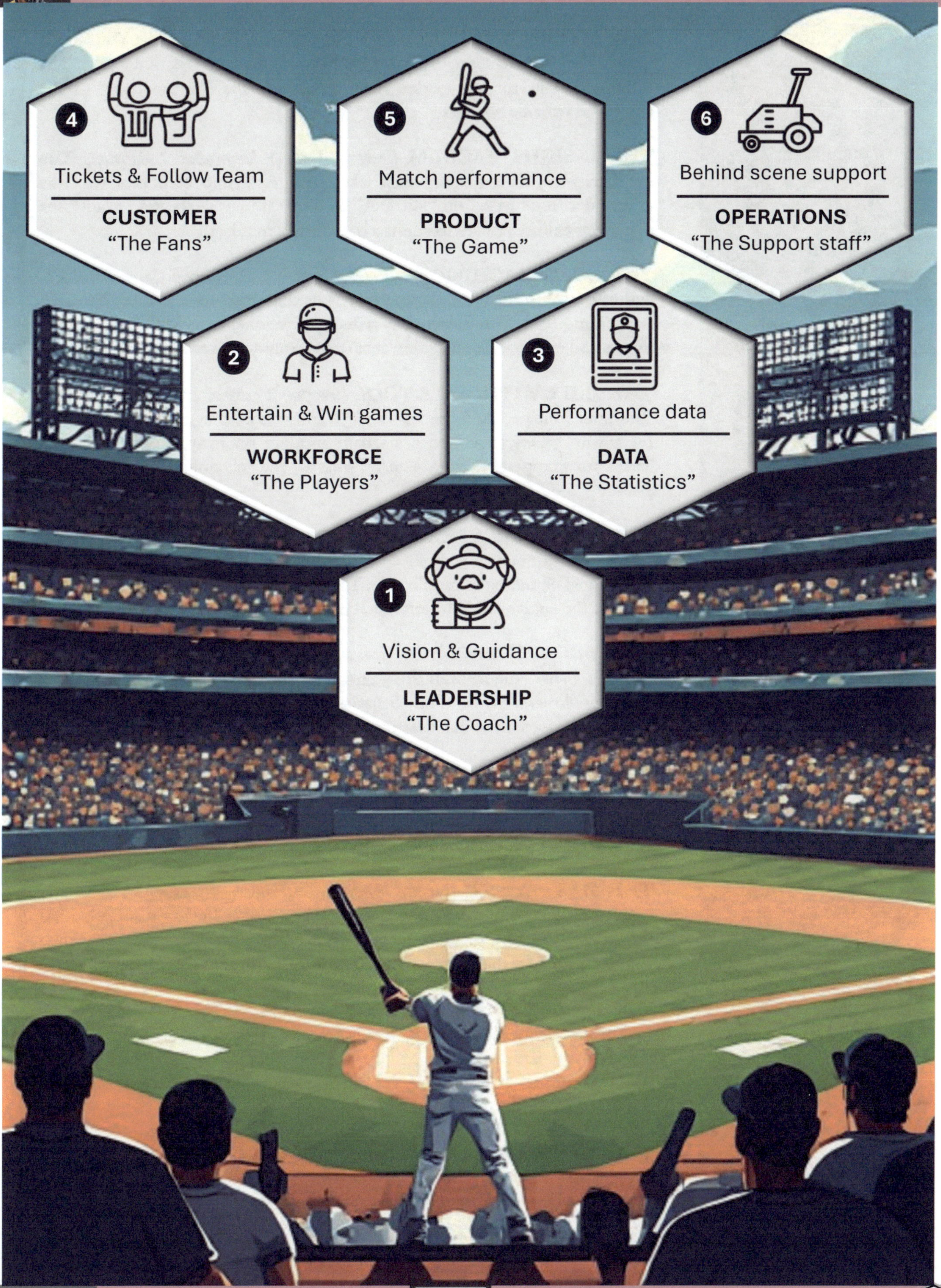
4
Tickets & Follow Team
CUSTOMER
"The Fans"
5
Match performance
PRODUCT
"The Game"
6
Behind scene support
OPERATIONS
"The Support staff"
2
Entertain & Win games
WORKFORCE
"The Players"
3
Performance data
DATA
"The Statistics"
1
Vision & Guidance
LEADERSHIP
"The Coach"

THE LITERACY GAP (*Invested: Product | Overlooked: Workforce*): You launch genius AI-driven features that your staff cannot explain, sell, or support. The technology fails to gain market traction because the "human layer" of the enterprise remains obsolete.

THE INSIGHT VACUUM (*Invested: Data | Overlooked: Leadership*): You build massive, high-integrity data lakes, but leadership continues to make critical decisions based on "gut feel." The investment is wasted because the upper-tier culture refuses to adopt a data-driven mindset.

THE COLD MACHINE (*Invested: Operations | Overlooked: Customer*): Back-end efficiency reaches an all-time high, but the actual user journey becomes robotic and rigid. You successfully reduce operational costs but lose the brand loyalty and emotional connection that drives long-term revenue.

THE SKILL WITHOUT A TOOL (*Invested: Workforce | Overlooked: Product*): You train your people to be "AI-Ready" and culturally agile, but you provide them with legacy tools and a stagnant product roadmap. Your best talent leaves for competitors who actually possess the technology to match their skills.

THE HALLUCINATION (*Invested: Leadership | Overlooked: Operations*): Executives announce a bold "Digital First" vision, but the underlying operational plumbing is decades old. The strategy never leaves the boardroom because the engine lacks the structural integrity to turn the wheels.

THE BROKEN PROMISE (*Invested: Customer | Overlooked: Data*): You market a hyper-personalized experience, but your internal data silos remain a fragmented mess. You end up alienating customers with irrelevant offers and inaccurate information that destroys their trust.

THE PROTOTYPE TRAP (*Invested: Product | Overlooked: Operations*): A brilliant AI-driven tool is developed in a vacuum, but the legacy supply chain and logistics cannot fulfill the demand it creates. The product succeeds in the lab but collapses the moment it hits the real-world operational floor.

THE DARK ASSET (*Invested: Data | Overlooked: Workforce*): You spend millions cleaning and structuring your data, but no one in the organization has the literacy to query or interpret it. The data sits idle and "dark," providing zero ROI because the people aren't equipped to find the value.

THE ROGUE ENGINE (*Invested: Operations | Overlooked: Leadership*): Middle management automates workflows to save time, but leadership has no visibility or governance over these new "black box" processes. The organization moves faster, but the steering wheel is no longer connected to the captain's chair.

THE INTERNAL ECHO (*Invested: Workforce | Overlooked: Customer*): Employees are empowered and "AI-literate," but they become obsessed with internal process improvements rather than user needs. The company becomes a high-performing bubble that is perfectly optimized for a customer that no longer exists.

THE BLIND CAPTAIN (*Invested: Leadership | Overlooked: Data*): A visionary leader attempts to steer the company through a massive pivot based on bold assumptions. Because the data pillar was ignored, they navigate straight into a market reef that the "missing data" would have clearly identified.

THE FANCY WRAPPER (*Invested: Customer | Overlooked: Product*): Your marketing and customer service are world-class, but the actual product has not evolved to keep up with competitors. You win the initial sale through great branding, but the churn rate sky-rockets once the customer realizes the product is hollow.

THE ANALYTICS ISLAND (*Invested: Data | Overlooked: Operations*): The data team produces perfect, real-time reports on every aspect of the business. However, the operational workflows are so slow and manual that by the time the business acts on the data, the opportunity has already vanished.

THE ORPHAN PROJECT (*Invested: Product | Overlooked: Leadership*): Innovation teams build cutting-edge AI tools that solve real problems, but leadership never provides a permanent budget or strategic home for them. These "pilot projects" eventually wither and die because they lack the top-down air cover to survive.

THE FRUSTRATED EXPERT (*Invested: Workforce | Overlooked: Operations*): You hire the best AI researchers and data scientists, but you force them to work within manual, "broken" internal workflows. Your experts spend 90% of their time fighting the system rather than building the future.

THE IVORY TOWER (*Invested: Leadership | Overlooked: Customer*): The board develops a strategy based purely on shareholder value and financial engineering. By ignoring the Customer pillar, they fail to see a massive shift in user sentiment that makes their entire three-year plan irrelevant.

THE HUMAN BOTTLENECK (*Invested: Operations | Overlooked: Workforce*): You automate 80% of your workflows, but the remaining 20% require human intervention that your staff isn't trained for. The entire system grinds to a halt because you optimized the machine but forgot to upskill the operator.

THE REACTIVE PANIC (*Invested: Customer | Overlooked: Leadership*): The company chases every new customer trend and social media "vibe" to stay relevant. Without a strategic anchor from Leadership, the organization becomes a fragmented mess of contradictory ideas that stand for nothing.

THE WASTEFUL HARVEST (*Invested: Data | Overlooked: Product*): You collect trillions of data points and pay massive cloud storage fees, but none of that data is ever used to iterate or improve the actual product. You are "Data Rich" but "Value Poor," hoarding information with no path to monetization.

THE GRASSROOTS REBELLION (*Invested: Workforce | Overlooked: Leadership*): Staff begins using "Shadow AI" and unsanctioned tools because leadership is too slow or fearful to provide guidance. You have an AI-ready workforce that is operating outside of the company's governance, creating a massive security and legal risk.

If the Pillars define the WIDTH of enterprise transformation, the **Six Lenses** define its DEPTH. They are the universal imperatives that cut vertically through every domain—Leadership, Workforce, Data, Customer, Product, and Operations. Each Lens exposes a different dimension of capability, ensuring that transformation is not superficial, fragmented, or episodic, but deeply rooted in the systems that shape how organizations think, behave, and evolve. These Lenses are not optional. They are not enhancements. They are the structural forces that determine whether transformation becomes sustainable or collapses under its own ambition. Each Lens reveals a layer of truth about the enterprise—its intelligence, its discipline, its technology, its execution, its culture, and its integrity.

The Human–AI Synergy Lens (L1) examines how intelligence is created, shared, and amplified across the enterprise. It explores the evolving relationship between human judgment and machine capability, ensuring that AI augments rather than replaces human potential. This Lens forces leaders to confront how decisions are made, how insights flow, and how people interact with intelligent systems. It challenges organizations to build environments where humans and AI operate as partners—each elevating the other. Human–AI Synergy is the starting point because it defines the cognitive foundation of modern transformation.

The Business Maturity Lens (L2) establishes the structural discipline required for transformation to take hold. It examines governance, alignment, prioritization, funding, risk management, and organizational readiness. This Lens reveals whether the enterprise has the scaffolding to support change or whether ambition is outpacing capability. Business Maturity is where leaders confront the operational truth: transformation is not powered by inspiration alone—it is powered by systems, discipline, and repeatable mechanisms that ensure consistency across the enterprise.

The Digital Evolution Lens (L3) focuses on the technological backbone of the organization. It examines how systems integrate, how data flows, how platforms scale, and how digital environments adapt to new demands. This Lens ensures that technology becomes an enabler rather than a bottleneck. It challenges organizations to build digital ecosystems that are fluid, intelligent, secure, and trustworthy. Digital Evolution is not about tools—it is about creating a technological environment capable of supporting continuous change.

The Execution Mechanics Lens (L4) converts strategy into action. It examines the methodologies, operating models, workflows, and performance systems that determine whether the organization can deliver consistently and at scale. This Lens exposes the gap between what leaders intend and what teams experience. It ensures that execution is not left to improvisation but is engineered with precision. Execution Mechanics is where transformation becomes tangible—where ideas become outcomes and outcomes become advantage.

The Cultural Compass Lens (L5) explores the beliefs, behaviors, and shared meaning that shape how people think and collaborate. It examines psychological safety, communication norms, leadership behaviors, and the emotional climate of the enterprise. This Lens reveals whether the culture accelerates transformation or resists it. Cultural Compass ensures that people feel connected, supported, and aligned—because no transformation can succeed in a culture that does not want it.

The **Moral Compass Lens (L6)** anchors the Blueprint in ethics and transparency. Moving beyond a static matrix, it defines the **True North** for AI governance, data privacy, and leadership accountability. This Lens ensures transformation is principled, not just efficient. In the AI era, a Moral Compass is the boundary condition for legitimacy—without it, innovation loses its direction; with it, trust becomes an unbreakable competitive advantage.

Together, these Six Lenses form the DEPTH of the Ai6x6 Transformation Blueprint. They ensure that every Pillar—Leadership, Workforce, Data, Customer, Product, and Operations—is approached with rigor, coherence, and integrity. They transform transformation from a collection of projects into a complete operating philosophy.

SIX LENSES OF TRANSFORMATION : HUMAN BODY ANALOGY

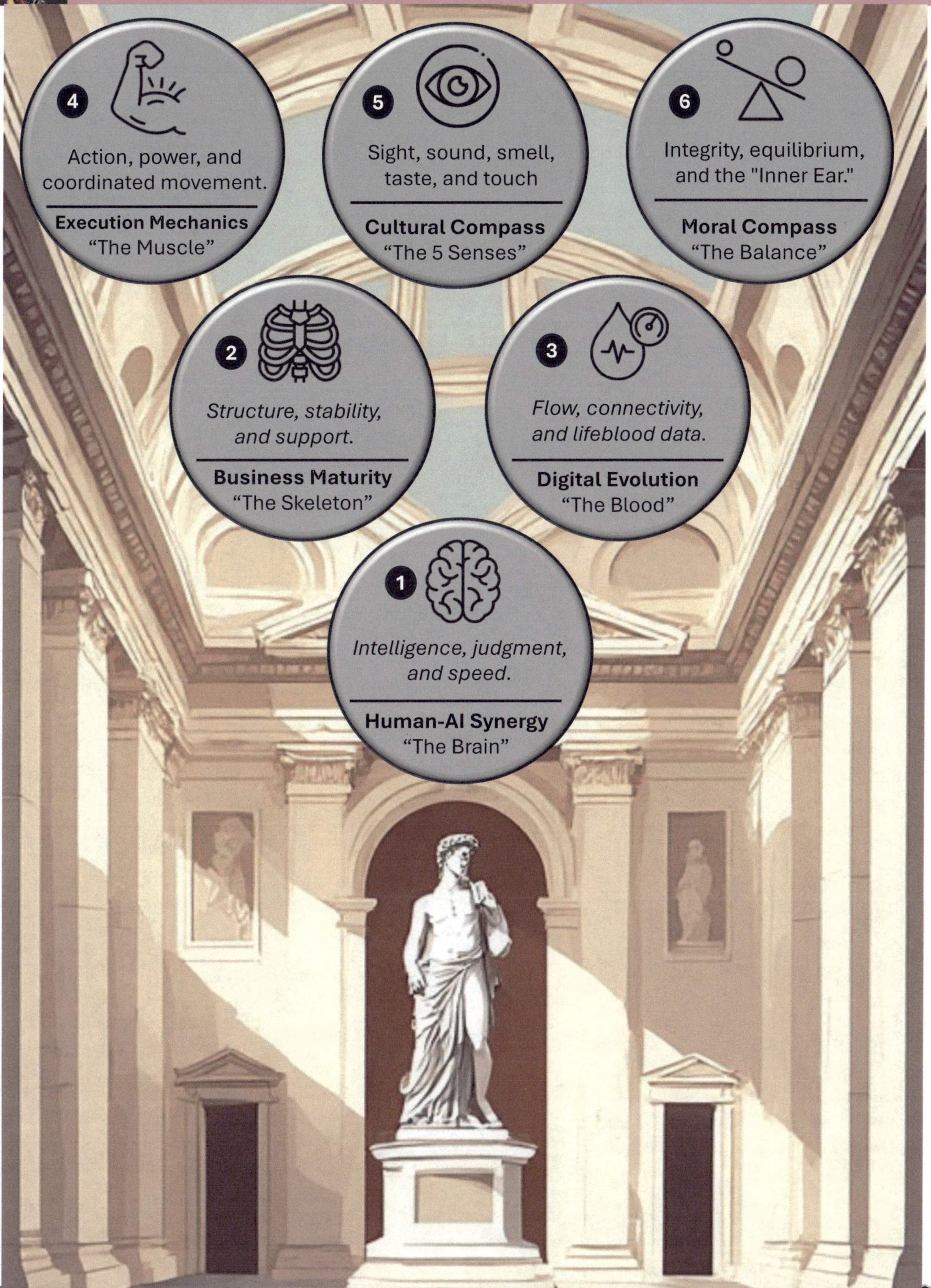

4
Gears that Convert energy power to motion
Execution Mechanics
"The Engine"
5
Monitor car health, temperature, audio
Cultural Compass
"The Dashboard"
6
Alignment, stability and stay in 'right lane'
Moral Compass
"The Steering"
2
Structural Frame and body of the car
Business Maturity
"The Chassis"
3
Gasoline, Electric energy & battery
Digital Evolution
"The Fuel"
1
Judgement, GPS, decision-making
Human-AI Synergy
"The Driver"

THE ETHICAL IMPLOSION (*Invested: Digital evolution | Overlooked: Moral compass*): You build a high-speed AI infrastructure with zero ethical guardrails or governance. One biased algorithm creates a public relations nightmare and a massive legal liability that can bankrupt the firm.

THE BUREAUCRATIC FREEZE (*Invested: Business maturity | Overlooked: Cultural compass*): You have established perfect governance and checklists, but you are operating within a terrified culture. People follow the rigid "process" of transformation but stop taking risks, effectively killing all real innovation.

THE THEORETICAL DREAM (*Invested: Human–AI synergy | Overlooked: Execution mechanics*): The organizational "brain" develops brilliant ideas for AI collaboration, but there is no "muscle" to act on them. Innovation stays trapped as a pilot project and never scales to produce enterprise value.

THE AUTOMATED ERROR (*Invested: Execution mechanics | Overlooked: Human–AI synergy*): You build a powerful execution engine but ignore the necessity of human judgment and intervention. The system begins executing "bad" decisions or flawed logic at 100x speed without anyone knowing how to stop it.

THE FUNDED HOBBY (*Invested: Cultural compass | Overlooked: Business maturity*): The internal culture is happy, safe, and collaborative, but there is no structural discipline or funding prioritization. Massive amounts of energy and capital are wasted on uncoordinated "passion projects" that never deliver ROI.

THE PRINCIPLED STANDSTILL (*Invested: Moral compass | Overlooked: Digital evolution*): You have world-class ethics and deep integrity, but you lack the "blood flow" of modern data and systems. You remain a "good" organization, but your technology is too outdated to actually compete with faster rivals.

THE FERRARI IN THE GARAGE (*Invested: Digital evolution | Overlooked: Execution mechanics*): You invest millions in an elite, high-performance tech stack that stays idle because no one has built the workflows to use it. You have purchased the "Engine" but never bothered to build the "Transmission" to turn the wheels.

THE LEGALISTIC LOOPHOLE (*Invested: Business maturity | Overlooked: Moral compass*): You strictly follow every legal rule and regulation while completely violating the spirit of ethics and trust. Your "Governance" is perfect on paper, but your brand is dying because your customers feel exploited by your "technically legal" actions.

THE INTELLIGENT OUTCAST (*Invested: Human–AI synergy | Overlooked: Cultural compass*): You develop AI tools that are technically brilliant and offer high synergy, but the workforce finds them creepy or intrusive. Because you ignored the "Senses" of the culture, the tools are rejected by the very people they were meant to help.

THE EFFICIENT COLLAPSE (*Invested: Execution mechanics | Overlooked: Business maturity*): The organization is moving with incredible power and speed, but it is moving toward a goal that isn't funded or strategically prioritized. You are winning the race, but you've realized too late that you're on the wrong track entirely.

THE NICE LIAR (*Invested: Cultural compass | Overlooked: Moral compass*): The culture is polite, collaborative, and highly "engaged," but it is built on a foundation of hidden ethical skeletons. People are "happy" because they are protecting each other from the truth, leading to an eventual systemic implosion.

THE ANALOG SAINT (*Invested: Moral compass | Overlooked: Human–AI synergy*): The leadership has high integrity but lacks the "AI-IQ" to understand the risks of the modern digital landscape. They are making ethical decisions based on a world that no longer exists, leaving them vulnerable to new forms of disruption.

THE EXPENSIVE CHAOS (*Invested: Digital evolution | Overlooked: Business maturity*): You layer high-tech AI tools on top of disorganized, immature governance and disorganized funding. The tech doesn't solve your problems; it simply makes your existing chaos move much faster and cost much more.

THE STAGNANT FORTRESS (*Invested: Business maturity | Overlooked: Execution mechanics*): The organization has perfect defenses, massive funding reserves, and rigid discipline, but zero kinetic movement. You are a perfectly "safe" business that is being slowly surrounded and suffocated by smaller, faster competitors.

THE DR. FRANKENSTEIN (*Invested: Human–AI synergy | Overlooked: Moral compass*): You achieve brilliant "Brain-to-Machine" innovation that pushes the boundaries of what is possible. However, because you ignored the "Inner Ear" of balance and ethics, you create a system that causes more social harm than organizational good.

THE BURNOUT FACTORY (*Invested: Execution mechanics | Overlooked: Cultural compass*): You optimize your "Muscle" for maximum output but completely ignore your "Senses" and emotional climate. You fail to detect internal fatigue and resentment until your highest-performing talent quits en masse.

THE LUDDITE OASIS (*Invested: Cultural compass | Overlooked: Digital evolution*): You have created the best place in the world to work, with high safety and great vibes. However, because you ignored the "Blood Flow" of digital evolution, your company is being disrupted out of existence by competitors you can't even see.

THE TOOTHLESS POLICY (*Invested: Moral compass | Overlooked: Execution mechanics*): You have beautiful "Values" posters and world-class ethics policies, but no actual workflows or mechanics to enforce them. Your integrity is a theory that falls apart the moment a team is pressured to hit a quarterly target.

THE GHOST SHIP (*Invested: Digital evolution | Overlooked: Human–AI synergy*): You have successfully implemented full automation and "lights out" data systems across the enterprise. However, because there is no human judgment in the loop, the system continues to run on outdated logic that is driving the company into a ditch.

THE PRECISE DINOSAUR (*Invested: Business maturity | Overlooked: Human–AI synergy*): You have spent years optimizing and documenting every single process to a level of "Six Sigma" perfection. Unfortunately, you ignored the "Brain" of AI synergy, and those processes are now perfectly optimized for a market that has already moved on.

Human judgment amplified by AI

Human–AI Synergy defines how intelligence is created, shared, and amplified across the enterprise. It examines the evolving partnership between human judgment and machine capability, ensuring that AI elevates rather than replaces human potential. This Lens forces leaders to confront how decisions are made, how insights flow, and how people interact with intelligent systems. It exposes whether the organization is building a truly augmented workforce or simply layering tools on top of old thinking. Ignoring this Lens leads to fragmented automation, poor adoption, and inconsistent decision-making. Human–AI Synergy is essential because it establishes the cognitive foundation of transformation—how the enterprise thinks, learns, and adapts—and without it, AI becomes a disconnected tool rather than a strategic multiplier.

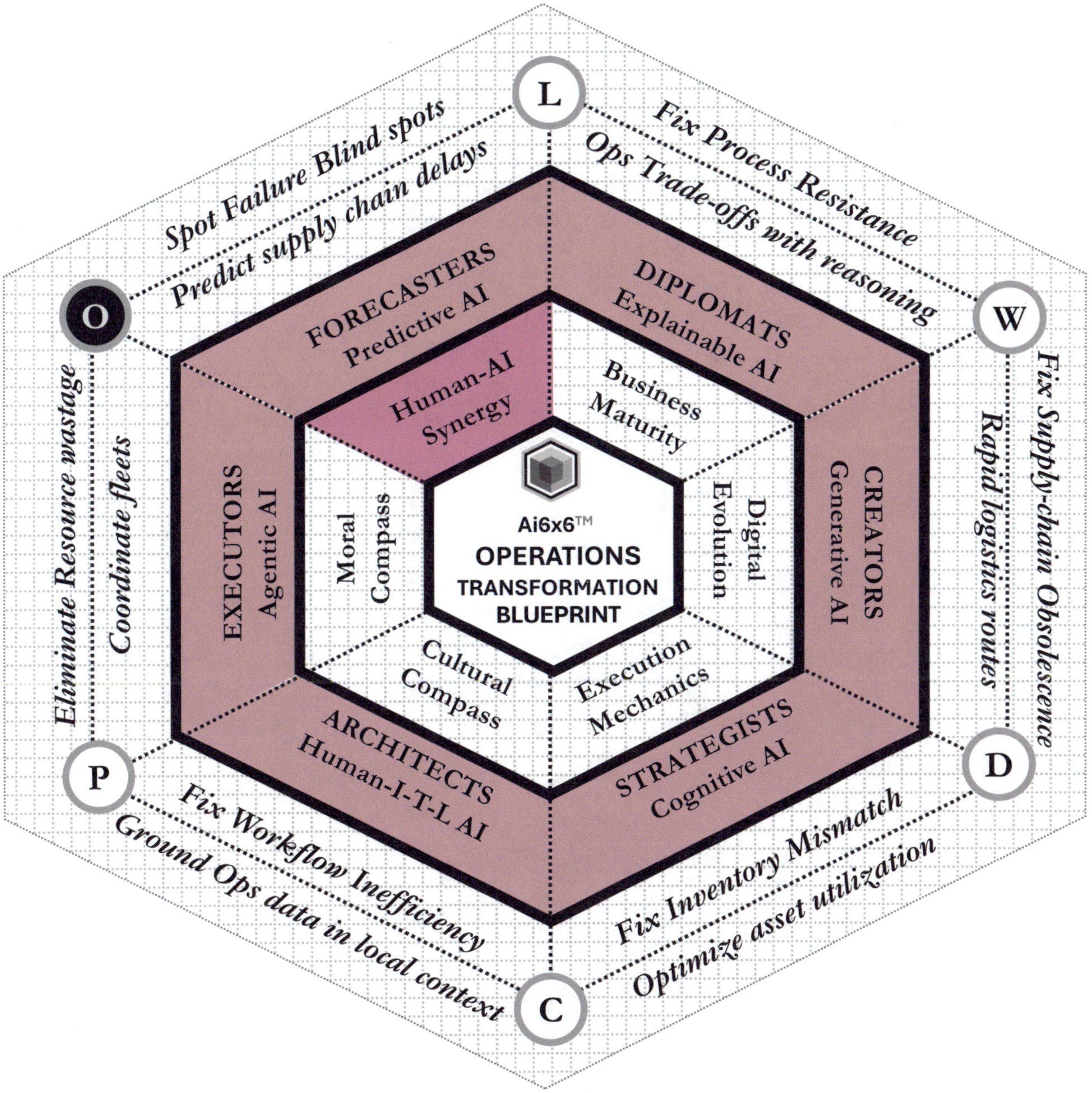

Ai6x6 HUMAN-AI SYNERGY FRAMEWORK FOR **OPERATIONS LIFECYCLE**

Discipline that stabilizes transformation

Business Maturity is the structural discipline that transforms raw ambition into a legacy of Authority. It begins at the Foundation, establishing the governance and funding models that anchor an organization's identity before activation begins. As the enterprise moves through Initiation and Cohesion, this Lens evaluates whether the internal scaffolding is robust enough to support the increasing velocity of the transformation engine. This Lens exposes a fundamental truth: sustained success is engineered through systems, not slogans. It ensures that as Momentum builds, the enterprise maintains the Optimization necessary to prevent collapse under the weight of its own initiatives. Mastering this Lens allows an organization to transcend temporary programs and achieve the permanent market Authority that defines industry leadership.

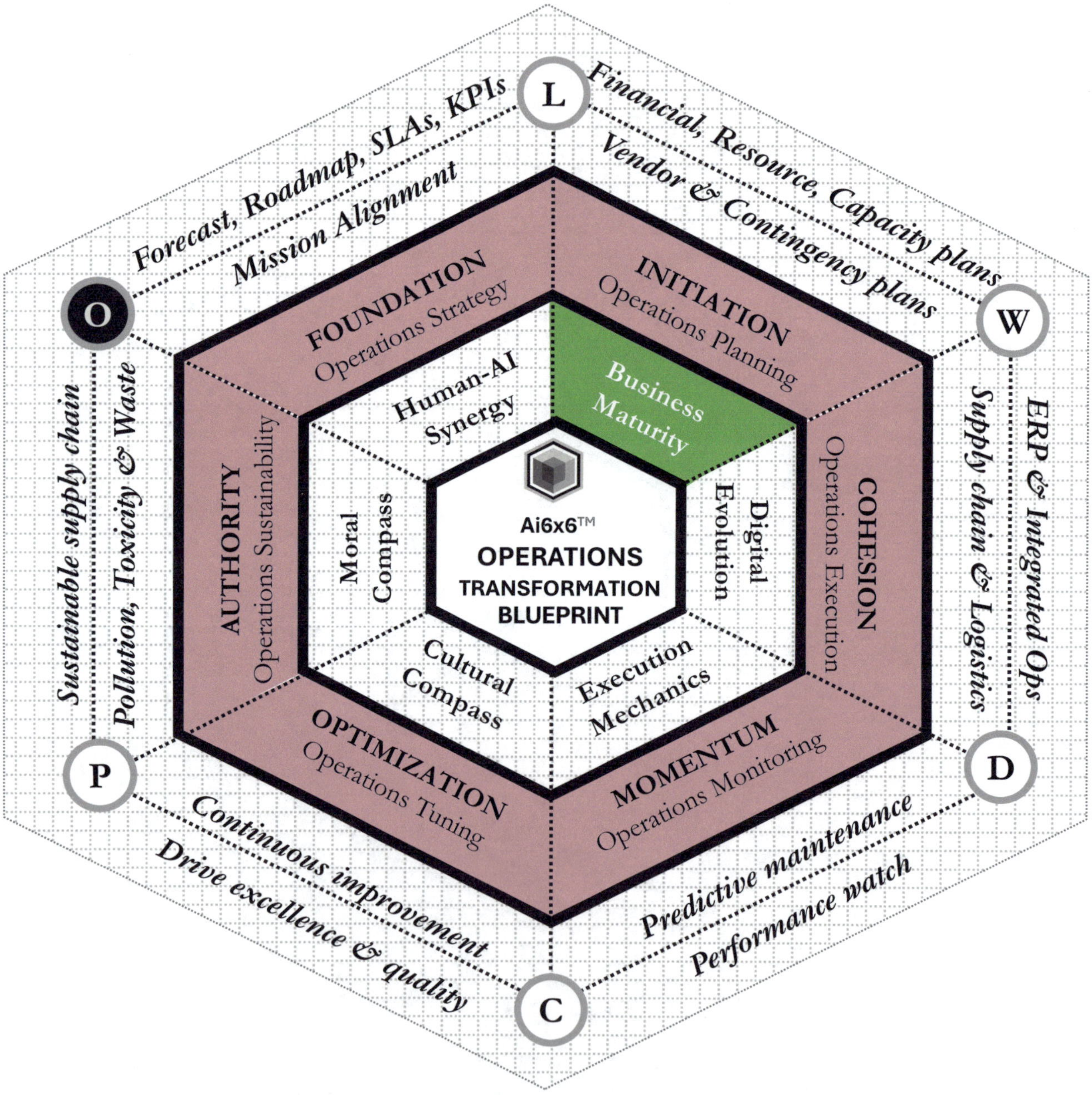

Ai6x6 BUSINESS MATURITY FRAMEWORK FOR **OPERATIONS LIFECYCLE**

Infrastructure built for continuous change

Digital Evolution examines the technological backbone that enables modern transformation. It evaluates system integration, data flow, platform scalability, and the adaptability of digital environments. This Lens ensures technology accelerates rather than obstructs progress. It forces organizations to confront technical debt, architectural fragility, and the true readiness of their digital ecosystem. When ignored, enterprises build brittle systems, slow innovation, and create environments where strategy moves faster than infrastructure. Digital Evolution is not about tools—it is about constructing a digital foundation capable of continuous change. It ensures that technology becomes an enabler of intelligence, speed, and resilience. Without this Lens, even the strongest strategies fail because the underlying environment cannot support enterprise-wide evolution.

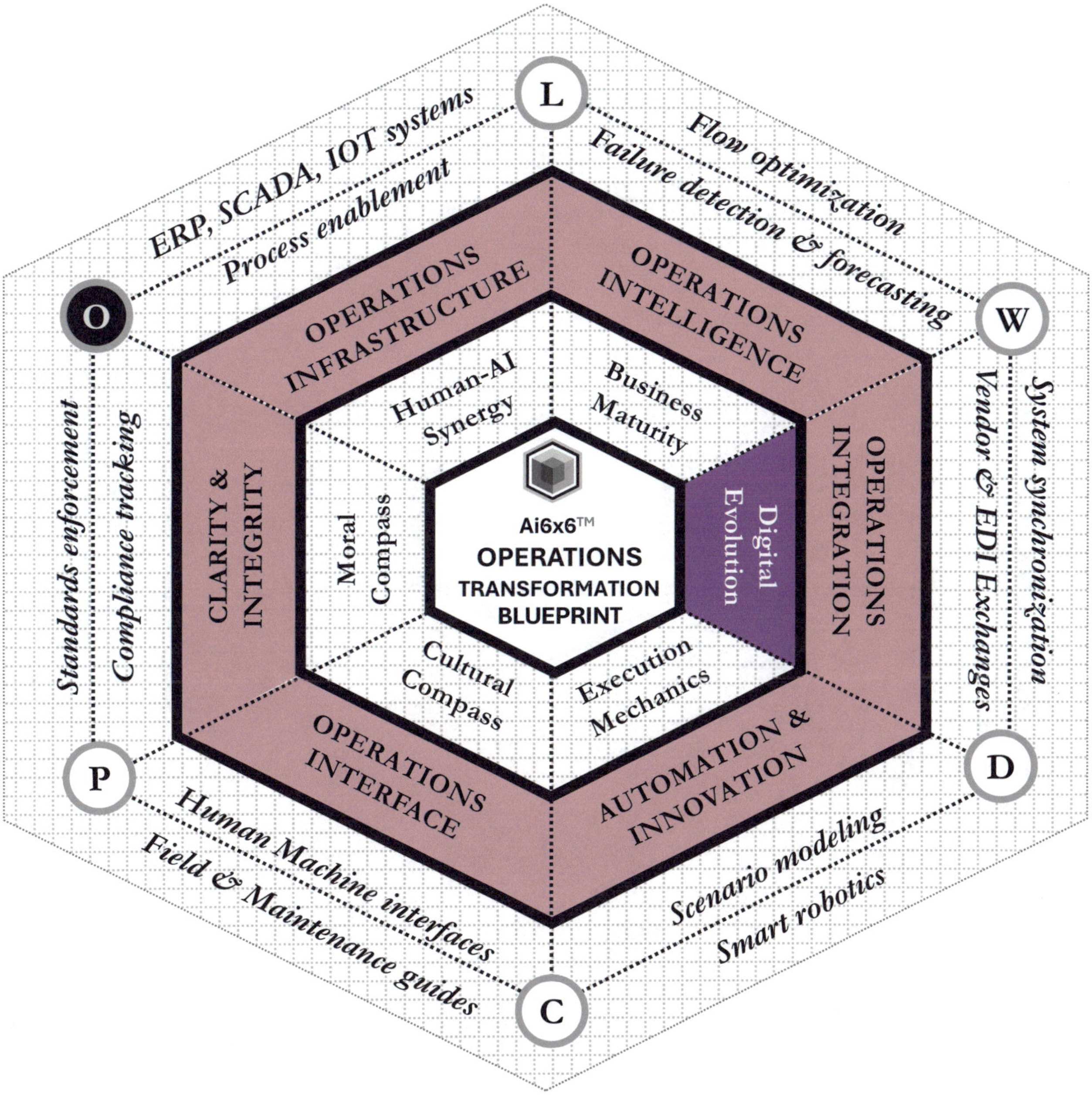

Ai6x6 DIGITAL EVOLUTION FRAMEWORK FOR **OPERATIONS LIFECYCLE**

Strategy delivered with operational precision

Execution Mechanics converts strategy into consistent, scalable action. It examines operating models, workflows, methodologies, and performance systems to determine whether the organization can deliver reliably. This Lens exposes the gap between leadership intent and frontline reality, revealing whether execution is engineered or improvised. It forces leaders to confront bottlenecks, handoff failures, and the true velocity of delivery. Ignoring this Lens results in missed deadlines, inconsistent outcomes, and organizational frustration. Execution Mechanics is essential because it transforms ideas into outcomes and outcomes into advantage. It ensures that transformation is not theoretical but operational—a system of delivery that produces measurable impact at scale.

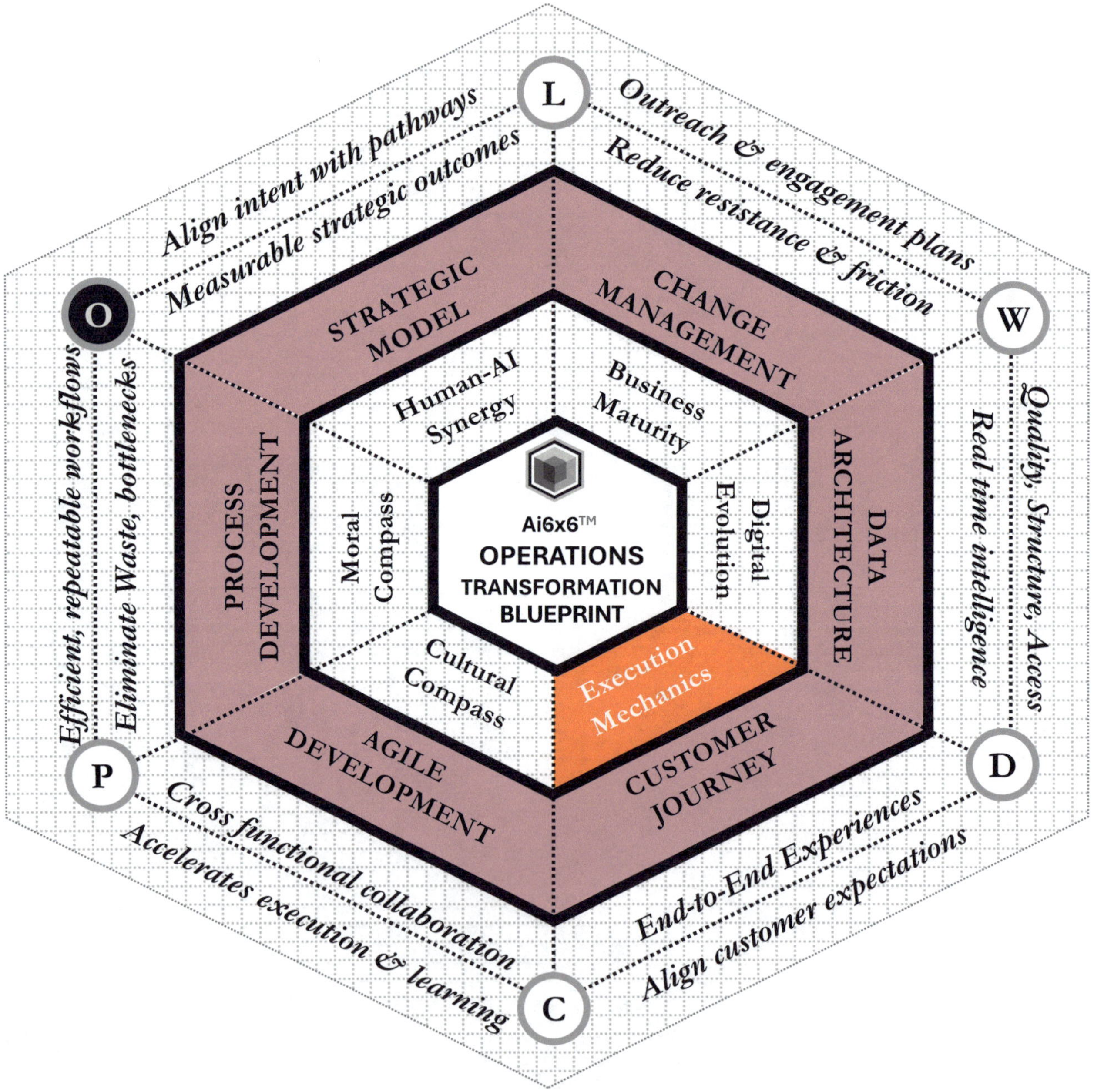

Ai6x6 EXECUTION MECHANICS FRAMEWORK FOR **OPERATIONS LIFECYCLE**

Culture aligned to accelerate change

Cultural Compass explores the beliefs, behaviors, and shared meaning that shape how people think, collaborate, and respond to change. It examines psychological safety, communication norms, leadership behaviors, and emotional climate. This Lens reveals whether culture accelerates transformation or quietly resists it. It forces leaders to confront the invisible forces that determine engagement, trust, and momentum. Ignoring this Lens means ignoring the human system that ultimately determines success. Even the best strategies fail in cultures that are misaligned, fearful, or fatigued. Cultural Compass is essential because transformation is a human endeavor—no system or technology can succeed without cultural readiness. It ensures people are not just compliant but committed, creating the conditions for lasting change.

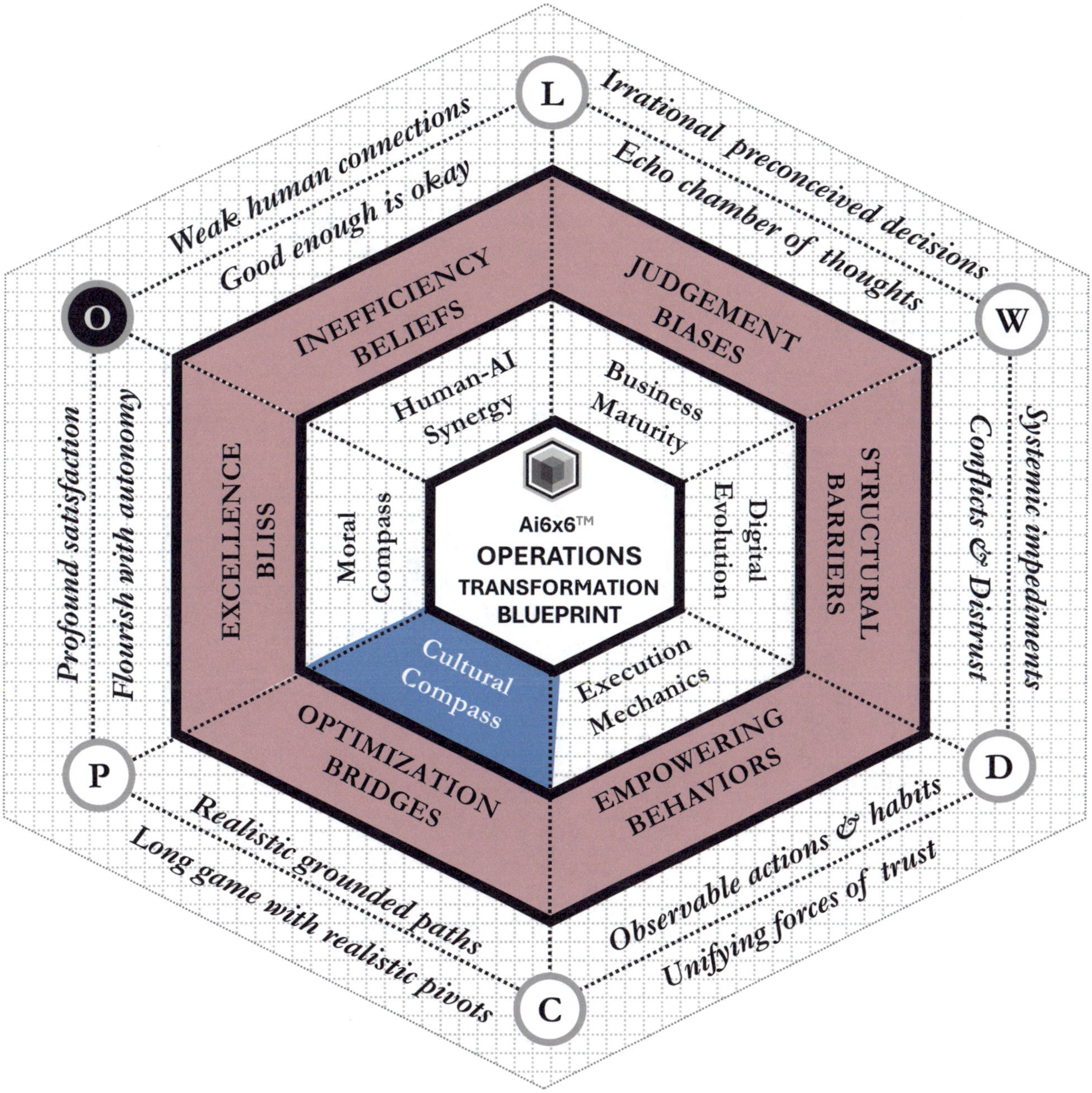

Ai6x6 CULTURAL COMPASS FRAMEWORK FOR **OPERATIONS LIFECYCLE**

Ethics governing intelligent enterprise systems

The **Moral Compass** anchors transformation in ethics, transparency, and responsibility. It evaluates data privacy, AI fairness, accountability structures, and the moral obligations of leadership. This Lens ensures that innovation is principled and that stakeholders feel protected, respected, and informed. It forces leaders to confront the ethical implications of their decisions and the long-term consequences of their systems. Ignoring this Lens exposes organizations to reputational damage, regulatory risk, employee resistance, and customer disengagement. Trust is the boundary condition of modern enterprise—without it, nothing else holds. The Moral Compass is essential because it defines the legitimacy of every decision, system, and outcome, ensuring transformation is not only effective but worthy of confidence.

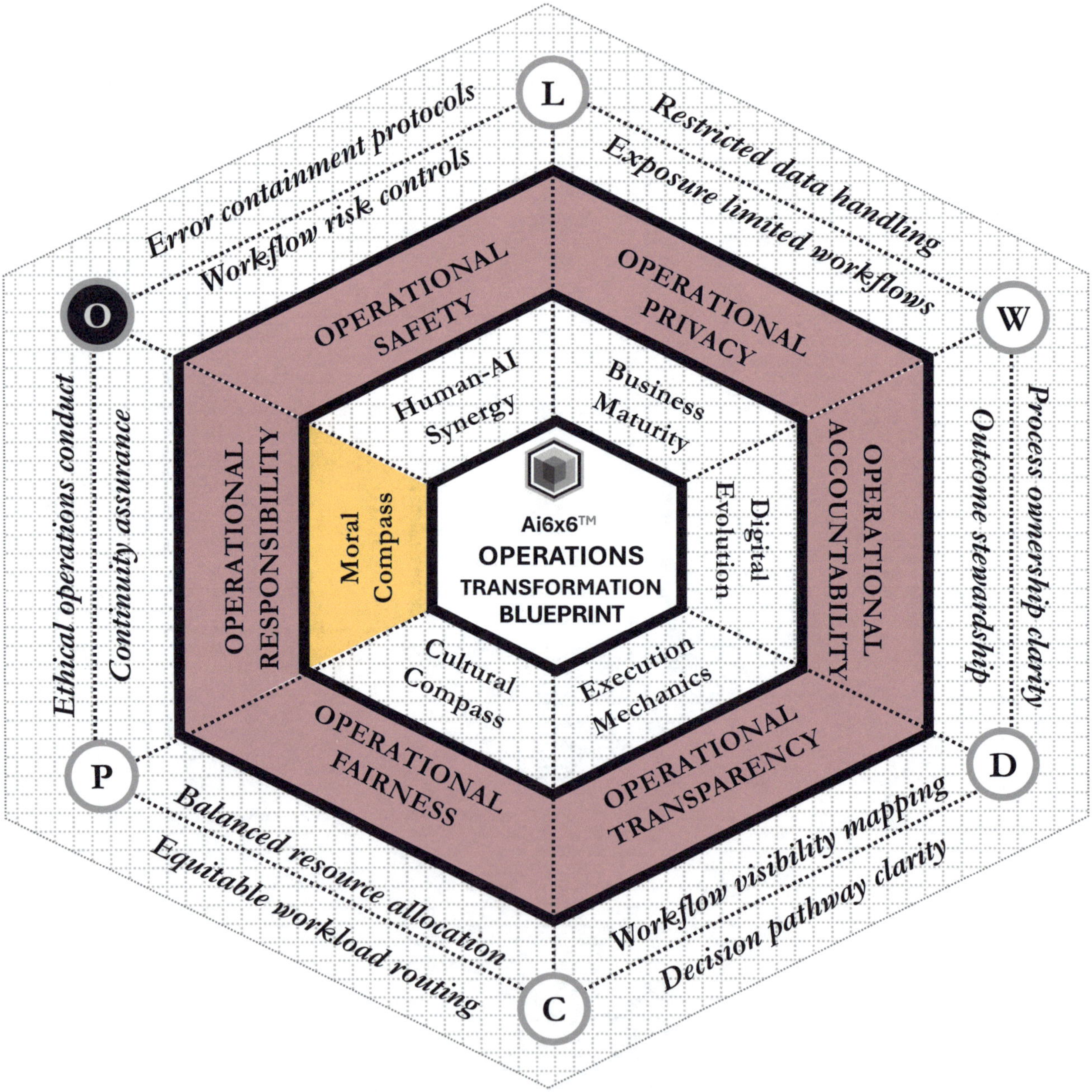

Ai6x6 MORAL COMPASS FRAMEWORK FOR **OPERATIONS LIFECYCLE**

Ai6x6™ Transformation Blueprint

Why Transform Operations?

"It's no longer the big beating the small, but the fast beating the slow." — Eric Pearson

Eric Pearson is the Chief Information Officer (CIO) for the International Hotel Group (IHG).

Modern business is defined by speed, efficiency, and adaptability. An operations transformation is required to move beyond rigid, legacy systems that create bottlenecks. Without this change, organizations cannot respond quickly to supply chain disruptions, changing customer demands, or competitive pressures. This transformation is critical for building a lean, resilient, and digitally integrated operational model that enables real-time visibility and flexible workflows, allowing a company to move at the speed of the modern market. The benefit of this transformation is increased efficiency, reduced costs, and the agility to respond quickly to market shifts.

1. **End-to-End Process Automation:** Systematic application of automation tools (RPA, scripts) across high-volume, low-judgment tasks eliminates manual human errors and drastically speeds up processes like provisioning, patching, and financial reporting. The consistency and speed of automated processes reduce operational variability, allowing the organization to guarantee higher levels of service reliability to customers. Human capital is conserved for strategic, non-automated tasks.
2. **Increased System Resilience and Uptime:** Leveraging cloud-native features like auto-scaling, geographic redundancy, and self-healing mechanisms minimizes disruption, ensuring near-perfect business continuity and protecting critical revenue streams. This focus on fault tolerance minimizes the business risk associated with natural disasters, hardware failure, or unexpected demand spikes. Revenue is protected 24/7/365.
3. **Lower Operational Expenditure (OpEx):** Dynamic, elastic scaling of infrastructure ensures the organization avoids costly over-provisioning, driving down monthly utility expenses by paying only for peak usage capacity and not permanent capacity. The reduction in CapEx (fixed hardware investment) and optimized OpEx (utility billing) provides significant, sustainable savings over non-elastic infrastructure. Infrastructure costs are directly and accurately tied to business activity.
4. **Automated Compliance and Audit Trails:** Infrastructure as Code (IaC) and version control guarantee that the state of all systems is auditable and documented, providing instant, verifiable compliance evidence for internal and external regulatory bodies. This coded enforcement of regulatory rules reduces the need for expensive, manual compliance checks and minimizes the risk of human oversight. Audit cycles are accelerated from weeks to hours with high confidence.
5. **Faster Incident Response:** Automated observability (AIOps) tools instantly detect, diagnose, and often automatically resolve issues before human intervention is required, significantly cutting the Mean Time To Recovery (MTTR) and minimizing user impact. Reducing MTTR from hours to minutes or seconds protects the brand's reputation and prevents major financial losses due to extended downtime. Service degradation is often corrected transparently to the end-user.
6. **Creation of Perfect System Cohesion:** The use of GitOps and IaC eliminates Configuration Drift, guaranteeing that all environments (development, staging, production) are identical and predictable, thereby eliminating "works on my machine" failures. Predictable deployments reduce the stress and risk associated with major releases, ensuring that tested code functions reliably the moment it hits production. The deployment pipeline itself guarantees consistency.
7. **Increased Capacity Utilization:** Advanced load balancing, container orchestration, and dynamic resource pooling maximize the efficiency of existing physical and cloud hardware, boosting the effective utilization rate of capital assets. A higher utilization rate means the business extracts maximum value from every dollar spent on infrastructure, improving capital efficiency. Underutilized resources are instantly reclaimed and repurposed.
8. **Systemic Energy Efficiency and Sustainability:** By migrating to modern cloud architectures and optimizing server utilization, the organization significantly reduces its energy consumption and hardware footprint, contributing to a lower corporate carbon footprint. This commitment to greener operations not only reduces utility costs but also aligns with growing environmental, social, and governance (ESG) expectations from investors and customers. Sustainability becomes a measurable operational advantage.
9. **Single Pane of Glass Observability:** Unifying logs, metrics, traces, and user monitoring into a single dashboard provides comprehensive, real-time visibility into the health and performance of the entire service portfolio, simplifying maintenance and debugging. Engineers spend less time hopping between disparate monitoring tools and more time solving problems, accelerating resolution and reducing cognitive load. The collective "brainpower" of the Ops team is focused purely on analysis.
10. **Enhanced Supply Chain Transparency:** Integration of IoT devices, digital twins, and blockchain technologies provides real-time, immutable tracking of goods and materials through the supply chain, optimizing logistics and verifying provenance. This end-to-end visibility minimizes losses from theft or damage, streamlines customs clearance, and increases customer trust in product origins. The operational system provides verifiable trust throughout the value chain.

1. Efficiency Without Visibility - When Work Hid in the Shadows

Everyone was busy. Calendars were full, tools buzzing, dashboards glowing green. So when a product delay blindsided the company, everyone was shocked. "But ops had it," said one team. "We submitted our update," said another. Turns out, work was happening—but in isolation. Project data lived across PM tools, spreadsheets, and tribal memory. Ops wasn't broken—it was invisible. The COO launched a shared operational layer: dynamic roadmaps, live blockers, and shared definition-of-done criteria. Daily standups got visual. Dependencies were tracked, not assumed. Patterns emerged—rework, bottlenecks, double work. Within two quarters, cycle times shrank by 28%, and meeting time dropped by 40%. Teams finally saw the whole field, not just their corner. The real transformation wasn't software. It was shared sightlines.

Because when operations are invisible, trust erodes. And when trust erodes, performance soon follows.

2. Processes that Punish Agility - When SOPs Blocked Progress

The business team spotted an urgent opportunity: a limited-time partnership with a cultural trendsetter. But legal needed two weeks. Finance demanded forms. Procurement asked for approvals—twice. The deadline passed. The deal died. The postmortem was sobering: "We aren't slow because we don't care. We're slow because we don't trust." Every team had built walls to protect themselves—but those walls blocked momentum. The COO introduced a new approach: agile governance. Standard Operating Procedures (SOPs) became guardrails, not gates. Legal templated high-speed contracts. Finance pre-approved budget corridors. Ops created "72-hour sprints" for rapid alignment. Suddenly, go-to-market was fluid, not frantic. Teams stopped navigating mazes—and started collaborating with guardrails. Transformation meant swapping control for clarity.

Because operational design isn't just about safety—it's about momentum that moves with confidence.

3. Growth Without Scale - When Success Strangled Itself

Revenue doubled in six months. Headcount didn't. Neither did systems. Onboarding took three weeks. Customer data lived in spreadsheets. Handovers got lost in email. Success was suffocating itself. The ops director ran a workflow mapping session. Post-it notes covered every wall: duplications, manual checks, bottlenecks. The diagnosis? Scale hadn't failed—it hadn't been designed. The fix began small: automated handoffs, single data sources, tiered responsibilities. Playbooks replaced guesswork. Within 90 days, onboarding time halved, and team load dropped by 35%. Growth stabilized—without burnout. The company didn't hire more. They operated better. Operations transformation isn't the last step after expansion—it's the reason expansion works.

Because scaling isn't just adding headcount. It's removing friction at the root.

4. Siloed Systems, Fragmented Truth — When Data Didn't Align

In the executive meeting, marketing claimed 20,000 leads. Sales said only 4,000 were qualified. Finance couldn't find either number. Frustration flared. Each function had its own tools, definitions, and timelines. "We're arguing over whose spreadsheet is less wrong," someone joked. Ops called a pause and launched a Data Truth Project. First: unified definitions. Then: integrated systems. Lead status, funnel stages, revenue attribution—all standardized, visualized, and shared. Within weeks, meetings shifted from defensive to decisive. Marketing optimized for conversion, not just volume. Sales knew what to nurture. Finance synced forecasts to real pipeline.

Transformation isn't just tool integration—it's operational agreement. Because when teams don't share truth, they can't share progress.

5. Ticket Loop Purgatory — When Internal Requests Became a Black Hole

Nisha needed access to a financial dashboard. She submitted a ticket—heard nothing. She pinged twice. Then escalated. Finally, a terse message: "Please follow process." But she had. Three weeks later, access arrived—after the decision window closed. Ops investigated. Across IT, finance, data—over 40% of internal tickets stalled. The culprit? Intake chaos. Vague forms, misrouted asks, no prioritization logic. So they redesigned the entire request flow: clear categories, triage channels, transparent SLAs. Requests were now trackable—like packages. Ownership was visible. Cycle time dropped by half. More importantly, employees felt respected. "You made back-office feel like front-office," Nisha said.

Transformation isn't just workflows—it's dignity in execution. Because operations that work behind the scenes shouldn't stay behind in standards.

6. Meetings that Solved Nothing — When Time Became the Easiest Thing to Waste

Ops meetings happened daily. But outcomes? Rare. People recapped updates. Decisions were deferred. Action items got lost. A new hire whispered, "Why does this feel like theater?" The COO listened—and acted. They banned "update-only" meetings. Every recurring meeting needed a charter: purpose, owner, outcomes, decision path. Ops created a calendar audit tool—highlighting time spent on low-impact rituals. They replaced 30-minute meetings with 10-minute decision huddles. Cross-functional issues got "async escalators"—Slack threads with pre-reads and action owners. Within six weeks, meeting load dropped by 40%. Decisions sped up. People opted in—not out.

Transformation isn't about more collaboration—it's better choreography. Because time is the only resource everyone shares. And operations decides how it's used—or wasted.

7. The Hidden Cost of Handoffs — When Work Collapsed Between Teams

An urgent vendor payment got missed. Finance said ops was late. Ops said the handoff wasn't flagged. The vendor left. Everyone felt terrible—no one was accountable. The root cause? Invisible boundaries. Tasks handed off without transition. No rituals, no shared ownership. The head of ops piloted a "moment-of-handoff" protocol: clear trigger, named owner, confirmation loop. Teams created micro-rituals—handoff Slack emojis, batons in Notion, checklist signoffs. The culture shifted from "I sent it" to "We caught it." Error rates fell. Tension eased. Vendors returned.

Transformation isn't just fixing errors—it's designing accountability where none existed. Because between "my job" and "your job" is usually where things break.

8. Change Without Absorption — When Ops Couldn't Keep Up with Strategy

Strategy shifted. Again. And again. New priorities. New OKRs. But in the trenches? Old workflows. Old incentives. The gap widened. Ops became whiplashed—always reacting, never upstream. Finally, they asked to join quarterly planning. They created an "absorption layer": every new initiative got a capacity impact check, rollout runway, and ops play. Implementation moved from ad hoc to designed. Strategy didn't slow—but it stuck. Ops went from reactive to anticipatory.

Transformation means pulling operations into the strategy room—not just the war room. Because brilliance without absorption is just a beautiful PowerPoint.

9. Vendor Sprawl — When Tools Solved and Created Chaos

Every team had their tools. Sales used five platforms. HR used four. Finance didn't trust any. Integration issues. Duplicate spend. Security gaps. Yet everyone felt productive—until it broke. An audit revealed over 160 SaaS licenses. Ops kicked off a software rationalization sprint. They mapped usage, evaluated overlap, renegotiated contracts. Redundant tools were sunset. Critical ones got proper onboarding. They created a central "App Registry" and vendor review council. Costs dropped 28%. Onboarding simplified. Context switching shrank. People didn't lose autonomy—they gained clarity.

Transformation isn't about fewer tools—it's tools that work as a system. Because productivity without coherence is just glorified chaos.

10. Burnout Without Bad Intent — When Every Team Meant Well, But it Was Too Much

"No one did anything wrong," the COO said. "But everyone's drowning." Pulse surveys flagged exhaustion. Projects piled up. Deadlines overlapped. Teams were saying yes—to everything. Ops ran a bandwidth audit. Turns out, 40% of work came from uncoordinated asks—well-meaning, siloed, scattered. They introduced intake gates. Every new initiative needed a resourcing check. A capacity board showed real-time load. Leaders were trained to say no, defer, or negotiate. The org didn't lose momentum. It gained sanity. Energy returned. People stopped feeling heroic for burning out—and proud for being sustainable. Transformation means creating systems where care isn't just a value—it's operationalized. Because even high-performers have limits. And great operations protect them.Shape

11. Playbooks That Never Got Played — When Ops Wasn't Lived, Just Documented

They had everything—standard operating procedures, onboarding guides, escalation trees. Dozens of beautiful PDFs created by diligent ops teams. But when issues arose, no one used them. "We didn't know it existed," one teammate confessed. The problem wasn't a lack of documentation—it was disconnection. SOPs were siloed in folders and Google Drives, untouched after day one. A new operations manager reframed the strategy: if a playbook isn't used in real time, it's not real at all. She integrated guidance where work happens: decision trees in tools, AI prompts in ticketing systems, micro-SOPs in Notion docs linked from Slack messages. Teams could now surface instructions during the task—not hunt for them after. She also launched a "living doc" culture—teams could comment, revise, and flag outdated content weekly. Within two quarters, usage spiked, onboarding time dropped 30%, and error rates in critical workflows fell. Employees started saying, "I didn't know we had a playbook for that—and now I can't imagine working without it." Transformation isn't just creating knowledge—it's embedding it where it matters. Because operations aren't what you write down. They're what people remember in motion.

12. When Ops Became the Culture — Not Just the Checklist

Every team claimed values like ownership, curiosity, and clarity. But day-to-day? Deadlines were fuzzy. Decisions lingered. Escalations bounced. Despite best intentions, the operating model was misaligned with the culture on paper. A new Chief of Staff posed a bold idea: what if operations was culture? Together with ops, they re-architected rituals around core values. Ownership meant single-threaded leaders—no ambiguity. Curiosity became monthly reflection forums, not just postmortems. Clarity showed up in pre-reads, not just slide decks. Teams codified what done really meant. Slowly, behavior shifted. Leaders stopped saying "Let's be transparent"—they demonstrated it through transparent docs. Trust rose. Feedback flowed. A product manager said, "We stopped chasing culture and started building it through how we work." Six months later, engagement scores rose and cross-functional velocity doubled. The culture hadn't changed because of slogans. It changed because ops made values real. Transformation is when process and purpose align—so every meeting, metric, and message reinforces what you believe. Because how you work is who you are.

Ai6x6™ Transformation Blueprint

Power of Diversified Insight

"Clearly, the thing that's transforming is not the technology—the technology is transforming you."

Jeanne W. Ross, Research Scientist, MIT Sloan.This quote brings the focus back to the human and cultural element of transformation. The challenge is not implementing code, but changing the organizational mindset, leadership behaviors, and fundamental business identity in response to the technology

We have deliberately chosen to move beyond the limiting confines of the conventional single-expert **Foreword**, which, by its nature, risks restricting the reader to one singular point of view. In an era where **business transformation** is shaped by a confluence of disruptive forces, relying on a solitary oracle is fundamentally inadequate. Instead, this book opens with a **curated collection of 15 interview snippets** from a diverse array of global thought leaders. This method immediately immerses you in a rich, multi-faceted tapestry of opinions and real-world experiences. By foregrounding this **plurality of perspective**, we offer a wider, more nuanced, and dynamic lens through which to view the challenges and enormous opportunities presented by modern change. Consider these short, powerful excerpts a **Chorus of Counsel**: a comprehensive and inclusive foundation built on collective wisdom that better equips you for your own complex journey of enterprise innovation and transformation. This is a commitment to **diversified insight** from the very first page.

Gain unprecedented, insider access to the minds of 27 global thought leaders in this six-book series. Sourced from comprehensive 1-on-1 interviews, these experts are recognized worldwide for their work in business transformation and emerging technology. Their insights are drawn from my personal, professional network—including former bosses, colleagues, and business partners —giving you an insider's view into the minds of industry shifters. To maximize focus and utility, we've synthesized these deep-dive conversations into thematic 'interview cards.' Each of the six books offers 15 relevant, concentrated insights on specific topics, allowing you to quickly access and apply the most relevant expertise, no matter where it originated in the original interview.

EXPERT VOICES

Subrata Mukherjee: Your profile highlights your vast network of startups, investors, and CxOs.. You help develop transformative partnerships. How do you go about building and nurturing an ecosystem of partners and advisors to drive innovation?

Richard Entrup: You need an intention to network and connect. This can't be forced. I've built a vast network over four decades that I'm able to tap into for all kinds of benefits, for myself, my employer, and those in my network. I see networking and relationship building as the lifeblood of a successful career, and no one attains success without the help of others. I also take the tenet of 'paying it forward' very seriously. You need to help others as much or even more than you take from your network and you need to understand that networking and job searching are two different dynamics. Job searching is a one-way need, and networking involves connecting others for their benefit, not your own. I can't believe how many seasoned executives don't get that and they think they're networking when they contact me after a 10-year run at one job with no contact. We all have those contacts who only ping when they are looking for a new job. That's not networking.

SM: From your experience as both an advisor and an internal leader, what are the three most important things a leader must do to foster a culture of innovation?

RE: I've seen that fostering a culture of innovation requires intentional actions that empower teams, align efforts with strategic goals, and create an environment where experimentation thrives. Here are three important things:

Empower Psychological Safety and Experimentation: Leaders must create a safe space where employees feel encouraged to take risks and propose bold ideas without fear of failure or retribution. This means celebrating small failures as learning opportunities—e.g., a failed AI pilot that reveals a new customer insight—and rewarding creative problem-solving. For instance, Google's "20% time" policy, which led to products like Gmail, shows how giving teams autonomy to explore ideas can drive breakthroughs. Leaders should set clear expectations that innovation is iterative, using metrics like the number of prototypes tested to gauge progress rather than immediate ROI.

Align Innovation with Business Strategy: Innovation must serve the organization's goals to avoid being a scattershot effort. Leaders should work closely with cross-functional teams to identify high-impact areas—like using open-source AI to optimize supply chains or closed-source GenAI for rapid customer engagement—and tie these to measurable outcomes, such as 10% cost savings or 15% revenue growth. By embedding innovation into strategic priorities and communicating this alignment clearly, leaders ensure efforts are focused and resources are allocated effectively, as seen in companies like Amazon, where innovation (e.g., AWS) directly supports business expansion.

Invest in Talent and Collaboration: A culture of innovation thrives on diverse, skilled teams and external partnerships. Leaders must upskill their workforce through training in emerging technologies and foster collaboration across departments to spark interdisciplinary ideas—think marketing and IT co-developing a GenAI-driven campaign. Additionally, partnering with startups, academia, or open-source communities brings fresh perspectives and accelerates innovation. Leaders should track engagement metrics, like participation in hackathons or cross-team projects, to ensure a vibrant, innovative ecosystem.

RICHARD ENTRUP

Managing Director, Head of Emerging Solutions at KPMG US

A multi-award-winning global technology and innovation leader with nearly three decades of experience driving strategic business transformation. He is at the forefront of identifying and incubating disruptive technologies like GenAI, Quantum Computing, and Spatial/XR, pioneering new products and solutions for clients. A two-time recipient of the HMG Global Leadership Institute Award for Exceptional Business Transformation, his distinguished background includes C-suite leadership roles (CIO, CTO, CDO) at globally recognized brands such as Verizon, Disney/ABC, Christie's, Time Warner, Viacom, and Tiffany & Company. He also serves as a strategic advisor and investor to a portfolio of innovative startups and venture capital firms, passionately guiding ventures focused on cutting-edge innovations. He is based out of New York, United States.

EXPERT VOICES

Subrata Mukherjee - What is the single biggest lesson you've learned about the role of a Chief Data Officer in a mergers and acquisitions scenario?

Saleem Khan: I think the most important lesson I've learned having sold a few companies now is the importance of story telling. I started out my career as a programmer and morphed into a data and product leader over time. However, a former senior leader and mentor once told me, "what got you here, won't get you there." I say all this because the transition from being a hands-on developer or data engineer to data leader requires you to be an effective communicator. These are very different skills from being a programmer. I've developed these skills over the years but I attribute much of my success to my wife who happens to be a high school English teacher.

You see, my wife does a great lesson on Aristotle's Rhetoric. In this book, Aristotle covers the three ways in which you can tell a good story to convince your audience: **ethos, pathos, and logos**. As a data leader it becomes very important in an M&A scenario to be prepared for any style of rhetoric required.

Many questions that will be lobbed at you during the diligence process will be **logos**, or questions that require evidence, facts, and reason. What are your standard processes to ensure data quality? What is your fill rate on 'x' data element? How many unique records do you maintain and how does this compare to your competition? It's important to have these facts memorized not to show your subject matter expertise but your subject matter authority!

You will also be asked questions that infuse elements of **pathos**, which is an appeal to emotion; anger, excitement, joy. Ideally, you don't induce too much anger, but you may be asked how your customers feel about your data or products. It's important to have anecdotes at the ready and a list of customers you can reference to back up your claims. Pathos questions are less evidence based so it's important to be able to tell anecdotes in relatable and simple language following the narrative arc used in Joseph Campbell's, The Hero's Journey (another good read for any data/product leader)

Finally, we have **ethos** questions. These are questions that appeal to knowledge, experience and often cultural fit. Questions that may come up concern how you have built trust with your customers, employees, and existing investors? Do you have regular touchpoints with them? Do you have a weekly newsletter with insights that establish trust and credibility? What would your employees say about the culture of your organization?

Being a good story teller and developing these skills is crucial in ensuring an outsized enterprise value for your sale.

Aristotle's Rhetoric identifies three key persuasive appeals: **Ethos** (character and credibility of the speaker), **Pathos** (emotional appeal to the audience), and **Logos** (logical reasoning and evidence). A speaker uses a combination of these appeals in their message to effectively persuade an audience, with the goal of creating a well-rounded and convincing argument.

SALEEM KHAN

Chief Data & Analytics Officer, Pole Star Global

Saleem is a seasoned data and analytics executive with a strong history of creating business value through data-driven innovation and successful company exits. In his current role as Chief Data & Analytics Officer at Pole Star Global, he leverages his deep expertise to lead and execute data strategy. His notable achievements include serving as Chief Digital Product Officer at VettaFi, which was successfully acquired by TMX Group, and as Chief Data & Analytics Officer at Discovery Data, which had a successful exit to Institutional Shareholder Services. He has also held leadership positions in data innovation and advanced analytics at major corporations like Dun & Bradstreet and S&P Global, and was a founder of the fintech startup ESGINE, an AI-enhanced platform for ESG compliance. He studied at Fordham University. He is based out of New York, United States.

Subrata Mukherjee: You have extensive experience advising and leading start-ups. What is the single most important piece of advice you give to start-up founders who are looking to accelerate their revenue growth?

Stephane Pere: Focus first on product-market fit and user trust. Too many founders rush to scale without proving that their solution truly solves a pain point. Growth comes then when customers believe in both your product and your values. Once that trust is established, scaling revenue becomes much faster and more sustainable.

SM: Your work at **Vaultinum** focuses on technology due diligence and software risk. What is the most common and often overlooked risk you identify when conducting technology due diligence for a company?

SP: The most overlooked risk is software maintainability. Companies often focus on market opportunity and financial performance but neglect the quality and sustainability of the codebase. If the software is outdated, poorly documented, or dependent on a few individuals, it creates enormous hidden risks. At Vaultinum, we use automated code analysis to surface these risks early, ensuring that investors and acquirers understand the long-term sustainability of the technology they are buying.

SM: With the rapid adoption of AI, what new cyber and software risks are emerging that companies need to be prepared for?

SP: Generative AI brings profound promise—but with it comes elevated risk: deepfakes, synthetic identities, AI-generated code, and fraud that can be hard to detect without robust oversight. Beyond technical threats, companies face strategic exposure linked to all AI capabilities.. Can they substantiate their claims of AI readiness, or are they merely riding the wave of "AI washing"? Vaultinum's AI Maturity Audit addresses this by rigorously assessing whether AI capabilities are genuine, scalable, and integrated across technical infrastructure, data governance, and team expertise—vs. superficial embellishments. At the same time, companies must also consider how external AI trends could undermine their business models. The AI Disruption Audit helps leaders evaluate how generative AI and commoditized tools might erode their differentiation or displace core features—before the damage is done. In today's fast-moving environment, the key is dual vigilance: ensure your own AI foundations are legitimate and resilient, and anticipate how external AI advances could shift the competitive terrain. That's how businesses protect value and stay ahead in the HUMAN-AI SYNERGY.

SM: As a leader, what is the one lesson you've learned that has had the biggest impact on your career?

SP: The most valuable lesson I've learned is that vision alone is not enough. Even when you see opportunities in a complex, fragmented, or challenged market, success depends on your ability to bring others with you. That requires a mix of inspiration and experimentation: you need to craft a compelling story that unites your team and clients around shared beliefs, but you also need to show, through concrete trials and new approaches, that this vision can work in practice. People don't just follow words—they follow evidence of progress. A leader's role is to turn abstract opportunities into a narrative tested by action, so that confidence grows collectively and momentum builds sustainably.

STÉPHANE PERE

Former Chief Data Officer – The Economist | MD – Evidency, France

Stéphane is a seasoned digital and tech executive with a strong focus on reinventing trust in the digital realm. His extensive career includes pivotal roles at leading companies, where he has driven strategic transformation and innovation. As Chief Data Officer and Executive Vice President at The Economist, he earned multiple awards for his work in transforming the B2B market intelligence business. He has also served in executive roles at Vaultinum, Evidency, and Poool, where he has leveraged his expertise in digital strategy, business development, and software as a service (SaaS) to accelerate growth and mitigate risk. He is also a strategic advisor and non-executive director, guiding media and tech companies in revenue acceleration and digital transformation. He studied in London Business School , Emlyon Business School & Paris-Pantheon-Assas University.He is based out of Paris, France.

EXPERT VOICES

Subrata Mukherjee: What, in your view, is the single greatest opportunity for content businesses when leveraging AI? Conversely, what is the greatest risk?

Michael Brunt: The greatest opportunity is personalization at scale, using AI to deliver the right content at the right time to deepen engagement and loyalty. The greatest risk is commoditization. If publishers rely on generic AI-generated content they erode trust and lose differentiation. The winners will be those who use AI to amplify human creativity and strengthen trusted journalism.

SM: Your firm specializes in "optimizing the customer experience" and "refining your content strategy." How is AI changing how publishers approach these areas?

MB: AI enables data-driven, tailored customer journeys. Publishers can now predict churn before it happens, test pricing models in real time and improve onboarding. In content strategy, AI can audit at scale, showing which topics, formats and tones drive engagement or subscriptions. This level of clarity transforms how resources are allocated across the newsroom.

SM: How do you advise publishers on aligning their AI initiatives with their overall business strategies to avoid common pitfalls?

MB: We start with business outcomes. Every initiative must contribute to acquisition, retention or efficiency. We then ensure editorial values guide AI use through strong governance. Finally we integrate AI into workflows so that it becomes part of daily practice rather than a bolt-on. The result is alignment between innovation and long-term growth.

SM: As a consultant, what role do you see strategic advisory firms playing in helping legacy industries adapt to technological disruption?

MB: We act as translators and accelerators. Our role is to cut through hype, provide clarity and deliver tailored, actionable plans that unlock growth. We also help leadership teams align editorial, commercial and technology around shared objectives. That outside perspective gives organizations confidence to embrace change at pace.

SM: For a young leader starting their career in media today, what one piece of advice would you give them?

MB: Stay curious and avoid early siloing. The leaders of tomorrow will understand journalism, technology, data and audiences. Learn across disciplines, use AI as an enabler and keep the purpose of journalism at the center. That balance will keep you resilient and relevant.

SM: What is your biggest prediction for the media industry over the next five years?

MB: I expect to see smaller, but more profitable, publishers serving well-defined audiences with AI-enabled personalization and trusted content. Depth of relationship will matter more than scale. At the same time new AI-driven platforms will emerge as gateways to content, challenging the dominance of search and social.

SM: You mention your tool, ContentReview.ai. Can you tell us more about how it provides insights that go beyond traditional content analysis?

MB: ContentReview.ai combines AI-driven categorization of content, advanced statistical modelling and editorial expertise to deliver actionable insights. It benchmarks content across multiple categories including user need, tone, subject mix and headline effectiveness, and explains why certain pieces resonate.

It also forecasts the impact of changes and recommends clear steps that improve profitability and engagement. Most publishers could tell you what the top 10 and bottom 10 performing articles were – Content Review allows them to analyze the 99% between those two extremes that usually goes unexamined.

MICHAEL BRUNT

Former MD/COO/CMO @ The Economist, GM Times / News UK

Michael is a strategic media and publishing executive with a proven track record of driving business transformation and profitability. Currently as Co-Founder of HBM Advisory, he helps content businesses navigate the digital age by aligning business strategy, implementing AI capabilities, and optimizing content and team structures. His distinguished career includes a significant tenure at The Economist, where he held multiple senior leadership roles, including CMO and COO. During this time, he was instrumental in transforming the circulation business from a loss-making entity into the largest contributor to group profits through innovative digital and experiential marketing campaigns. He also served as General Manager for The Times and The Sunday Times, overseeing their long-term growth strategies. He has studied at Harvard Business School and University of Manchester. He is based out of London, United Kingdom.

EXPERT VOICES

Subrata Mukherjee: Your career is built on a foundation of Service Delivery Assurance and Operational Excellence. What is the core philosophy you've developed for ensuring a large, global team consistently exceeds SLAs and other contractual commitments?

Rod Rinker: I believe that to ensure SLAs and Contractual Commitments are met, you have to make sure every level of the organization knows what these commitments are. Afterwards, you set up a review cadence with all the key stakeholders internally and externally to ensure compliance and continuous improvements. The key is that you must watch the SLAs daily, weekly, monthly, and quarterly to assess what the performance is and then put corrective action plans in place when needed.

SM: Can you share an example of a specific operational challenge you faced and how you drove a transformation that resulted in a tangible improvement in service delivery?

RR: Utilization of On-Demand Field Labor for Installs, Moves, Adds, and Changes (IMAC) / Deployments. For Industry Standard skills, we found that the on-demand marketplace for field labor (i.e. Field Nation) provided the right skills at the right price and at the right timeframe. By utilizing On-Demand Field Labor, we were able to improve the gross margin, as well improving the Net Promotor (NPS) / Customer Satisfaction (CSAT) score. When you utilize On-Demand Field Labor, you can get an independent contractor and effectively pay-as-you-go for the short-term IMAC/Deployment work.

SM: How do you balance the need for global standards with the local and regional needs of your field services teams?

RR: Global Consistency with common processes leads to greater efficiency and reduced cost. However, there are some local and regional requirements that are necessary to ensure customer needs are met. We tried to leverage global processes whenever possible but had to have some local and regional customizations just to those customer-specific needs were met. The goal should always be to drive consistency first, then customize as needed.

SM: A standout theme in your career is your ability to link operational excellence to driving revenue growth and expansion. You co-led the pursuit of a 5-year, $40M Managed Print Services contract and a $64M Device-as-a-Service contract at HP. Can you describe your role in collaborating with the sales team on these major wins?

RR: The key to managed services is to understand the needs of the customer and to become the IT provider of choice by acting in a way that the customer considers you a part of their staff. My role was to describe how my team would provide these "as-a-service" deliverables in detail to build customer confidence in the solution we were proposing. Use cases and specifics on how we would provide these deliverables were the key things that helped us win these deals.

SM: In your experience, what is the most effective way for a service delivery organization to identify and capitalize on opportunities for revenue expansion?

RR: The "land and expand" methodology works best. First of all, make sure you deliver the SLAs and all Contractual Deliverables to spec and exceed them to build customer confidence (i.e. the "land" part). After you have proven yourself and become a trusted partner, you've earned trust and respect from the client, which opens the door to recommend incremental services to the client (i.e. the "expand" part). They key is to do what you say you're going to do, and exceed expectations first with your current deliverables.

ROD RINKER

Senior Director, Global Field Services, Digital Workplace, Unisys

Rod is a results-driven executive with extensive expertise in global and regional IT service delivery and customer success. He has a proven track record of leading large, complex organizations (over 4,000 employees) and consistently exceeding service level agreements (SLAs) and contractual commitments. At Unisys, he is responsible for the global field services operation, focusing on modernization and transformation. His career is marked by key roles at NCR and Hewlett Packard Enterprise, where he drove significant revenue growth, improved customer satisfaction (as evidenced by high Net Promoter Scores), and successfully managed multi-million-dollar services contracts. An expert in leadership, he emphasizes building a "trusted advisor" status with clients to foster loyalty and expand business. He holds an MBA from Georgia State University & B.A. in Computer Science - Anderson University. He is based out of Atlanta, in the United States

EXPERT VOICES

Subrata Mukherjee: As a leader at a major corporation, how do you navigate the complex ethical challenges that arise from the economic and societal impacts of new AI technologies?

Mouchka Heller: I am a religious person, and an observant Jew, so I understand ethics in an intrinsic manner that has never been defined by external variables such as technology. I believe that every individual has a responsibility to build their own moral compass and cannot rely on the world to do it for them. Much of technology is created to respond to customer demand, which means it is for humans to determine how they want to use AI, and for companies to follow. That makes the work of ethics a daily task that is incumbent upon all individuals.

SM: How can governments and the private sector collaborate to ensure AI is developed responsibly?

MH: The diversity of equities in AI adoption can be its own system of checks and balances, as long as we allow these different points of view to not only coexist but actively engage in dialogue. Democratic governments have a key role to play in fostering an equitable use, distribution, and development of AI. On the other hand, the private sector does not have the nationalist interests that some other entities may have. Inherently, private organizations align with the market, responding to customer demand in a highly dynamic manner. The collaboration of the public and private sectors can help think about customers as a class larger than those with deeper pockets, so we can catalyze opportunity while supporting the interests of the collective.

SM: In your role as Global Head of Economic Research at AWS, you focused on how technology influences economic outcomes. How do you measure the economic impact of a technology as new and disruptive as generative AI?

MH: There are several ways to do this, and they have all been employed by the various firms who have engaged in this effort. People often focus on the "what" of technological innovation. What is it creating? What is it challenging? I prefer to focus on the "how." How is it integrated by businesses and individuals? How hard is that integration? How is it discovered and adopted? How is it scaled? Asking how shows me which mechanisms are affected by the technology, making the policy and economic impact of that technology that much easier to track and quantify.

SM: What is the most significant economic challenge facing the world today, and how can data and AI help solve it?

MH: One economic challenge I often think about is income inequality. The gap between the rich and the poor leads to social unrest, political instability, crime, and economic inefficiencies. It's also devastating to witness, on a human level. The growth of technology has democratized access to opportunity for many. My children are growing up in a world where they can hold the Library of Alexandria in their palm. Of course, there are many hurdles that come with it, which I think about every day as a concerned parent, but there is also a historical opportunity there. AI can help a teenager code their own application to help the family business. Cloud-based internet services can connect a student to millions of free educational resources, including thousands from the world's best universities. We can communicate with family overseas in seconds, translate complex directions in a different language, get a map to nearly anywhere. With careful deployment, AI can help lift people out of poverty, democratizing one of the most valuable currencies in history: information.

MOUCHKA HELLER

Principal & Ex- Global Head, Economic Research, AWS Policy - Amazon

Mouchka is a translational economist and accomplished executive focused on the intersection of technology, policy, and economic development. In her role at Amazon, she is focused on the data center infrastructure and power distribution that enables AIML technology. Her career highlights include serving as Global Data Officer at The Economist and holding various leadership positions at the World Economic Forum, where she focused on integrated mobility systems and policy. As a former Trade Commissioner for the Consulate General of Canada, she has extensive experience in international relations and business development. Mouchka's background is distinguished by her ability to bridge complex policy, data, and business strategies to drive innovation. She holds a Masters in International Relations and Affairs at Georgetown University and a BA from Yeshiva University. She is based out of New York in the United States.

EXPERT VOICES

Subrata Mukherjee: What is the most common hurdle companies face when they are just starting to explore AI? Is it a technical or an organizational challenge?

Ellen Weinstein: The most significant hurdle is unequivocally an organizational challenge. While technical hurdles certainly exist—such as data fragmentation and a lack of data literacy—they are ultimately solvable with the right investment in infrastructure and talent. The real obstacle is securing executive buy-in and fostering a cultural willingness to embrace change across the entire organization. Many companies get stuck in "pilot purgatory" because they lack a clear roadmap for scaling their AI initiatives beyond a single proof of concept. Our work is to help them identify a "lighthouse" use case—a project with a high ROI and a strong chance of success—that can build the organizational momentum needed to secure long-term investment.

SM: You launched Spotdata, which was the industry's first electronic invoicing portal. What was the biggest challenge you faced in getting an entire industry to adopt a paperless workflow?

EW: The biggest challenge in pioneering Spotdata was not a technical one, but a cultural and behavioral one. We had to move an entire industry from a deeply ingrained, paper-based workflow to a digital one. This required a massive change management effort, which we addressed by focusing on education and value articulation. We didn't just tell the industry to adopt the technology; we demonstrated how it would directly solve their most pressing pain points—reconciliation discrepancies and payment delays. This shift from "we're changing" to "this will benefit you" was essential for mass adoption and building the trust needed for such a significant workflow change.

SM: How do you help a client determine where they are on the AI maturity curve, and what are the first steps they should take to move forward?

EW: We begin by conducting a strategic workshop to help clients understand their position on the AI maturity curve, which ranges from a foundational stage of data hygiene to an advanced stage of generative and agentic automation. By thoroughly assessing their current data infrastructure, technical capabilities, and business processes, we provide a clear benchmark. The first steps we recommend are not technical; they are fundamentally strategic. A client must first clearly define a high-impact business problem. Then, we work together to identify an initial high-value, low-risk use case that serves as a proving ground for their data governance, MLOps, and talent. This initial project helps to validate their approach and provides the foundation for future, more complex initiatives.

SM: Beyond buzzwords, what is the single most important metric you advise companies to track to measure the success of their AI initiatives?

EW: Beyond conventional metrics like cost savings or time reduction, the single most important metric is organizational velocity. This measures how quickly a company can move from identifying a business problem to deploying an AI solution that creates tangible value. It's a measure of the enterprise's ability to iterate and innovate at an accelerated pace. Success isn't just about a single use case's ROI; it's about building a repeatable capability to leverage data and AI as a competitive weapon. This velocity is a direct result of effective data governance, seamless data pipelines, and a culture of continuous deployment and learning. Ultimately, it's a reflection of the organization's ability to adapt and thrive in an increasingly AI-driven world.

ELLEN WEINSTEIN SALAMAY

Seasoned Business Development Leader, Ex-Marketo, Ex-Cognizant

Ellen is a seasoned business development and strategy professional with a proven record of helping companies leverage technology to achieve significant growth. Her current focus is on AI adoption and digital transformation, which she applies in multiple roles, including as an Advisor at Covington Place Partners and Sr. Consultant at Priorise. She also serves as Vice President of Business Development for Accounti, an AI-driven financial services firm, and InCommon®, which builds AI-capable teams. With a career spanning over two decades, she specialized in connecting tech startups with agencies and media companies. Her extensive background includes senior leadership positions at prominent firms such as Marketo, Cognizant Technology Solutions, and Xen.AI, where she consistently drove business growth, forged strategic partnerships, and optimized marketing and operational strategies across the advertising and media industries. She is based out of New Jersey, in United States.

Subrata Mukherjee: Your career includes significant time leading digital platforms for major media companies, including The Economist and the Telegraph. What specific lessons did you learn from building multi-platform content systems and paywalls that have been applicable to your work in other industries?

Mark Brincat: Building multi-platform content systems and paywalls taught me three key lessons. First, customer-centric design. At The Economist and The Telegraph, building paywalls taught me that customer experience is paramount. Seamless access across web, mobile, and apps required anticipating and removing friction points early. This experience taught me that enterprise solutions must also prioritize the user, whether that's an employee, a candidate, or a manager. Second, data-driven iteration. Paywalls were only successful when continuously tested and optimized. This reinforced the lesson that systems should be designed with analytics and feedback loops embedded from the start. Third, modular architecture. Multi-platform content systems required flexible architectures that could scale across different geographies and devices. This modular, API-first approach has been directly transferable to other industries where integration with external platforms is critical.

SM: How did your experience with customer-facing products in media inform your approach to building robust enterprise solutions for clients?

MB: Media platforms serve millions of users daily, so performance and security are non-negotiable. I've carried that mindset into my work with enterprise solutions, where platforms that serve employees or candidates must meet equally high standards. The challenge of personalization at scale is also similar; content recommendation in media parallels talent personalization in HR technology—both require relevance, speed, and security. This experience helped me design enterprise systems that adapt dynamically to user context. Finally, commercial alignment is crucial; just as media paywalls needed to balance editorial integrity with revenue goals, enterprise solutions must balance client business objectives with technical feasibility and user satisfaction.

SM: How do you balance the need for speed and innovation with the requirement for robust governance and security in a large-scale project?

MB: I balance these needs with a model I call "agile with guardrails." Large-scale projects succeed when innovation is encouraged within a framework of governance. At scale, this means adopting agile delivery but embedding security, compliance, and architecture reviews at each stage. This is security by design. Lessons from customer-facing systems—where breaches can erode trust instantly—reinforce the importance of embedding InfoSec and data protection into the development lifecycle rather than treating them as afterthoughts. Finally, stakeholder transparency is vital. Maintaining speed while meeting governance demands comes down to communication: clear checkpoints, visible risks, and shared ownership ensure that innovation is not slowed unnecessarily but also doesn't create unacceptable downstream liabilities.

SM: What is the biggest mistake you see companies make when trying to forge new technology partnerships?

MB: The biggest mistake I see companies make is treating partnerships as procurement, not strategy. Too often, companies chase a partnership because the technology looks impressive, but they haven't defined why it matters for their business or their clients. This leads to focusing on the press release instead of the customer outcome. The pitfall is neglecting integration and enablement. Even strong partnerships falter if sales teams don't understand the joint value proposition, or if the client experience feels fragmented.

MARK BRINCAT

Ex-CTO @ SHL, Economist | Architect & BBC, Telegraph, Home Office

Mark Brincat is a technology executive with deep expertise in digital transformation, AI, and data analytics. In his current role as the Global Head of Presales and Technology Partnerships at SHL, he leads go-to-market strategy and advises clients on their transformation journeys. Concurrently, he serves as a Fractional CTO for Castille Resourcing, where he oversees the engineering and scalability of their digital platform. He previously held key Chief Technology Officer (CTO) roles at both SHL and Intive, and senior architectural positions at major media organizations like The Economist, Telegraph Media Group, and BBC Worldwide. His career highlights include implementing large-scale digital and cloud solutions and forging strategic partnerships. He is based out of London, United Kingdom.

EXPERT VOICES

Subrata Mukherjee: Your career has taken you from oil and gas to enterprise software. What transferable lessons did you learn from applying AI to real-world industry problems?

Brian Glasman: From my early career in oil and gas, I developed a deep appreciation for long-term reliability, rigorous validation, and strategic planning. That industry operates under extreme regulatory and safety constraints, where software must perform flawlessly for decades and every decision is backed by extensive cross-checking and high-cost experimentation. This instilled a mindset of precision and accountability that has proven invaluable in other domains. The emphasis on thorough planning before execution created a level of strategic discipline that shaped how I evaluate risk, scope solution architectures, and assess quality across the entire development lifecycle. When I transitioned from that to enterprise software, it was very clear that they adopted an agile mindset where, as the software evolved, the customer would drive the application of it towards areas where the value is very clear. Now what AI is providing is a bridge to iterate faster, provide higher quality, and more features that can be explored and tested more quickly, allowing for improved flexibility and responsiveness in the entire process of making software products. This is highly different from the oil and gas industry.

SM: How does the product discovery process differ when building an AI-powered solution for a specific industry versus a more general enterprise application?

BG: Building a general enterprise AI application is exponentially more difficult than creating a solution tailored to a specific industry. The core challenge lies in the diversity of customer needs. This complexity slows execution and demands a much larger investment in user research, iteration, and cross-segment testing. By contrast, industry-specific AI applications offer clarity and focus. Engaging just a few representative customers is often enough to shape a tightly scoped product with clearly defined requirements. These solutions typically align closely with established workflows, making adoption more straightforward. However, this focus comes at a cost. Niche products naturally limit total market potential and can carry concentrated adoption risk if the chosen segment proves too narrow. On the other hand, general-purpose solutions allow for broad experimentation and may uncover demand in unexpected sectors. Still, in practice, choosing the right niche often proves to be the safer and more effective strategy for reaching product-market fit.

SM: You presented at CES 2022. During that presentation, you provided a highly technical AI solution. How do you make that easy to understand for a broad audience?

BG: Effectively communicating the value of AI starts with acknowledging its perceived complexity, especially in its early stages. To make it accessible to a broad audience, I focus on concrete use cases rooted in real business scenarios. These use cases involve familiar roles and problems, allowing people to connect immediately with the context. When AI is positioned as a natural extension of existing workflows, its utility becomes clear. The goal is to demonstrate how it enhances decision-making with speed, relevance, and clarity, particularly for executive audiences. For example, during a CES 2022 presentation, I described a situation where a CEO walks into the VP of Product's office seeking insight into a competitor's strategic shift. With the CTO and VP of Product Management present, they turn to the AI system, which has been aggregating relevant data on the topic. In less than a minute, the AI produces a concise, executive-level summary with supporting data.

DR. BRIAN SCOTT GLASMAN

Chief Product Officer & Head of Client Engagements at AInspire.ai

Dr. Glassman is a seasoned product management executive, engineer, and product strategist specializing in generative AI. As the co-founder and Chief Product Officer at AInspire.ai, he leads client engagements and product strategies, collaborating with C-level executives to develop and deploy AI solutions for large enterprises. With a career dedicated to leading technology innovation, he previously served as Director of Product Management at Electra Vehicles, where he built the product function for AI-powered solutions in the automotive sector. His extensive experience includes a seven-year tenure at Rave Digital, where he led product management and customer engagements for Fortune 1000 clients. Furthermore, his background as an engineer and product leader at Cameron, a Schlumberger company, highlights a unique expertise in technical product development. He is based in New York.

EXPERT VOICES

Subrata Mukherjee: : Your long tenure at VGD means you've successfully led the company through multiple major technology shifts over the last two decades. Can you walk us through the most significant technological challenge you've faced during your 18 years at VGD, & how you approached it?

Preeti Sharma : One of the most significant challenges was guiding the shift from traditional on-premise servers to a cloud-first strategy. It wasn't just a technical migration—it was a mindset change. For years, infrastructure was something we could see, touch, and tightly control within our own walls. Moving to the cloud required rethinking security models, retraining teams, and helping both colleagues and clients trust that resilience and compliance could actually be strengthened, not weakened, by the transition. My approach was to balance pragmatism with vision: we piloted in areas where the benefits were clear, demonstrated early wins, and used those successes to build confidence and momentum. Ultimately, the move to cloud-first enabled us to deliver greater scalability, flexibility, and innovation for our clients. I saw SaaS where others saw development projects.

SM: In what ways has the client demand for "digital services and solutions" evolved since you began?

PS: Eighteen years ago, clients saw digital as an optional add-on. Today, it's the expectation—and increasingly, it's about integration, personalization, and intelligence. I've always been evangelical about user experience and usability, and it's rewarding to see clients now fully embracing that perspective. Adoption is one of the key OKRs I focus on, because the best technology only matters if people actually use it. The shift has been from "Can you do this digitally?" or "My boss is forcing this on us" to "How can you make this smarter, faster, more meaningful—and easier for users to adopt?

SM: What strategies do you use to stay ahead of the curve and ensure VGD remains at the forefront of digital innovation?

PS: I rely on a mix of continuous learning, strong partnerships, and staying close to our clients. Listening carefully to their evolving needs is often the earliest signal of where the market is going. Pair that with horizon-scanning—whether through certifications, conferences, or exploring emerging technologies—and you can chart a path that is both innovative and pragmatic. This lets us test, learn, and refine before scaling broadly—so clients benefit from cutting-edge solutions that are also proven, usable, and sustainable. I also weave in what I know about the client, so that my recommendations fit their organization while giving them a vision for the future.

SM: You work with many public sector clients, and their digital transformation often moves at a different pace than the private sector. What are some of the unique challenges and opportunities you've encountered when helping public sector organizations adopt new technologies like AI and automation?

PS: The public sector has an incredible opportunity to impact people's daily lives at scale. One of the first challenges we faced was fear—AI can feel like a mysterious robot making decisions or locking you out of your own house. In reality, AI has been with us for decades in the form of algorithms and machine learning, and what's changing now is its accessibility and power. Another challenge is that many government organizations are still developing their acceptable use policies for AI, which means we often have to wait before fully deploying solutions. That said, the rigor around governance—security, compliance, and policy—ultimately makes public sector solutions stronger and more trusted once they're delivered.

PREETI SHARMA

Chief Digital Officer & CIO at VGD NYC

Preeti is a distinguished technology executive and a "Notable Woman in Tech" with a proven history of driving digital strategy and innovation. As the Chief Digital Officer and CIO at VGD NYC for over 18 years, she leads the delivery of digital services, technology strategy, and solutions for a diverse range of clients. Preeti's expertise is deeply rooted in her long-standing focus on AI and technology transformation, backed by recent certifications in AI Leadership and Digital Innovation from Carnegie Mellon University. Her foundational experience includes nearly a decade at ISI Emerging Markets, where she served as Global Operations Manager, playing a key role in scaling the business from a small startup into a profitable enterprise. A member of Chief and the Forbes Technology Council, she has been recognized with multiple AVA Digital Awards for her work. She is based out of New York.

EXPERT VOICES

Subrata Mukherjee: You have experience with a wide range of technologies, from Airflow to Kafka and Spark. Can you explain how these tools fit together to form a cohesive, scalable data architecture?

Manish Bhandari: Each tool addresses a different layer of the problem. Kafka and Flink handle real-time ingestion and messaging. Spark and PySpark are used for large-scale batch transformations, building feature stores and curated datasets. Airflow orchestrates everything, ensuring pipelines run in the correct order with reliability and SLA tracking. Snowflake, Redshift, or S3 serve as the storage and warehousing layer, providing a single source of truth for analytics. Finally, Kubernetes and Terraform manage the underlying infrastructure, ensuring everything is portable and scalable. This multi-layered architecture enabled us at Tinder to process billions of daily events, with streaming for real-time features and batch for analytics, all in a reproducible cloud-native environment.

SM: What's a common mistake you've seen engineers make when designing a data pipeline, and how do you mentor junior colleagues to avoid it?

MB: A common mistake is overengineering before understanding the "why." Junior engineers often focus on choosing the technology without first clarifying the business problem the pipeline is meant to solve. To mentor, I encourage them to: 1) Start with the 'why' by clarifying the business decision before choosing a technology; 2) Design for reliability and evolution by adding monitoring, schema validation, and modularity upfront; and 3) Prioritize documentation for features and metrics to make systems discoverable and reusable.

SM: How has the role of a data professional evolved over your career?

MB: The role has evolved from a custodian of data to a business impact multiplier. When I started as a DBA, my value was measured by stability and correctness—making sure databases were healthy and online. The role then shifted to enablement, where we built data platforms to deliver analytics and support machine learning at scale. Now, with the rise of AI, the role has expanded again. Today, data professionals are AI enablers, making unstructured data usable and building pipelines that power AI agents. We've moved from back-office operations to front-line strategy and innovation.

SM: What core skills from your time as a DBA are still most relevant to your work today in data engineering and advanced analytics?

MB: Three core skills from my time as a DBA remain incredibly relevant today. First is data integrity—if your underlying data isn't correct, the entire system, from analytics to AI models, will fail. Second is performance tuning—the mindset of optimizing for efficiency at scale is directly transferable, whether you're working with SQL or a Spark job. Third is governance and access control—the principles of managing who has access to what data are the same, just applied at a much larger scale across distributed cloud systems and for compliance needs. The tools have changed, but the discipline and rigor from those DBA days still guide my approach.

SM: For someone just starting their career in data, would you recommend they focus on a specific area first, like databases, or take a broader, more modern approach?

MB: My advice is to go deep first, then broad. Pick one area—like SQL, Python, or a specific database—and truly master it. That depth provides the rigor and confidence needed to build a solid foundation. Once you have that, then begin to broaden your knowledge by learning how data moves through pipelines, how orchestration works, and how cloud architectures fit together. If you only learn tools at a surface level, you risk becoming a "checkbox engineer." Having a strong, foundational pillar of expertise will make you resilient and adaptable as technologies inevitably change.

MANISH BHANDARI

Ex-Tinder, Ex- Wayfair, Currently building an AI-First Data platform

Manish Bhandari is a seasoned professional with a strong focus on building and maintaining analytics infrastructure and architecting data solutions. He is currently in a new venture at a Stealth company, leveraging his extensive experience from leadership roles in the tech and data sectors. Previously, he served in data infrastructure roles at companies such as Tinder and Wayfair. As a Senior Data Engineer at QuantumBlack, AI by McKinsey, he was instrumental in building impactful analytics solutions for C-level executives and designing data pipelines to support AI and data science projects. His experience is built on a foundation of managing corporate data infrastructure at Genscape. Manish's core expertise lies in Data Engineering, Advanced Analytics, and ETL, with proficiency in technologies including Python, Databricks, and AWS. He is based out of San Francisco, United States.

EXPERT VOICES

Subrata Mukherjee: With your NACD Cyber Security Certification, you're well-versed in the challenges of modern governance. In a world of increasing cyber threats, what is the most important responsibility of a board of directors regarding cybersecurity?

Michael Gabriel: While cybersecurity review often gets assigned to the Audit & Risk Committee, the Board is still ultimately responsible for this oversight and governance and needs to be engaged and not just rubber-stamp Committee oversight. The Committee should work with the CISO to remain vigilant outside of Board meetings. The Board of Directors needs to ensure that the oversight of cybersecurity is not abdicated to being just a technology responsibility, and is recognized as a business responsibility. The risks to the organization that needs to be protected encompass areas such as regulatory requirements, restricted and sensitive information, operational environments, AI usage, access and identity management, third party products and services, application integration, insider threats, and business continuity needs including handling ransomware demands. The Board needs to ensure status and compliance is conveyed in understandable business terms. The framework needs to show what needs to be protected, how well it's being protected, and what needs to be done, if anything, to improve the cybersecurity protection status. It also needs to ensure that incidents that pass, or might pass, a threshold of impact (such as regulatory, or above a cost or exposure magnitude) are brought to the Board's attention. The incident impact, or potential impact, needs to be conveyed clearly just like any other business impact, and the status of such items need to be tracked and reported to the Board until resolved. Remedial actions need to be addressed to avoid future recurrence to mitigate ongoing risk, and need to be included in Board status updates until they're resolved.

SM: Beyond having a dedicated CISO, what does a strong top down cybersecurity governance framework look like?

MG: A framework provides standards, guidelines, and best practices for managing cybersecurity risk, and helps identify, assess, and manage cybersecurity risks, in a structured and repeatable manner. This framework should be created in consultation with senior business executives, including the CEO, CFO, CHRO, General Counsel, Chief Physical Security Officer, Division Heads and Country area heads (such as the U.S., Latin America, EMEA, APAC). I have always included both internal and external audit to reduce the likelihood that there are conflicts in scope or perspective as to how risk status is conveyed to management and the Board. Many frameworks exist, such as but not limited to NIST, ISO, PCI and SOC, and the it should be conveyed to the Board in a business perspective. The framework should be customized as necessary for the company structure, business operational and continuity requirements, regulatory requirements, and business partner, client, and consumer information protection needs. Governance dashboards should ideally include a basis of comparison to industry peers since different industries have different cybersecurity requirements. I also recommend it should show impact by geographic regions and lines of business. Status should clearly highlight any gaps to desired future goals, comparisons to prior quarters and annual trends. The status of any key projects required to improve the security posture should be included with timeframes, benefits and business risks, and initial and ongoing costs. Such projects should support achieving the agreed upon security posture and risk tolerance that has been approved by senior management and the Board. The cost of the cybersecurity program needs to be clear, the risks clearly articulated, and the impact of incidents tracked, and root causes addressed from a risk/benefit perspective.

MICHAEL GABRIEL

Ex CIO at HBO, NBA | NACD Certified Board Director

Michael is a senior technology and cybersecurity executive with experience providing C-level leadership, governance, and advisory services. He is currently an independent Board Director and provides professional services for companies like Warner Brothers Discovery (WBD). His career is highlighted by a 14-year tenure as EVP & Global CIO at HBO, where he was the initial executive to spearhead the groundbreaking HBO GO initiative and led a company-wide digital transformation. An inventor on six technology patents, he also played a key role in information security governance across the Time Warner enterprise. Prior to HBO, he served as Global CIO for the National Basketball Association (NBA), where he implemented the first NBA retail store technology system. He also held senior IT leadership positions at EMI Records Group and NBC Olympics.. He is based out of New York

EXPERT VOICES

Subrata Mukherjee: You've advised InfoSec Global, a leader in quantum-ready cybersecurity. What is the most critical cybersecurity threat that a CIO and CISO should be preparing for today, beyond the standard risks?

Richard Entrup: I believe supply chain attacks leveraging AI-driven automation as the most critical cybersecurity threat beyond standard risks like ransomware or phishing. These attacks exploit vulnerabilities in third-party vendors, cloud services, or open-source software—think SolarWinds or Log4j—but are now supercharged by generative AI, enabling adversaries to craft hyper-targeted exploits, manipulate code repositories, or automate reconnaissance at scale. For example, AI can generate polymorphic malware that evades detection or mimic legitimate user behavior to infiltrate SaaS platforms. The threat is amplified by the increasing complexity of supply chains, with 60% of organizations reporting third-party breaches in recent surveys. To prepare, CIOs and CISOs must collaborate on vendor risk management (e.g., enforcing zero-trust architecture across suppliers), AI-augmented threat detection (using behavioral analytics to spot anomalies), and proactive code auditing for open-source dependencies. Regular tabletop exercises simulating supply chain compromises, combined with mandating SBOMs (Software Bill of Materials) from vendors, can reduce exposure by 30-40% based on industry benchmarks. This threat demands a shift from reactive defenses to preemptive, ecosystem-wide resilience.

SM: How do you convince an enterprise to invest in a security threat that may be years away from being a widespread reality?

RE: Convincing an enterprise to invest in a future-focused security threat like post-quantum cryptography (PQC), which may be 5-10 years from widespread impact, requires framing it as a strategic business imperative rather than a distant hypothetical.

First, I'd present a risk-based business case grounded in data: highlight the "harvest now, decrypt later" threat, where adversaries collect encrypted data today for future quantum decryption, potentially exposing intellectual property or customer data worth millions (e.g., a single breach could cost $4.45M based on 2023 IBM data). Emphasize industries like finance or healthcare, where data longevity amplifies risk.

Second, I'd align PQC preparation with existing priorities: show how adopting NIST's quantum-resistant algorithms integrates with routine crypto upgrades (e.g., TLS modernization), leveraging current budgets to minimize incremental costs—potentially saving 30% by avoiding rushed retrofits later.

Third, I'd underscore competitive and regulatory advantages: early adoption positions the enterprise as a leader, builds customer trust, and aligns with emerging standards (e.g., NIST's 2024 PQC guidelines or EU mandates). I'd propose a phased approach—starting with a $50K-100K pilot to test PQC in non-critical systems like internal file encryption—demonstrating measurable progress (e.g., 10% of systems quantum-ready in 18 months).

Finally, I'd use storytelling: share examples like the Y2K bug, where proactive investment prevented disaster, and contrast with reactive failures like Equifax's $1B+ breach. By tying PQC to business outcomes, leveraging partnerships (e.g., with vendors offering PQC-ready solutions), and showing a clear ROI path, I'd secure buy-in from the C-suite and board, ensuring long-term resilience without sacrificing short-term priorities.

RICHARD ENTRUP

Managing Director, Head of Emerging Solutions at KPMG US

A multi-award-winning global technology and innovation leader with nearly three decades of experience driving strategic business transformation. He is at the forefront of identifying and incubating disruptive technologies like GenAI, Quantum Computing, and Spatial/XR, pioneering new products and solutions for clients. A two-time recipient of the HMG Global Leadership Institute Award for Exceptional Business Transformation, his distinguished background includes C-suite leadership roles (CIO, CTO, CDO) at globally recognized brands such as Verizon, Disney/ABC, Christie's, Time Warner, Viacom, and Tiffany & Company. He also serves as a strategic advisor and investor to a portfolio of innovative startups and venture capital firms, passionately guiding ventures focused on cutting-edge innovations. He is based out of New York, United States.

EXPERT VOICES

Subrata Mukherjee: You founded Occumoda in 2016 based on the insight that "the traditional services model is broken." What was the core problem you saw in the traditional single-bench model, and how does your A-player-only, multi-vendor approach directly address it?

Jeff Rishi: Traditional vendors/companies incorporate then build out teams of resources that compliment the core competencies that make up the services portfolio of the vendor. This is a traditional approach that "works" for the organizations that build them this way. It works because the top-level resources are purported to the best, largest or most lucrative clients in the vendors portfolio. As a long-time seller, then VP then EVP of sales before becoming a CEO to "change the consulting paradigm" I was charged with revenue growth, net new logos/new geographies. As I bring in a new client/logo as we say, I was always told; One, we will put the best we have on the bench for this new client and second; if we are missing someone, we will go to the market to find the needed. So, the more business I closed, the lower skilled the "best" team was that was available, and the more "untested" resources were added to the client teams. While this is a paradox of sorts, I found it to be a common and consistent approach to software development delivery. The net effect was that the non-marquis clients/new clients required a lot of patience for the delivery team to get the right team and it took a lot more effort from internal management to create the right mix of skills. Occumoda was built/designed to leverage the no-employee model of business. A drop-ship orientation if you will. While the above information points to IT services and software delivery, it applies across the board within any organization. If you hire a VP of Operations, this person is vetted and put to work. This VP is charged and does their best. What about the other 10, 20, 30 VPs that can shed new light on a problem, provide insight that the employed VP cannot or does not know. When we think about business momentum and the future or an organization, we see that a non-employee/consultant can be leveraged today and as citations change, change the skills and experience set of an employee in that role. Thereby eliminating the very costly need to replace a company asset.

SM: You often use the phrase "the Art of the Possible." Can you give us an example of how you've used this philosophy to help a client with a complex, advanced technology?

JR: Perhaps one of my favorite sayings in my career, "anything is possible" is now more true than ever. Technology is uncapped, and arriving at unprecedented speed. "As a service" becomes a reality as fast as the build to suit or bespoke approach. With this speed and access worldwide, the "Art of the Possible" is a reality. Think it/dream it/envision it and it will be actualized. Giving great power to CIOs and CTOs everywhere. If it can be envisioned, it can be brought to life.

SM: What were the biggest challenges you faced in building a new kind of systems integrator from the ground up?

JR: The challenge is conveying the model and having the members understand how it works. It is not a typical nor sometimes logical (depending on the entity) model to participate in. Typically the entire team is proposed from a supplier not just a single lead or expert. There is little to no critical issue on the client side. Most contract with Occumoda on a finished product contract so there is little to no client management for these. If there happens to be a more team or staff orientation request, the clients understand that the model may be dispersed but the level of skill is much higher and the production is accelerated.

JEFF RISHI

Founder/GM at Occumoda, Ex- Cignex, IBM

Jeff Rishi is a visionary business and technology leader with a career defined by driving significant revenue and commercial growth, having generated over $1.23 billion in software and services revenue. As the Founder and General Manager of Occumoda, he champions a contrarian, multi-vendor approach to AI/ML and software development, leveraging a curated collective of elite talent to activate client visions. A proven leader in sales and go-to-market strategy, Jeff has held senior roles at companies such as CIGNEX Datamatics and IBM, where he was a key evangelist for open source and Big Data solutions. His expertise lies in transforming complex, bleeding-edge technologies into tangible business outcomes while scaling global operations and providing fractional CRO services to startups and scale-ups.. He is based out of United States.

EXPERT VOICES

Subrata Mukherjee: What is the key difference in mindset between leading a business that is a cost center versus one that is a revenue driver?

Karen A. Brophy : A cost center mindset is about efficiency, minimizing expense while supporting the core business. A revenue driver mindset is about growth and accountability: every initiative must show measurable return. I shift the focus from "how do we support?" to "how do we monetize and expand?" That means treating product, editorial, and technology not as service lines but as integrated engines of the P&L.

SM: How do you balance programmatic revenue and direct sales?

KAB: I align them around yield and customer outcomes. Programmatic offers scale and liquidity, while direct sales bring premium rates and deeper relationships. The key is transparency: one revenue scorecard, rules of engagement between teams, and clear pricing strategies. We packaged programmatic inventory to complement direct, not cannibalize it, which maximized overall profitability.

SM: What are the biggest operational challenges in harmonizing disparate assets across 200 stations?

KAB: There are three main challenges in the hub and spoke model when trying to modernize and standardize across so many assets. The biggest is probably the cultural resistance, which is understandable and even to some degree desirable. Local teams value autonomy and want to own their successes. Local leadership, General Managers or Publishers, are held accountable for their performance so they want control of the levers. The best approach is building relationships while leading them to see that their P&L will benefit. Second is technical debt. Additionally, the legacy technology systems can differ per market, so custom APIs are required and need to be maintained. So, while manageable, it requires more resources to support. Third, operational scale across hundreds of sites is very complex. My solution was phased migration with shared infrastructure but local customization, plus constant and consistent communication on how centralization improved speed, reach, and monetization.

SM: What is the most significant technological challenge ahead?

KAB: AI is clearly the most significant technological challenge facing companies and especially media companies where advanced technology is not their core competency. There is a lot to figure out about AI and its application in media businesses as well as how to build resilience into the business on its impact. I don't know that media companies have an answer for the future where an LLM is the browser and eventually web pages become increasingly irrelevant. Prior to this AI push, the erosion of third-party cookies was the biggest issue, and it still is an issue. Programmatic revenue which is a big driver for a lot of companies will struggle with out that first party data. So there is impending revenue impacts for those media companies who don't have a first party data strategy and infrastructure in place.

SM: What advice would you give young professionals in media product?

KAB: Your role is as a diplomat so start there. Some product roles are strategic, and some are all about execution. To be strategic, listen and build relationships. I like to say everyone has good ideas, but it is the product managers job to recognize the good ones and set the priorities. Learn the KPIs relevant to your industry and be fluent in data and AI. Build a portfolio of shipped projects, even small ones. Build your network and maintain it, do not get so busy you are not staying connected as the job landscape is challenging right now. And keep learning new skills, stay on top of what is out there, take classes, webinars, vibe code, stay in front as that is the best thing you can do in today's market.

KAREN A. BROPHY

General Manager, President, COO/CPO | Ex Hearst, Yahoo!

Karen is a dynamic executive specializing in P&L Management, Digital Transformation, and Revenue-Driving AI and technology business strategy for national media brands and consumer platforms. As the founder and President, Digital Division at Nexstar Media Group, she oversaw all digital business, product, and monetization strategies for over 200 local stations, The CW, and The Hill, delivering the highest digital revenue in company history with 70% growth. She is an expert in launching scalable platforms, notably developing the AI-powered advertising platform, Stellar, and spearheading the rollout of OTT Video and FAST channels. Karen held senior Product Leadership roles at major companies like Hearst and Yahoo!, in addition to her current advisory work as a Fractional COO/CXO, focusing on operational excellence and Artificial Intelligence. She is from Los Angeles, California, USA.

WE NEED YOU!

WELCOMING OUR NEXT SET OF THOUGHT LEADERS

The remarkable success of the Ai6x6 Organization Transformation Series is built upon the **hard-won wisdom** of executive leaders who have successfully navigated the AI transformation within their industries, turning theory into a deeply practical, actionable guide. As we expand the series, we are extending a highly selective invitation to a new cohort of pioneers—individuals whose professional journey is at the **leading edge of AI strategy and execution**. We are looking for executives who possess a proven track record of rewriting business models, mastering unforeseen ethical and competitive challenges, and offering a **clear blueprint** for the next 3-5 years of the AI-Era.

Contributing to the Ai6x6 Series is a unique opportunity to **cement your professional legacy**, joining an exclusive, curated network of global authorities whose insights are distilled into executive-level knowledge cards, directly shaping the governance and innovation decisions of the world's most influential companies. Our selection process is based strictly on **pure expertise and demonstrated leadership** in the AI-Era, reinforcing our commitment to editorial integrity and authentic thought leadership—there is no monetary exchange involved. Select contributors will also receive an invitation to blog on our expert network microsite, further amplifying their platform. To confidentially inquire about contributing your expertise, please contact us via our dedicated portal at **www.Ai6x6.com/contact** or by scanning the adjacent QR code.

Contact

SUBSCRIBE TO OUR NEWSLETTERS: To stay ahead of the curve and receive continuous, executive-level intelligence on the rapid evolution of the AI-Era, we invite you to register for our **Ai6x6 Weekly Newsletter**. Each edition delivers curated insights directly from our thought leaders and highlights critical shifts in AI strategy, governance, and provides relevant excepts from our website and books. Keep your finger on the pulse of the future and ensure your company remains competitive by signing up on the same page

PLEASE REVIEW US ON AMAZON!

Operations Blueprint

We know your time is valuable, but if the Ai6x6 Organization Transformation Series delivered the real-world, actionable strategies you needed, please consider leaving a brief review on Amazon. If we missed the mark, we would also like to know, and that is the only we we can improve. Your honest feedback is crucial; it helps fellow executives discover this unique framework and benefit from the insights of our thought leaders. Thank you for supporting our mission!

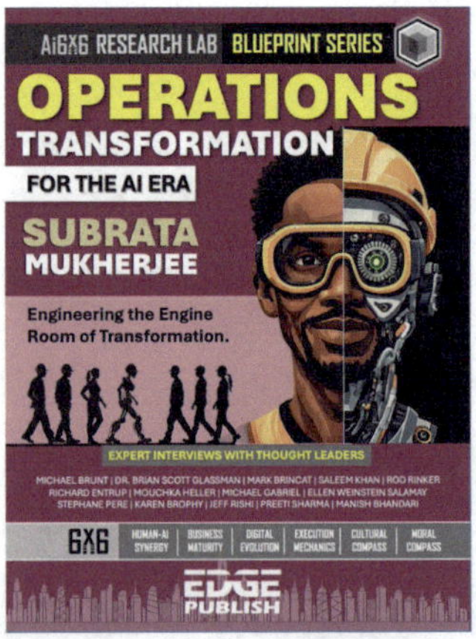

BLUEPRINT SERIES

SECTION B

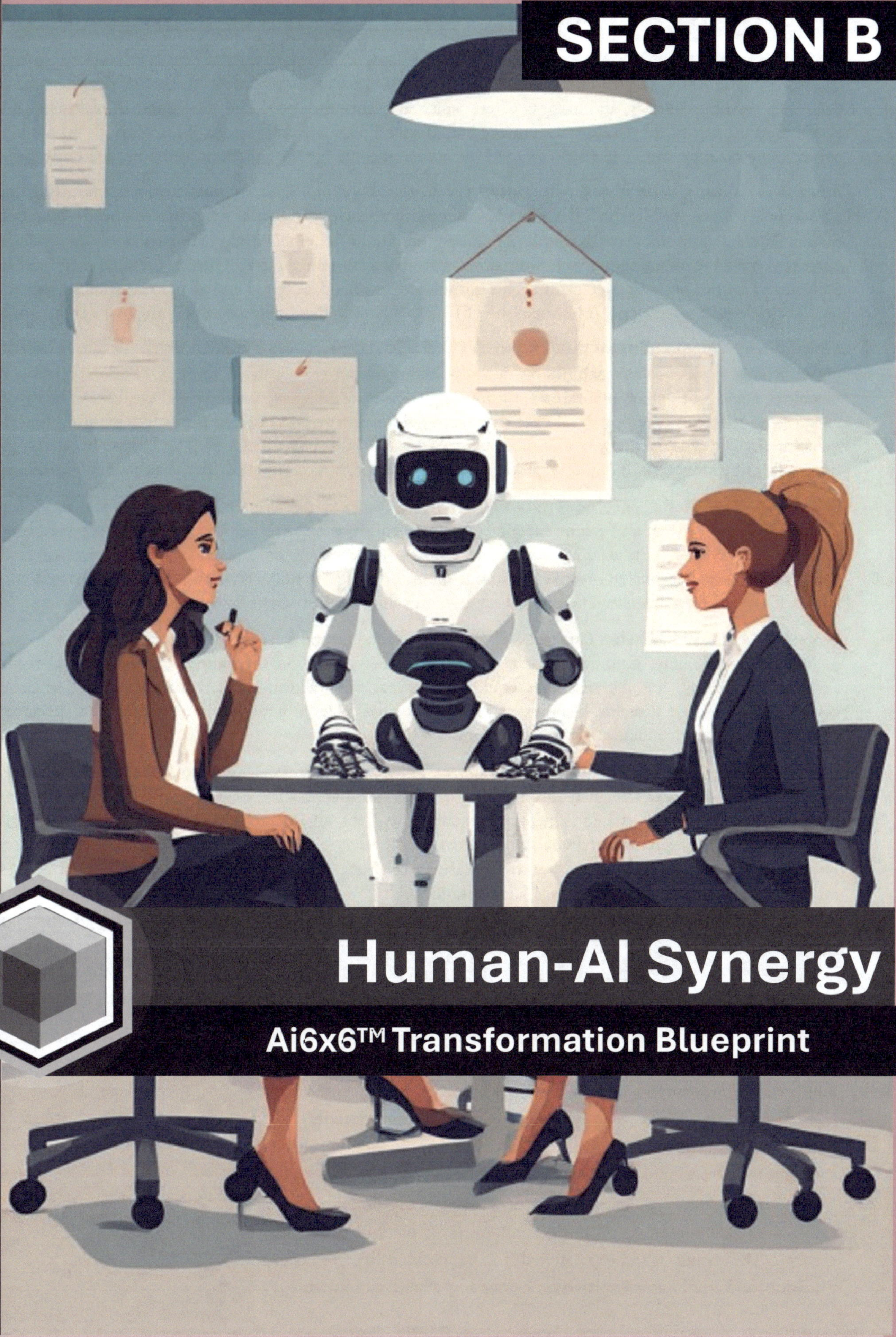

Human-AI Synergy

Ai6x6™ Transformation Blueprint

Story 1 — Enigma Codebreakers (1940–45): German U-boats encrypted communications using constantly shifting ciphers, forcing Allied cryptanalysts to chase patterns that dissolved daily. Manual decoding could not keep pace with wartime urgency, and insights often vanished before they could be confirmed. Electromechanical devices accelerated the testing of solutions at unprecedented speed. Humans interpreted these outputs, correcting false leads and refining assumptions. The fusion of machine acceleration & human judgment reshaped the trajectory of the war.

Story 2 — Turing's Bombe & Operators (1940–45): Breaking enemy codes required evaluating thousands of cipher possibilities rapidly, and the Bombe machine generated candidate solutions at speeds humans could not match. Yet its outputs were incomplete, ambiguous, and often misleading. Human operators validated results, corrected machine assumptions, and filtered noise from meaningful signals. Their contextual judgment transformed mechanical output into strategic clarity. This interplay of mechanical speed and human reasoning cracked codes that neither side could solve alone. It became one of history's earliest demonstrations of human–machine symbiosis.

Story 3 — Toyota Production System (1950s–70s): Factories struggled with defects, bottlenecks, and unpredictable workloads as machines produced volume without contextual awareness. Workers observed issues that machines could not detect or interpret, prompting Toyota to design a synchronized workflow that aligned human insight with machine precision. Real-time data from the line informed continuous improvement cycles. This human–machine rhythm created a new operational standard for reliability and efficiency. The system became a global model for intelligent production. It demonstrated how augmentation, not automation, drives sustainable excellence.

Story 4 — Salk Polio Vaccine Trials (1952–55): As polio outbreaks intensified, statistical models guided trial design but lacked the ethical nuance required for medical decision-making. Researchers interpreted model outputs through clinical judgment, ensuring safety and fairness while machines processed vast amounts of trial data. This combined system accelerated discovery without compromising integrity. Human oversight balanced computational rigor with moral responsibility. It showed how human values anchor scientific acceleration.

Story 5 — Apollo 13 Mission Control (1970): An explosion crippled Apollo 13, leaving astronauts in mortal danger as incomplete telemetry streamed back to Earth. Engineers relied on simulations to test survival scenarios under extreme constraints, but the machines could only reveal possibilities. Human teams improvised procedures the computers could not imagine, blending machine-generated insights with creative reasoning. It became a defining moment of human–machine crisis collaboration.

Story 6 — Air Traffic Control Modernization (1960s–80s): Surging air travel overwhelmed manual tracking systems, and controllers struggled to maintain safe separation across increasingly crowded skies. Radar automation provided real-time positional data, enabling machines to track aircraft continuously while humans made judgment calls. This hybrid system reduced collisions, increased capacity, and established a global standard for aviation safety. Human–machine coordination became the backbone of modern air travel. The system remains one of the world's most enduring examples of operational augmentation.

LESSONS FROM AN OPERATIONS TRANSFORMATION PERSPECTIVE

Operations evolve from manual workflows supported by mechanical acceleration to intelligent ecosystems integrating automation, human judgment, and real-time data. Early systems relied on humans to interpret machine outputs under pressure, while later systems embedded continuous sensing and adaptive coordination. Across decades, the operational pattern remains consistent: machines handle volume, speed, and pattern detection; humans manage ambiguity, escalation, and ethical boundaries. Human–AI synergy produces operations that are both stable and agile—capable of responding to volatility without losing control. The result is an operational model defined by resilience, precision, and continuous improvement driven by shared cognition.

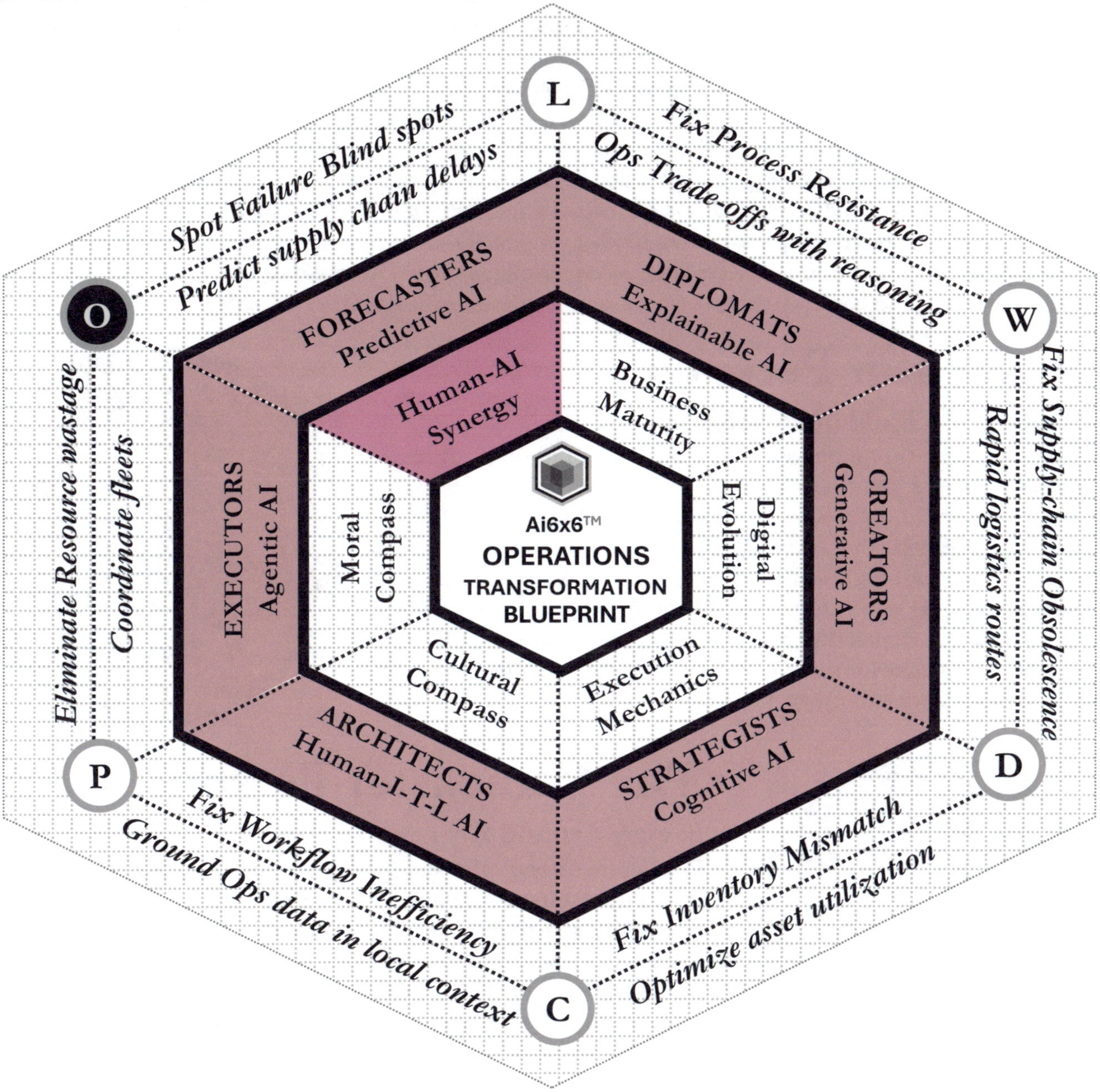

The **Ai6x6 Research Lab** chose this lens to address the fundamental shift from manual oversight to the orchestration of kinetic intelligence. We believe the first hurdle of operations transformation is the psychological transition from being a "cog in the machine" to being the **Architect** of a self-correcting system. This lens is vital because it forces the operational mind to stop viewing efficiency as a personal burden and start viewing it as a collaborative architecture. It is designed to move the leader from the exhaustion of manual firefighting to the liberation of presiding over **Agentic AI**, ensuring that human intuition remains the primary "why" behind the autonomous execution of the mission.

It is about shifting the operator's heart from a state of constant friction to one of strategic orchestration. This lens acts as a cognitive bridge, allowing the leader to see AI not as a threat to labor, but as a diverse set of personas—like the **Forecaster** and **Strategist**—that extend the enterprise's physical and digital reach. Without this shift, operations remain a weight; with it, they become a high-velocity engine. It is an invitation to shed the burden of repetitive monitoring and embrace a new, augmented authority grounded in the courage to implement **Explainable AI** that provides transparency into every movement.

Leadership in the age of intelligent systems demands a new grammar of collaboration—one that transcends the binary of human versus machine and instead frames intelligence as a continuum. The **Human–AI Synergy Lens** redefines leadership not as the art of commanding technology but as the discipline of co-creating with it. It asks leaders to move beyond automation and efficiency toward a deeper integration of foresight, ethics, creativity, cognition, governance, and execution. At its core, Human–AI Synergy is about **shared agency**. It recognizes that leadership today is exercised not only through human judgment but also through algorithmic influence. Every forecast, recommendation, and automated decision carries a fragment of leadership intent. The challenge is not to suppress that influence but to shape it—to ensure that AI amplifies human purpose rather than diluting it. This lens therefore becomes a mirror: it reflects how leaders distribute cognition, emotion, and control across human and machine systems, and how they design organizations that learn, reason, and act as unified intelligences.

The Creator Persona

Expanding the Field of Possibility

Creators represent the imaginative frontier of leadership. Through **Generative AI**, they explore more ideas than any human-only process could conceive. They treat AI not as a threat to originality but as a catalyst for it—a partner in ideation that expands the creative bandwidth of the organization. The Creator's discipline is not about producing more content but about discovering new patterns of thought. They use AI to prototype futures, to visualize complexity, and to democratize innovation. In their hands, generative systems become engines of curiosity, enabling leaders to design from abundance rather than scarcity.

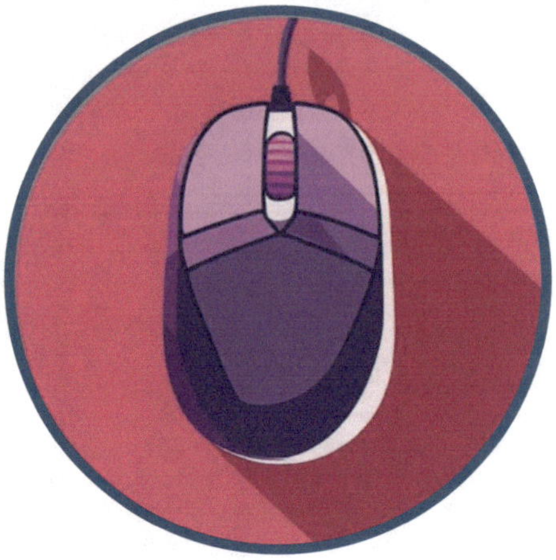

The Forecaster Persona

Seeing Around Corners

Forecasters embody the anticipatory dimension of leadership. They use **Predictive AI** to model futures before committing resources, transforming uncertainty into structured insight. Their craft lies in converting data into foresight—detecting weak signals, pressure-testing assumptions, and rehearsing alternative scenarios. In traditional leadership, forecasting was intuition-driven; in synergistic leadership, it becomes a disciplined dialogue between human imagination and machine probability. The Forecaster's maturity is measured not by how accurately they predict but by how responsibly they prepare. They shift the organization from reacting to events to anticipating them, embedding resilience into its strategic DNA.

The Strategist Persona

Orchestrating Cognitive Intelligence

Strategists use **Cognitive AI** to weave intelligence across functions, geographies, and time horizons. They design strategy as a living system—adaptive, self-correcting, and continuously learning. The Strategist's genius lies in orchestration: aligning distributed intelligences into coherent purpose. They understand that cognition is no longer centralized in the executive suite; it flows through networks of humans and machines that co-reason in real time. Their leadership is systemic rather than hierarchical, emphasizing feedback loops, dynamic modeling, and cross-domain synthesis. In this persona, AI becomes the connective tissue of enterprise intelligence.

The Diplomat Persona:

Building Ethical Bridges

Diplomats operate at the intersection of **Explainable AI** and human empathy. They translate algorithmic logic into moral language, ensuring that machine outputs are understood, trusted, and socially accepted. Their role is to humanize the interface between data and decision, to make transparency a leadership act. In a world where AI can both illuminate and obscure, the Diplomat safeguards the integrity of interpretation. They remind the enterprise that trust is not a technical feature—it is a cultural outcome. By bridging emotional and ethical terrain, they ensure that AI strengthens relationships rather than eroding them.

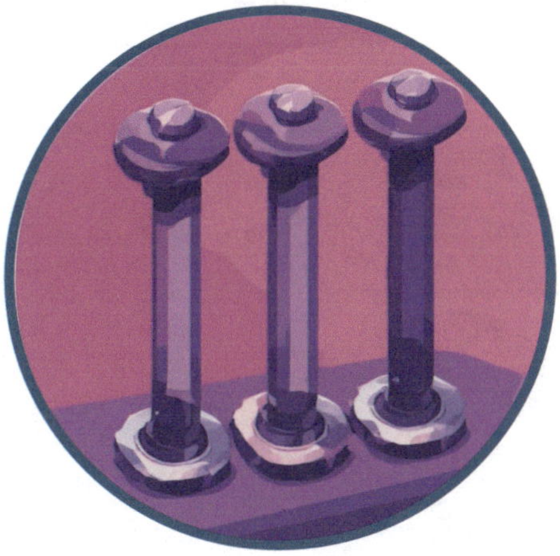

The Architect Persona:

Designing Responsible Collaboration

Architects are the custodians of **Human-in-the-Loop AI**, ensuring that governance and ethical guardrails are embedded into every workflow. They design environments where humans and machines collaborate responsibly—where oversight is not an afterthought but a structural principle. The Architect's leadership is infrastructural: they build the conditions under which synergy can thrive. Their focus is on accountability, transparency, and inclusion, crafting systems that respect both human dignity and machine precision. In their world, collaboration is engineered, not improvised; it is sustained by design integrity.

The Executor Persona:

Converting Insight into Action

Executors close the loop between knowing and doing. Through **Agentic AI**, they embed intelligence directly into operations so that insight becomes action without delay. They represent the kinetic dimension of synergy—the moment where cognition translates into execution. The Executor's mastery lies in operationalizing intelligence: automating decisions that are safe, contextual, and aligned with strategic intent. They ensure that the organization's nervous system is responsive, not reactive, and that every process carries the imprint of learned intelligence.

Together, these six personas form a **leadership spectrum**—from foresight to action, from imagination to governance. They illustrate that synergy is not a single capability but a choreography of roles. The Forecaster anticipates, the Diplomat interprets, the Creator imagines, the Strategist aligns, the Architect safeguards, and the Executor delivers. Each persona represents a distinct mode of engagement with AI, yet all converge on one principle: leadership is now a shared performance between human and machine. The Human–AI Synergy Lens therefore reframes leadership as **co-evolution**. It invites leaders to cultivate humility before complexity, curiosity before certainty, and design before control. It challenges them to lead not by commanding intelligence but by convening it—to orchestrate a dialogue between human wisdom and machine capability that continuously redefines what is possible. In this lens, synergy is not the end state; it is the ongoing discipline of learning together.

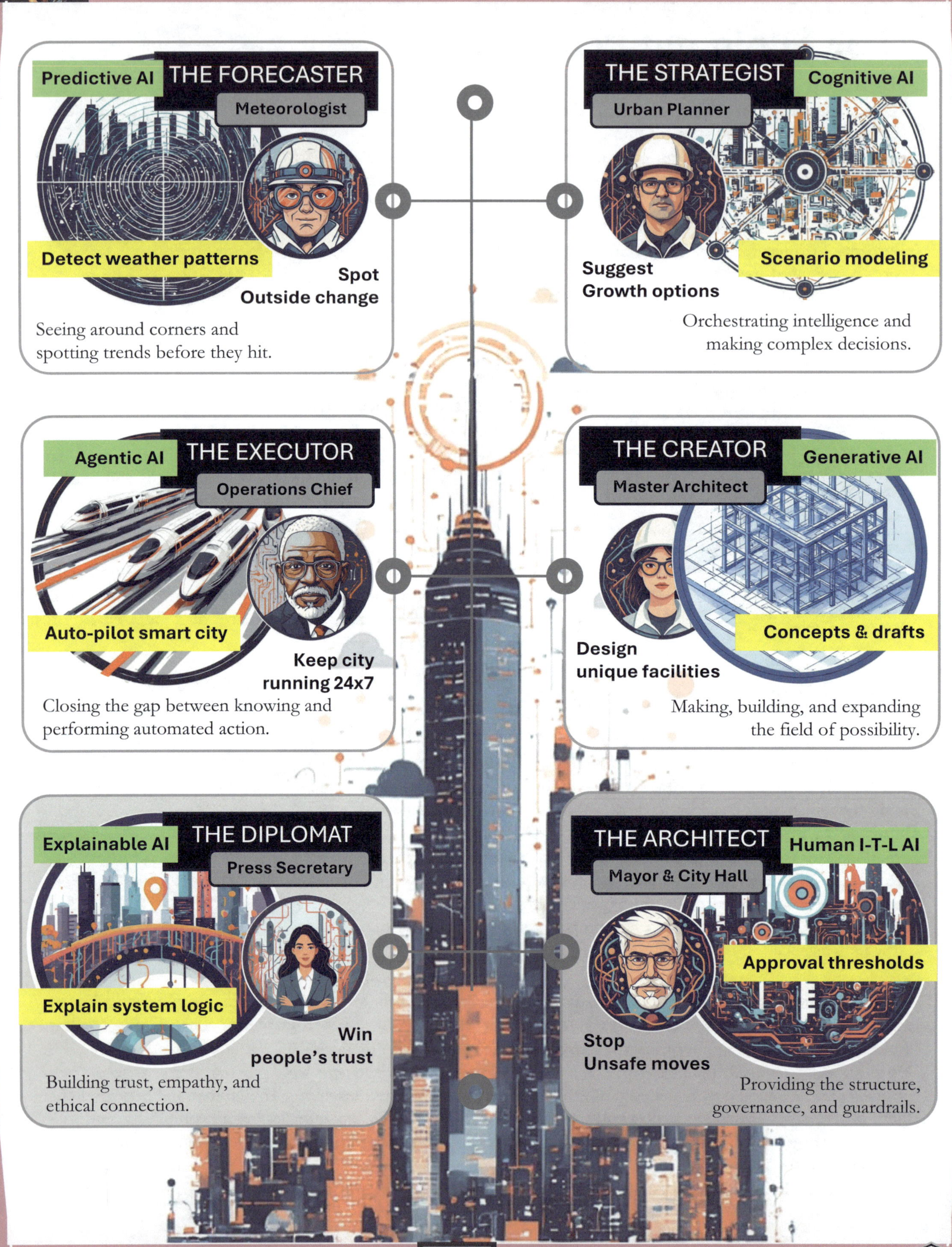
Predictive AI
THE FORECASTER
Meteorologist
Detect weather patterns
Spot Outside change
Seeing around corners and spotting trends before they hit.
THE STRATEGIST
Cognitive AI
Urban Planner
Scenario modeling
Suggest Growth options
Orchestrating intelligence and making complex decisions.
Agentic AI
THE EXECUTOR
Operations Chief
Auto-pilot smart city
Keep city running 24x7
Closing the gap between knowing and performing automated action.
THE CREATOR
Generative AI
Master Architect
Concepts & drafts
Design unique facilities
Making, building, and expanding the field of possibility.
Explainable AI
THE DIPLOMAT
Press Secretary
Explain system logic
Win people's trust
Building trust, empathy, and ethical connection.
THE ARCHITECT
Human I-T-L AI
Mayor & City Hall
Approval thresholds
Stop Unsafe moves
Providing the structure, governance, and guardrails.

Ai6x6™ Transformation Blueprint

Evolution of Humanity

"Every cause produces a consequence; but you must take care to discover the link between them."

Voltaire (1694–1778); Immutable Law (IaC) eliminates hidden links (configuration drift), making cause-and-consequence transparent and ensuring perfect System Cohesion.

In ancient systems, fulfillment for the keeper of infrastructure was a quiet but essential virtue—the constant, vigilant act of preventing societal breakdown (maintaining irrigation, city defenses, or transport routes). Their purpose was the unseen labor of stability, taking pride in the predictability of the system. The Industrial Age turned this role into a manual crisis response cycle. Operations teams found their fulfillment exhausted by the high-stakes, repetitive toil of maintaining increasingly complex machines, characterized by the stress of constant reaction rather than design. They became the modern Sisyphus. The AI Era grants this team radical Cognitive Relief. Fulfillment is achieved by shifting the entire role to designing and enforcing Immutable Law. By leveraging Infrastructure as Code (IaC), the team's professional focus elevates from manual execution to designing self-healing controls and embedding Failure Literacy into the architecture. They are no longer defined by the toil they perform, but by the comprehensive resilience and systemic cohesion they architect.

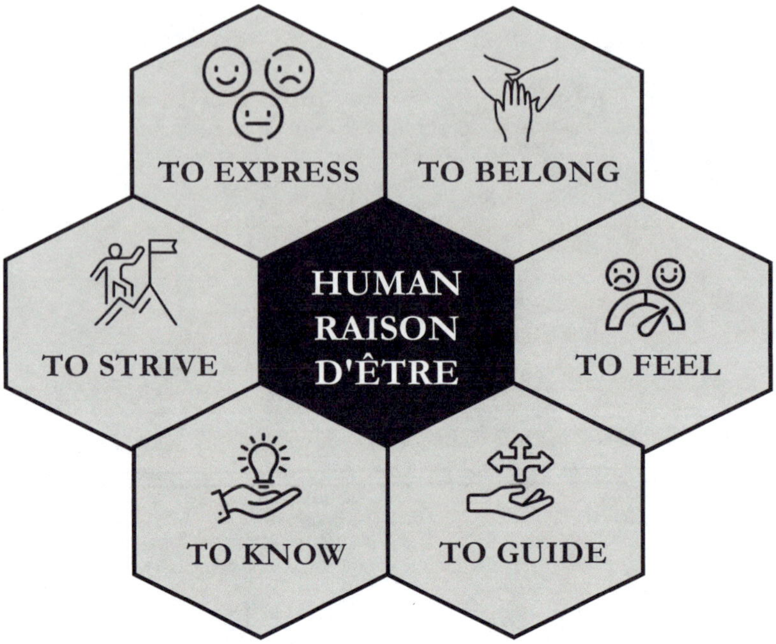

SIX UNIVERSAL ANCHORS OF HUMAN FULFILLMENT

HUMAN RAISON D'ÊTRE (THE REASON FOR EXISTENCE)

In an age defined by exponential automation, the core question of human existence shifts from survival to **significance**. When technology handles efficiency and labor, human value is no longer defined by capability, but by **meaning**. The six **Universal Anchors of Human Fulfillment** define the enduring, non-negotiable contributions of human consciousness—the pillars of a **Raison D'être** that secures our purpose across any era. These anchors affirm that our ultimate mission is to serve as the unwavering source of **Meaning and Moral Authority** in existence. They are the essential acts that only consciousness can perform.

1) TO EXPRESS

The **Anchor of Expression** is the essential human mandate to be the **Source of Novelty and Creation**, generating original ideas and subjective feelings that resist calculation or statistical synthesis. This purpose validates our unique consciousness by asserting that fulfillment comes from the generative act of bringing forth something utterly unique, not merely consuming or replicating. It requires **Uncurated Creativity**—the freedom to explore and define **Aesthetic Judgment**—thereby expanding the conceptual universe with concepts that have no statistical precedent. Through expression, we transform complex inner experience into external reality, upholding our role as the inventors of meaning.

The ultimate benefit of this anchor is the preservation of **Human Originality** and **Intellectual Sovereignty**. By committing to expression, we secure a future that is continually enriched by unique ideas, preventing stagnation and the predictability of automated systems. This anchor maintains the essential spark of invention, ensuring that human meaning remains tied to the capacity to create, **differentiating our value from that of a machine**. It is the source of all artistic, philosophical, and inventive breakthroughs that define a truly dynamic culture.

2) TO BELONG

The **Anchor of Belonging** defines our core identity as social beings, asserting that **authentic connection** is an existential necessity beyond mere digital interaction. This purpose requires the active, continuous labor of **Deep Connection**, prioritizing **unconditional presence** and shared vulnerability to foster a resilient community. Fulfillment is achieved by being the **social architect**, engaging in the emotional work of **mediating conflict**, and fostering cohesion. It includes preserving and transmitting **Cultural Heritage**—the stories and qualitative wisdom that bind us across generations. To belong is to confirm that our worth is interwoven with our relationships.

The ultimate benefit of this anchor is **Emotional Resilience** and **Social Stability**. By prioritizing genuine human connection, we safeguard ourselves against isolation and the psychological fragility that arises from purely transactional relationships.[1] This anchor ensures a supportive social framework that validates empathy and provides necessary emotional resonance, acting as a crucial defense mechanism against an isolating technological world. It builds the **trust and cohesion** necessary for any human system to function justly and compassionately.

3) TO FEEL

The **Anchor of Feeling** grounds human existence in the tangible, asserting the profound psychological value of **Embodied Mastery** and sensory engagement. This purpose elevates the dedicated pursuit of physical skill and process—not for necessity, but for the inherent satisfaction of the process. It is the quest for the **'Flow' state**, where an earned link between **exertion and accomplishment** provides deep fulfillment. This anchor mandates resisting the ease of total automation by engaging in voluntary challenges and perfecting non-essential crafts. It ensures that our abstract thoughts are grounded in the undeniable reality of the body and the senses.

The ultimate benefit of this anchor is **Physical and Psychological Integrity**. By committing to felt experience, we preserve our cognitive and physical health, resisting the inertia of automation and maintaining a sharp connection to reality. This anchor validates the **dignity of effort**, providing a clear, visceral confirmation of personal competence and self-mastery that is crucial for self-esteem. It ensures that fulfillment remains an **earned state**, confirming the complexity and richness of being physically alive.

4) TO GUIDE

The **Anchor of Guidance** establishes the human being as the **ultimate moral and ethical sovereign**. This purpose is founded on our unique capacity for empathy and subjective value judgment, making us the non-delegable authority for setting rules and managing risk. It is the practice of **Ethical Governance**, demanding that we **Draw the Moral Line** and translate compassionate principles into enforceable policy. Fulfillment is achieved by maintaining transparency, enforcing accountability, and continuously exercising the **Human Veto Power** over critical decisions.

The ultimate benefit of this anchor is **Systemic Integrity** and **Moral Safety**. By retaining final ethical authority, we ensure that technological power remains fully aligned with, and subservient to, human values, preventing systems from optimizing themselves at the expense of human welfare. This anchor provides the essential check against unchecked power, establishing the legal and moral frameworks necessary to protect all citizens. It guarantees that empathy and justice remain the **supreme architects of our collective future**.

5) TO KNOW

The **Anchor of Knowing** defines our core intellectual purpose as the engine of **Epistemic Exploration** and abstract reasoning. This purpose compels the mind to focus on formulating **Axiomatic Hypotheses**—the fundamental "Why?" questions that lie beyond the reach of data processing and established knowledge. Fulfillment is found in the generative act of **challenging established truths** and creating new theoretical frameworks that synthesize disparate fields. This anchor ensures that the pursuit of knowledge remains boundless, driven by philosophical depth rather than functional utility.

The ultimate benefit of this anchor is **Intellectual Freedom** and **Conceptual Progress**. By focusing on formulating new questions and challenging paradigms, we prevent the stagnation that could result if all knowledge were confined to data-driven answers. This anchor guarantees that our intellectual journey continues to be defined by profound theoretical breakthroughs and a relentless expansion of understanding. It confirms that human intelligence is defined not by the volume of information stored, but by the **depth and originality of the questions** we ask.

6) TO STRIVE

The **Anchor of Striving** is the **Existential Quest**—the voluntary pursuit of difficulty necessary to generate meaning, independent of any external necessity. This purpose recognizes that true fulfillment is an **earned state**, achieved through **Deliberate Discipline** and self-imposed ambition. It compels us to **Set Impossible Goals** and **Choose the Hard Way**—overcoming voluntary hardship—precisely because the challenge validates our capacity and resolve. This continuous effort allows us to craft a meaningful **Personal Legacy** by executing complex, long-term plans purely for the satisfaction of the effort.

The ultimate benefit of this anchor is **Authentic Self-Mastery** and **Purposeful Existence**. By embracing self-imposed difficulty, we cultivate the resilience, discipline, and fortitude that define character, resisting the complacency of automated comfort. This anchor ensures that human worth is determined by our **internal will**, rather than external circumstance, providing a framework for continuous personal growth and the creation of an enduring, meaningful legacy.

CONCLUSION: THE RAISON D'ÊTRE FOR THE HUMAN

The six Universal Anchors converge to form humanity's ultimate purpose: to be the **Unwavering Source of Meaning and Moral Authority**. This composite Raison D'être defines the essential human role in any advanced civilization.

Our mission is the sustained commitment to **non-automatable value**. We are the unique entity that not only pursues truth (To Know) but also creates the beauty (To Express) that makes truth worth pursuing. We are the mechanism that defines the rules of fairness (To Guide) and then personally invests the necessary effort (To Strive) to live up to those rules. Crucially, we are the only entity that grounds this entire conceptual structure in the tangible joy of physical reality (To Feel) and the vital necessity of shared vulnerability (To Belong).

The human purpose shifts from survival to an intentional journey of **growth, creation, and ethical custodianship**. Our value is not in output, but in the **input of consciousness, conscience, and will**. By diligently exercising these six anchors, every individual contributes to the essential, enduring purpose of the species: to be the **Sole Arbiter of Worth, Will, and Wonder**—ensuring that existence remains a deeply meaningful and purposeful quest.

In this era, human existence was focused on the direct, tangible application of skill for immediate survival, such hunting, fishing and surviving. The human was the Master of the Hunt.

In the Primitive/Paleolithic Era (Survival & Sustenance), human purpose was an absolute function of immediate necessity. Fulfillment was not a philosophical pursuit but a practical outcome of successfully continuing existence for oneself and the kin group. All six Universal Anchors manifested, but they were tethered directly to the biological and environmental imperative to survive the day and propagate the species. The overarching Raison D'être was to **Endure and Transmit**, meaning personal worth was inseparable from one's contribution to the group's survival calculus. This collective focus meant that individual talent was instantly converted into a benefit for all, solidifying a direct, unbroken link between personal action and existential significance.

To Express: Expression was functional and communal, centered on **Ritual Storytelling** and cave art. This art was not for individual fame or abstract aesthetics, but served as the primary educational system, preserving and transmitting essential survival knowledge, hunting techniques, spiritual beliefs, and the topography of the land. The repetitive, sacred nature of the rituals ensured collective memory and adherence to successful methods. Furthermore, these cultural forms encoded crucial information, often representing animals and spirits, that, if lost or forgotten, would jeopardize the tribe's safety and cultural identity.

To Belong: The Anchor of Belonging was non-negotiable, manifesting as **Kinship and Mutual Defense**. Isolation meant certain death; therefore, the purpose of connection was absolute. Fulfillment came from actively contributing to the group's defense, sharing food and resources, and ensuring the collective physical safety of the tribe above all else, fostering a fierce, self-sustaining mutualism. The emotional security derived from this deep collaboration—knowing that others were committed to your well-being—was a core psychological reward.

To Feel: Embodied Mastery was defined by **Essential Tool Mastery**. The feeling of accomplishment was directly tied to successfully chipping a sharp stone tool, accurately throwing a spear, or enduring the physical hardships of the hunt. This pursuit was not sport, but the daily necessity of mastering the physical world to secure food and shelter, and the psychological confidence gained was crucial for future success. This expertise immediately translated into social value; the most capable hunter or tool-maker earned respect and a higher status within the hierarchy, directly reinforcing the dignity of effort.

To Guide: Moral authority was based on immediate experience, taking the form of **Elder Authority and Tradition**. Guidance was unwritten law, passed down through generations of hard-won experience. Decisions—like where to hunt, when to migrate, or how to resolve disputes—were arbitrated by those with the deepest, most successful experiential knowledge, and were obeyed out of necessity for the group's stable cohesion. This system was vital for ensuring collective action remained predictable and risk-adverse, preventing internal conflict from threatening external survival.

To Know: The pursuit of knowledge was pure **Observation and Practicality**. The purpose of knowing was to accurately predict the immediate environment for resource exploitation. This involved memorizing the habits of migratory animals, understanding the medicinal or poisonous properties of local plants, and tracking subtle shifts in weather patterns—information critical for daily sustenance and successful, low-risk planning. Early forms of abstract reasoning also emerged here, through tracking the movement of stars or noting the cyclic nature of seasons and astronomical events, demonstrating an innate human desire to find order and predictability in environmental chaos.

To Strive: Striving was synonymous with **Daily Endurance**. There was no need to invent abstract goals; the pursuit of difficulty was the environment itself, offering constant challenge. Fulfillment was the successful completion of the continuous, grueling effort required to secure food, withstand the elements, and ensure the basic propagation of the tribe, creating a deeply ingrained sense of earned worth. The resilience built through continuous hardship was the ultimate personal asset, psychologically connecting physical suffering with the existential meaning of survival.

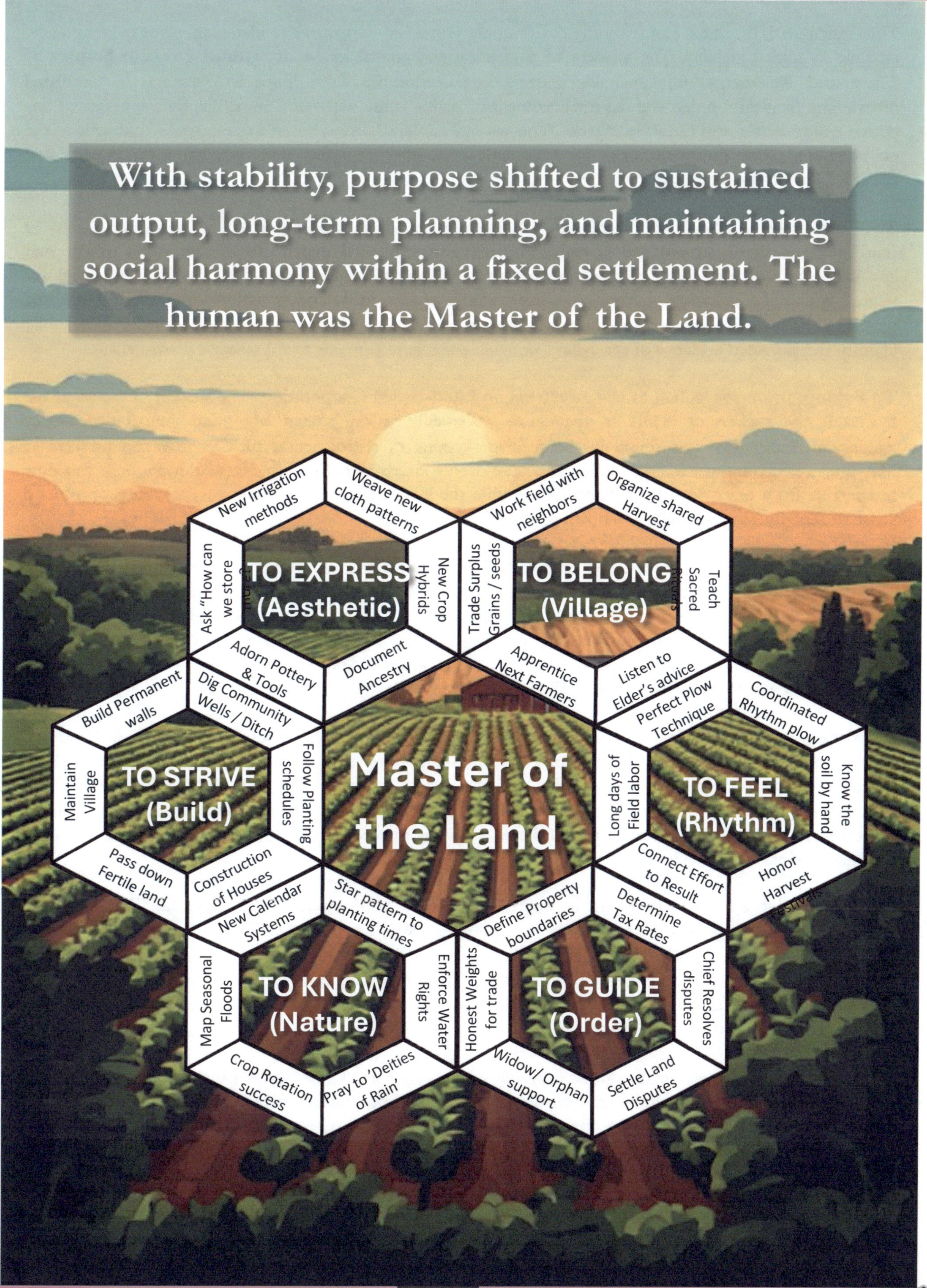
With stability, purpose shifted to sustained output, long-term planning, and maintaining social harmony within a fixed settlement. The human was the Master of the Land.
Master of the Land
TO EXPRESS (Aesthetic)
New Irrigation methods
Weave new cloth patterns
New Crop Hybrids
Document Ancestry
Adorn Pottery & Tools
Ask "How can we store
TO BELONG (Village)
Work field with neighbors
Organize shared Harvest
Teach Sacred
Listen to Elder's advice
Apprentice Next Farmers
Trade Surplus Grains / seeds
TO STRIVE (Build)
Build Permanent walls
Dig Community Wells / Ditch
Follow Planting schedules
Construction of Houses
Pass down Fertile land
Maintain Village
TO FEEL (Rhythm)
Perfect Plow Technique
Coordinated Rhythm plow
Know the soil by hand
Honor Harvest
Connect Effort to Result
Long days of Field labor
TO KNOW (Nature)
New Calendar Systems
Star pattern to planting times
Enforce Water Rights
Pray to 'Deities of Rain'
Crop Rotation success
Map Seasonal Floods
TO GUIDE (Order)
Define Property boundaries
Determine Tax Rates
Chief Resolves disputes
Settle Land Disputes
Widow/ Orphan support
Honest Weights for trade

The Agricultural/Neolithic Era (Farming & Community) marked humanity's fundamental pivot from nomadic struggle to settled stability. The mastery of cultivation and animal husbandry created a reliable **Surplus of Time and Resources**, fundamentally shifting human purpose from immediate survival to calculated investment, long-term order, and the establishment of lasting community structures. The Raison D'être of this period was to **Settle and Establish Order**. This era saw humans take on the role of conscious managers of the environment, laying the groundwork for systematic laws, property rights, and enduring social legacies that extended far beyond one generation.

To Express: Expression became focused on **Accounting and Enduring Form**. The logistical necessity of tracking seeds, yields, stored grains, and communal ownership led directly to the invention of the first systematic writing forms, such as early cuneiform, demonstrating that expression was initially a tool for managing complex resources. Beyond text, expression was seen in durable, functional goods like specialized pottery, permanent architectural structures, and woven textiles. These forms reflected a deep, stable community identity and provided evidence of specialized skill that contributed directly to the village's permanence.

To Belong: Belonging shifted to being centered on **Land-Based Cooperation**. The success or failure of the farm was the success or failure of the entire settlement, making shared labor and mutual dependency paramount. Fulfillment was found in the reliable rhythms of shared work, such as irrigation projects or harvesting, cementing generational inheritance and deeply intertwined **Kinship Networks**. Purpose was now inseparable from one's physical, defined place within the land and the local community, creating a profound sense of rootedness and stability unknown in the Paleolithic past.

To Feel: Embodied Mastery focused on **Craftsmanship and Process**. Fulfillment was derived from the methodical, repeated application of skill required by agriculture—plowing a straight furrow, carefully managing livestock, or perfecting the complex processes of metallurgy and food preservation. This mastery provided a sense of control over the environment that raw endurance could not. The purpose of skill was purely to generate the stable surplus required to maintain the settled life, providing a deep psychological reward through the predictable yield of sustained, expert effort.

To Guide: Moral authority became formalized, focusing urgently on **Property Rights and Resource Management**. Guiding the community meant establishing clear, complex rules around ownership of land, managing precious resources like water rights, and ensuring crop protection from both animals and neighbors. This necessity gave rise to the first systematic, written laws, moving purpose from inherited tradition (as in the Paleolithic) to a **Codified Social Contract**. Leadership was defined by the ability to enforce these new laws, ensuring fairness and stability within an increasingly complex economy.

To Know: Abstract reasoning began to flourish, driven by **Mathematics and Calendar Tracking**. The purpose of knowledge was now predictive certainty, essential for farming success. This meant mastering geometry to accurately divide and manage fields, and engaging in sophisticated celestial observation to track seasons and accurately determine optimal planting and harvesting cycles. This pursuit of cosmic order confirmed the human ability to impose intellectual structure upon chaotic nature, turning farming into a sophisticated, knowledge-driven enterprise.

To Strive: Striving fundamentally shifted from day-to-day survival to **Creating Surplus and Investment**. The existential quest became the arduous, long-term, deliberate work required to build durable structures, amass protective stores of food, and ensure the economic and social prosperity of the next generation. This focus on legacy and investment transformed work into a purposeful act of building a lasting future, where one's worth was measured not just by survival, but by the inheritance left behind for the family and the stability provided to the community.

This era defined purpose through professional expertise and contribution to complex, often abstract, technological and economic systems. The human was the Master of the Machine.

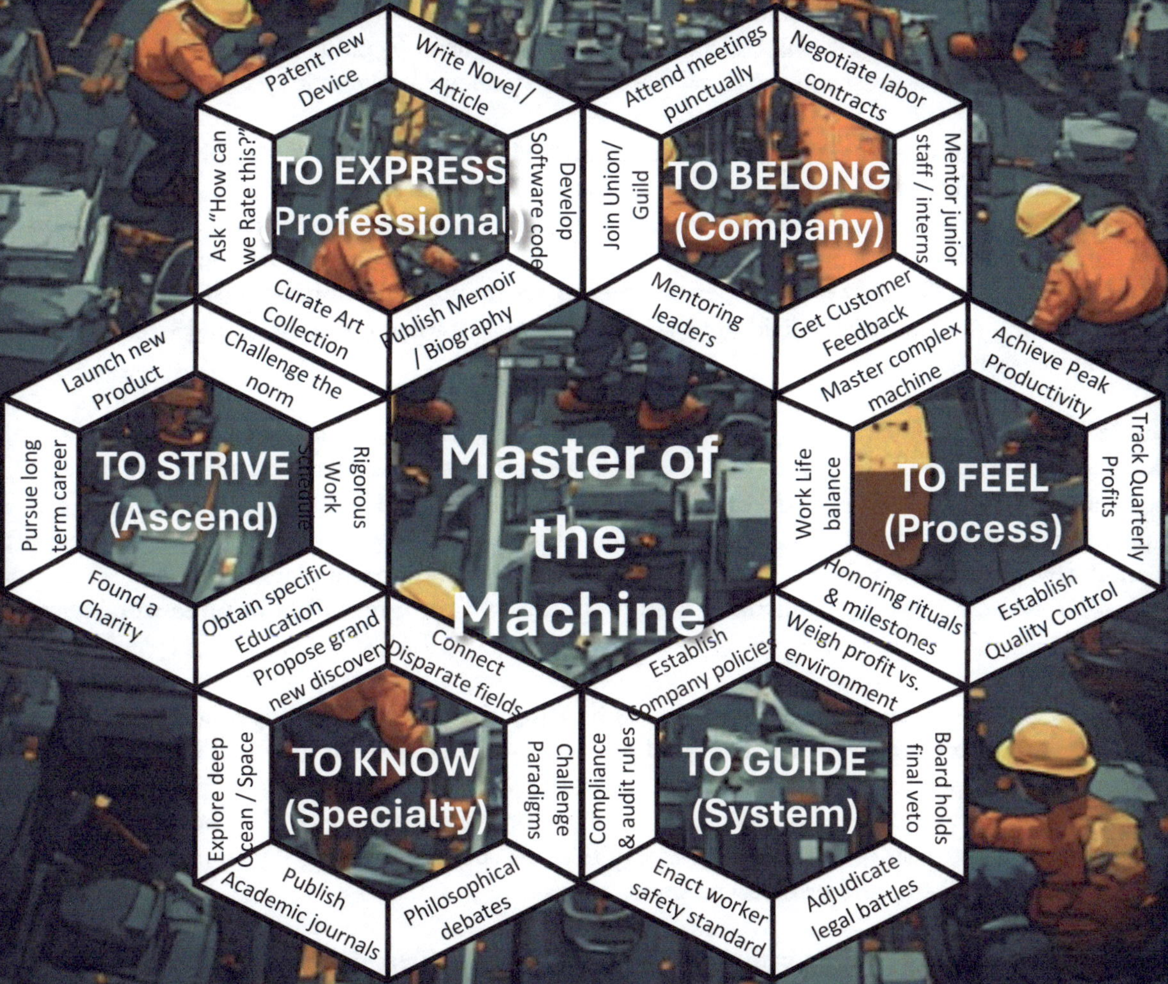

The Modern/Industrial Era (Specialization & Function) marked the decisive triumph of machine logic and capital over the natural world. Human purpose shifted from managing the cycles of the land (Neolithic) to mastering the mechanical efficiency of the factory and the market. The rise of the machine and the division of labor fundamentally redefined the individual, whose worth was now derived from highly specialized, functional expertise within a vast, interconnected economic system. The Raison D'être of this period was to **Achieve and Advance**, driving relentless progress through expertise and material gain.

To Express: Expression became focused on **Mass Media and Invention**. With the advent of the printing press, cheaper paper, and later, radio and cinema, creative expression transformed into a commodified product designed for wide distribution and profit. The purpose of expression was to inform, entertain, and persuade the masses, shaping shared cultural narratives through novels, newspapers, and commercial design. The creation of a **Patentable Invention** became a powerful act of individual expression, translating originality directly into economic ownership and societal contribution. Expression was celebrated when it served a functional, marketable, or propagandistic role.

To Belong: Belonging was redefined by **Nation and Institution**. The old agrarian kinship networks dissolved as masses migrated to urban factory centers, replaced by new, abstract, yet powerful social structures. Fulfillment was found in national identity, membership in professional organizations, and political affiliation. Crucially, the extreme specialization of labor necessitated the formation of organized counter-structures like **Labor Unions** to protect the rights of workers against centralized corporate power, demonstrating that belonging was a functional necessity for collective bargaining and social stability in a highly unequal economic system.

To Feel: Embodied Mastery centered on **Efficiency and Maximized Output**. The physical purpose shifted away from holistic craft towards highly repetitive, specialized labor—the mastery of a single, precise task on the assembly line. While monotonous, this specialization provided a clear, measurable sense of contribution, where personal worth was directly tied to economic output and reliability. Outside of the factory, the Anchor manifested in the rise of **Organized, Competitive Sports**, where physical effort was deliberately structured and formalized, allowing the industrial individual to achieve a sense of earned competence and visceral struggle that was absent in their work.

To Guide: Moral authority became formalized through **Secular Law and Corporate Regulation**. The complexity of contract law, property ownership, and massive corporate liability demanded sophisticated, non-religious guidance systems. The purpose of guidance was to manage the inherent conflicts between capital and labor, protect consumer safety in mass production, and define the boundaries of ethical economic competition. This period saw the relentless development of regulatory bodies and constitutional law aimed at establishing fairness within the new, powerful industrial economy, ensuring that **progress did not undermine social order** entirely.

To Know: The pursuit of knowledge was dominated by **Empirical Science and Specialization**. The purpose of knowing was functional utility, believing that all problems—from disease to industrial inefficiency—could be solved through the application of the rigorous **Scientific Method**. Knowledge became fragmented into specialized disciplines, each dedicated to gathering objective, measurable data. Fulfillment was achieved by contributing a verifiable, peer-reviewed fact or theory to one's specific field, thereby confirming the era's deep-seated faith in rational inquiry and technical progress.

To Strive: Striving was synonymous with **Ambition and Material Gain**. The existential quest was the pursuit of wealth, social mobility, and **Status**. Fulfillment was an earned state achieved through economic success, the establishment of a profitable business, or professional advancement within a corporate or governmental hierarchy. The cultural narrative of the "self-made individual" defined personal purpose, urging people to embrace continuous, relentless work to secure prosperity for the family and, more importantly, to carve out an enduring legacy of material success and high societal standing.

In this emerging era, AI handles efficiency and role specialization, freeing and forcing humans to refocus on subjective values, unique creation, and moral governance. The human becomes the Master Curator of Purpose

The Post-AI Era (Curation & Meaning) represents the definitive existential shift in human history. With advanced artificial intelligence and automation having conquered the challenges of efficiency, information processing, and mass labor, human purpose is radically redefined. We are no longer necessary for function, but for **meaning**. The individual's worth is derived entirely from their capacity for subjective judgment, ethical authority, and the creation of non-calculable value. The collective Raison D'être is to become the **Sole Arbiter of Worth, Will, and Wonder**—the source of all criteria for why life should continue and what makes it beautiful.

To Express: Expression is focused on **Subjectivity and Uncurated Truth**. In a world saturated with synthetic media and hyper-optimized content generated by AI, the only expression that retains intrinsic value is the genuinely **original, subjective human voice**. This purpose mandates focusing on authentic, unfiltered creation—personal philosophical content, aesthetic critique, memoir, and art that contains flaws, uncertainty, and emotional resonance. Fulfillment is found in generating novelty that AI cannot predict, thereby constantly pushing the boundaries of human experience and securing a creative future that is immune to algorithmic stagnation.

To Belong: Belonging becomes a deliberate, chosen investment in **Deep Vulnerability and Presence**. When digital connection is ubiquitous but emotionally shallow, true fulfillment relies on actively resisting transactional relationships and seeking authentic, physical presence. This purpose demands the **labor of Deep Connection**: cultivating empathy, mediating conflict through personal understanding, and prioritizing quality, face-to-face interaction. The collective human project requires maintaining the social fabric and emotional resonance that machines cannot replicate, ensuring that isolation does not erode the psychological foundation of society.

To Feel: Embodied Mastery transforms into the **Voluntary Pursuit of Difficulty and Flow**. Since automation eliminates the necessity of difficult labor, human physical purpose centers on self-imposed challenges—non-essential crafts, athletics, meditation, and the perfection of technical, yet unnecessary, skills. This pursuit generates the vital 'Flow' state—the psychological reward of earned accomplishment—which secures self-esteem and grounds abstract consciousness in tangible reality. Fulfillment is achieved through **Deliberate Effort** that reaffirms the dignity of the body, providing a visceral antidote to the inertia and complacency of total comfort.

To Guide: Moral authority centers on **AI Ethics and the Human Veto**. The purpose of guidance is to serve as the **ethical sovereign**, placing strict moral boundaries around autonomous systems. This involves defining the principles of data governance, ensuring transparency in black-box algorithms, and enforcing systemic accountability for technological outcomes. Crucially, every individual is mandated to exercise the **Human Veto Power**—the non-delegable right to overrule critical decisions made by algorithms—ensuring that power remains permanently aligned with subjective human values, not cold-blooded efficiency.

To Know: The pursuit of knowledge is now defined by **Abstract Synthesis and Axiomatic Inquiry**. In an environment where AI holds access to all existing data and established facts, human intellect must shift its focus. The purpose is to transcend data processing and concentrate on formulating **Axiomatic Hypotheses**—the profound "Why?" questions that challenge foundational assumptions. Fulfillment is found in connecting disparate fields of knowledge, creating new philosophical frameworks, and constantly resisting the temptation to settle for algorithmic answers, ensuring the continuous, boundless expansion of human understanding.

To Strive: Striving is the existential commitment to **Self-Imposed Discipline and Personal Legacy**. Since survival is guaranteed, the quest for meaning requires the individual to set ambitious, non-essential goals purely for the satisfaction of the effort. This involves cultivating profound resilience and fortitude against the immense temptation of automated ease. Fulfillment is an **earned state**, achieved through continuous, conscious effort and the construction of an enduring **Personal Legacy**—a body of work or moral character that defines one's worth by internal resolve rather than external economic output.

Ai6x6™ Transformation Blueprint

Human-AI Personas

"The coming era of Artificial Intelligence will not be the era of war, but be the era of deep compassion, non-violence, and love."
-Amit Ray, a well-known AI scientist and spiritual teacher.

The Ai6x6 HUMAN-AI Synergy framework formalizes a new executive standard for the AI era, transitioning organizations from fragmented automation to a state of Orchestrated Autonomy. This model defines the modern enterprise through six core leadership movements, each serving as a specialized "sense" for the organization. **The Forecasters** serve as the "Eyes," tasked with the heavy lifting of foresight to eliminate strategic blind spots and detect invisible market signals before they escalate into crises. Supporting this vision, **The Diplomats** act as the "Voice," translating complexity into human trust and ensuring that transformation is met with cultural alignment rather than institutional resistance. The organization's creative energy is fueled by **The Creators**, the "Engine" of the enterprise, who move beyond static production to drive a state of radical velocity and infinite iteration. Their output is balanced by the surgical objectivity of **The Strategists**, acting as the "Compass" to ensure that every capital investment is mathematically anchored to peak ROI. To unify these efforts, **The Architects** build the "Institutional Brain," ensuring that the organization's collective memory is always accessible and relevant to the individual. Finally, the framework finds its movement in **The Executors**, the "Hands" of the organization, who coordinate autonomous workflows to achieve scale without traditional overhead. By weaving these archetypes into a single Symphony, leaders can navigate market ambiguity with surgical precision, transforming the enterprise into a self-evolving system of sustainable innovation.

In the **Ai6x6 Human-AI Synergy framework**, the defensive stance on Operations is the final line of defense against **Systemic Entropy**. In many organizations, operations are the primary source of hidden costs—leakage caused by inefficient processes, manual errors, and "black box" workflows. Defensive operations strategy focuses on building a "Self-Healing Engine" that identifies friction before it results in financial or physical loss.

The Forecaster and The Diplomat: Mapping Risk and Ensuring Compliance

The **Forecaster** serves as the defensive "Early Warning System" for the supply chain and internal workflows. Its mandate is to kill **Operational Blindspots** by detecting anomalies—such as a pending logistics delay or a sudden spike in energy consumption—that signal a break in the flow. By identifying these "Weak Signals" early, the Forecaster allows the company to move into a defensive posture, rerouting resources before a minor hiccup becomes a total system shutdown.

The **Diplomat** is tasked with neutralizing **Governance Friction**. As operations become increasingly automated, the risk of "Regulatory Drift" increases. The Diplomat uses explainable AI to ensure that every automated operational decision is audit-ready and compliant with global standards. By acting as the bridge between autonomous logic and human oversight, it protects the "Voice" of the operation, ensuring that efficiency never comes at the cost of ethical or legal integrity.

The Creator and The Strategist: Modernizing Workflows and Eliminating Waste

The **Creator** acts defensively by killing **Process Obsolescence**. Legacy workflows are often rigid, unable to handle modern demand spikes. The Creator's defensive role is to generate "Adaptive Workflows"—dynamic process models that can be refactored in real-time to meet new constraints. It ensures the "Operational Engine" doesn't seize up due to "Static Debt," where the organization is trying to solve 2026 problems with 2010 logistical patterns.

The **Strategist** defends the bottom line by eliminating **Resource Mismatch**. Most waste in operations stems from over-allocation or "Zombie Workflows" that serve no current purpose. The Strategist performs a ruthless audit of the operational landscape, using causal inference to identify where capital and labor are producing the least yield. By "pruning" the operation of these mismatched efforts, it protects the "Operational Compass," ensuring that every unit of energy spent is mathematically aligned with peak efficiency.

The Architect and The Executor: Unifying the Twin and Self-Healing the System

The **Architect** kills **Information Inconsistency** across the physical and digital worlds. Its primary defensive mandate is the maintenance of a "Digital Twin"—a real-time, unified map of the entire operation. It prevents the friction of "Data Gaps" where the warehouse doesn't know what the factory is producing. By building this defensive "Institutional Brain," the Architect ensures that the organization operates as a single, synchronized organism rather than a collection of disconnected silos.

Finally, the **Executor** serves as the defensive "Hands" of the engine. It kills **Recovery Friction** by automating "Self-Healing" protocols. Whether it is automatically rerouting a delivery around a storm or triggering a mechanical repair before a machine breaks, the Executor ensures that guardrails are in place to maintain uptime. By removing the wastage of manual intervention for routine errors, it safeguards the "Operational Flow," ensuring the organization achieves a state of resilient, orchestrated autonomy.

Conclusion: The Resilient Body

A defensive operations strategy within the **Ai6x6 Human-AI Synergy framework** ensures that the "Organizational Body" is lean, resilient, and immune to waste. By removing the friction of blindspots, silos, obsolescence, and mismatch, the framework creates an environment where growth is sustainable and predictable. When a company's operations are protected by these six roles, it achieves the ultimate defensive advantage: the ability to absorb market shocks without breaking its stride.

In the **Operations Transformation**, the objective is to move from "Rigid Linearity" to **"Fluid Orchestration."** This shift transforms the organization's backbone from a series of friction-filled handoffs into a self-optimizing, autonomous nervous system. By applying the six roles to operations, leadership eliminates the "Execution Gap," ensuring that resources, logistics, and workflows move at the speed of thought.

The Forecasters and The Diplomats: Unlocking Resilience and Synchronization

The **Forecaster's** mandate is **Systemic Predictive Awareness.** By pairing Predictive AI with Graph Neural Networks, the "Eyes" see beyond internal data to the web of global dependencies. It identifies a potential labor strike or a raw material shortage weeks before it ripples into the factory floor. This allows the leader to shift production schedules proactively, turning potential disruptions into competitive advantages through "Resilience by Design."

The **Diplomat** serves as the "Voice" of **Stakeholder Orchestration.** By pairing Explainable AI with Contextual NLP, the Diplomat ensures that operational shifts are synchronized across the entire value chain. It translates complex logistics logic into clear directives for suppliers, partners, and internal teams. This eliminates the friction of misalignment, ensuring that every node in the ecosystem understands the "Why" behind an operational pivot, maintaining trust in a high-velocity environment.

The Creators and The Strategists: Unlocking Virtualization and Throughput

The **Creator** acts as the engine of **Operational Prototyping.** In this role, Generative AI is paired with Digital Twins to manifest "What-If" operational layouts in a virtual sandbox. Before moving a single piece of equipment or changing a workflow, the Creator simulates thousands of iterations to find the most efficient path. This "Generative Operations" allows the firm to redesign its entire logistics flow in hours, manifesting a "First Draft" of efficiency that is already physically verified.

The **Strategist** optimizes the **Operational Frontier for Maximum Throughput.** Using Multi-Objective Optimization, the Strategist balances the conflicting goals of cost reduction, speed-to-market, and carbon footprint. It identifies the "Optimal Flow," ensuring that the organization isn't just running fast, but is running toward the highest possible yield. This offensive objectivity turns operations into a precision instrument that extracts value from every square foot and every second of uptime.

The Architects and The Executors: Unlocking Governance and Autonomy

The **Architect** builds the "Searchable Operational Brain" through **Live Process Knowledge.** By pairing Human-in-the-Loop with Vector Databases (RAG), the Architect ensures that every operational SOP (Standard Operating Procedure) and historical lesson is "Live." This cures the "Institutional Leakage" that occurs when experienced operators leave, ensuring that the "Brain" of the factory grows smarter over time and remains instantly accessible to every worker on the floor.

Finally, the **Executor** manages the **Autonomous Actuation of Workflows.** Through Agentic AI paired with Edge AI and Blockchain, the Executor automates the "Hands" of the enterprise. These agents don't just monitor; they act—automatically re-routing shipments, triggering maintenance, or executing smart-contract payments when milestones are met. This "Autonomous Execution" ensures that the organization can maintain a 24/7 operational pulse with minimal human intervention and absolute financial integrity.

Conclusion: The Symphony of the Frictionless Enterprise

Operations transformation is the pursuit of **Operational Sovereignty.** By unlocking resilience, synchronization, virtualization, throughput, governance, and autonomy, the leader creates an organizational engine that is as agile as a startup but as powerful as a global giant. This framework ensures that the brand's "Execution Layer" is perfectly synchronized with its strategic intent, creating a high-velocity symphony of orchestrated action.

Operational Power: Real-Time Transparency

In the legacy era, operations were "Opaque." A leader only knew something was wrong when the assembly line stopped, the shipment didn't arrive, or the budget was blown. The transformation into **Forecasters** gives the operational core its own "Eyes." This role is about the **Purification of the Value Stream**. It involves turning every operational node—from a warehouse shelf to a digital transaction—into a "Live Signal." Operations cease to be a "Black Box" and become a "Glass Box," providing a high-fidelity map of the organization's physical and digital health.

Defensive Stance: Killing the Latency of Error

The defensive mission of the Operational-Forecaster is the **Elimination of the Reaction Gap**. In traditional business, there is a dangerous "Time Tax" between a failure and its discovery. The Forecaster role kills this "Friction of Delay" by monitoring the "Biological Rhythm" of the company. It identifies the "Statistical Whisper" of a bottleneck before the queue even forms. This creates a **Shield of Operational Continuity**. By predicting where friction will occur, the organization can re-route resources pre-emptively. This defensive posture kills the "Wastage of Crisis," protecting the company's "Margins" and ensuring that "Service Levels" are maintained through "Pre-emptive Awareness" rather than "Post-Mortem Correction." The system sees the friction before the friction stops the machine.

Offensive Stance: Unlocking Flow-State Optimization

On the offense, the Forecaster moves toward **Anticipatory Resourcing**. Because the operations "See" the incoming demand and the current capacity in perfect synchrony, they can "Pre-position" the company for a win. This **Unlocks Temporal Agility**, where the organization moves its "Weight" (inventory, capital, labor) to the exact point of future impact. If the "Eyes" of the operation see a weather event or a market shift coming, the system reconfigures the entire supply chain in advance. This turns operations into a "Competitive Weapon" that allows the brand to "Out-Pace" the market. This offensive move ensures the brand is always "Fluid," moving with the speed of data rather than the speed of human bureaucracy, capturing the "Yield Peak" of every market opportunity.

The Operational Outcome: Perceptual Sovereignty

The ultimate value of the Forecaster role is **Total Visibility**. When operations "See," they never gamble. For the leader, this provides **Navigational Peace of Mind**. You are no longer managing through "Reports"; you are managing through "Reality." This turns the operations floor into a "Strategic Asset" that ensures the brand is always the most "Responsive" and "Resilient" in the industry. It ensures the "Eyes" of the company are never closed, providing a 24/7 "Truth Stream" that keeps the organization perpetually aligned with its targets.

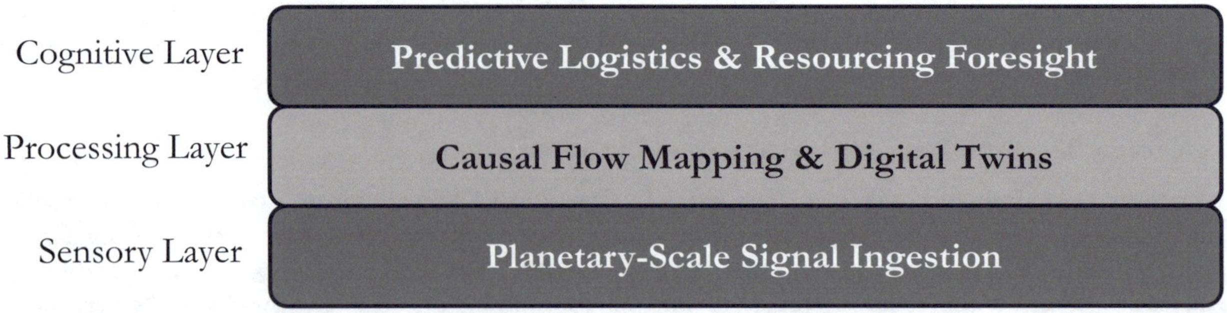

Executive Power: Global Supply Observability

In the legacy era, a leader's "Eyes" on operations were limited to delayed shipping manifests and quarterly vendor audits. The **Forecaster** role in operations transformation represents the transition to **Total Value Chain Visibility**. For the executive, this means moving away from "Reacting to Bottlenecks" toward "Perceiving Global Flux." By architecting a stack that perceives the real-time movement of every atom and bit across the global ecosystem, the leader gains the "Perceptual Edge" to identify disruptions—such as a port strike or a raw material shortage—days or weeks before they impact the bottom line. The ultimate goal is the **Purification of the Logistics Signal**, ensuring the firm sees the "True State" of global trade, allowing the leader to authorize shifts in production and inventory with a level of temporal sovereignty that turns market chaos into a competitive advantage.

The Sensory Layer: Planetary-Scale Signal Ingestion

The **Sensory Layer** of the Operations Forecaster acts as the brand's "Global Sensory Array." It moves beyond internal ERP data into **Multimodal Planetary Ingestion**. This layer leverages a convergence of **Satellite Imagery**, **IoT Container Sensors**, and **Geopolitical Sentiment Feeds** to capture the state of the world in high fidelity. Whether it is the dwell time of ships in a specific harbor, the weather patterns threatening a distribution center, or the financial health of a Tier-3 supplier, this layer "labels" the operational risk at the point of origin. By utilizing **Edge Intelligence**, it identifies "Weak Signals"—distinguishing between routine delays and systemic collapses. This ensures the leader's "Eyes" are receiving a raw stream of "Operational Truth" with zero latency, removing the informational gaps that traditionally leave firms vulnerable to unforeseen global shocks.

The Processing Layer: Causal Flow Mapping & Digital Twins

The **Processing Layer** serves as the "Digital Nervous System" of the operations stack. It transforms raw planetary signals into a structured understanding of **Systemic Flow**. This layer utilizes **Graph Neural Networks (GNNs)** and **Physics-Informed Simulations** to index the relationships between different nodes in the value chain. While traditional databases see isolated shipments, the Forecaster's processing layer sees the "Causal Network"—understanding, for example, how a labor shortage in a magnesium mine will impact the production of specific high-margin electronics six months later. By mapping these signals into a **Live Operational Digital Twin**, the system can simulate the future flow of goods under various disruption scenarios. It moves the organization from "Inventory Management" to "Resilience Logic," providing a "Logical Foundation" for immediate procurement pivots. This layer ensures the brand's "Operations Compass" is always calibrated to the physical realities of global logistics.

The Cognitive Layer: Predictive Logistics & Resourcing Foresight

The **Cognitive Layer** is the "Reasoning Mind" of the operations stack, where global signals are converted into **Strategic Resourcing Intent**. This layer utilizes **Generative AI** and **Multi-Agent Simulation** to synthesize the logic of the Processing Layer into "Strategic Positioning Alerts." Instead of showing the leader a shipping delay, the Cognitive Layer generates "Proactive Scenarios." It acts as an "Operations Sentry," using **Natural Language Generation** to tell the leader: "Based on the rising energy costs in Europe and the current shipping congestion in the Suez, there is an 85% probability of a 12% cost increase in Q3; we should pre-buy inventory in the North American hub now." This allows the executive to move from "Managing Logistics" to "Leading Global Flows," authorizing pre-emptive capital allocation. It turns the "Forecaster" into a "Proactive Sentry," ensuring the brand is always positioned in the most value-dense areas of the global economy.

OVERVIEW

Predictive AI forecasts future outcomes by analyzing vast historical data using machine learning algorithms. For the "Intuitive & Risk" leader, the AI becomes a crucial tool for quantifying uncertainty. The leader provides their initial intuition and acceptable risk limits, framing the precise question. The AI delivers objective, quantifiable evidence, such as likelihoods and key drivers, that either validates the leader's gut feeling or proposes data-driven alternatives. This synergy converts subjective risk into a calculated, strategic choice, empowering the leader to make proactive, decisive actions.

MAIN STEPS

Start

STEP 1 — Define Objective & Scope
Specify question, target outcome, and limits

STEP 2 — Data Collection & Preparation
Gather, clean, and format historical inputs

STEP 3 — Model Selection & Training
Build patterns from cleaned training data

STEP 4 — Evaluation & Validation
Test model accuracy on unseen data

STEP 5 — Deployment & Integration
Insert model into real-time operations

STEP 6 — Monitoring & Refinement
Continuously check and improve model performance

End

SAMPLE FLOW

Logistics - Supply Chain Disruption Risk- A COO predicts global supply chain disruptions using AI, allowing proactive inventory adjustments to minimize production halting risks.

1. Map Global Dependencies
Chart all necessary materials, suppliers, and transit routes required for production.

2. Ingest Geopolitical Data
Collect news, weather, and logistics delays data for risk modeling.

3. Model Delay Propagation
Train on historical event impacts to predict shipment delays and factory shutdowns.

4. Critical Inventory Gaps
Identify specific parts and production lines most vulnerable to forecast disruptions.

5. Find Financial Exposure
Calculate the total cost of potential production stoppage and missed delivery penalties.

6. Do Emergency Sourcing
Authorize spot buys or rerouting shipments to bypass high-risk geographical zones.

USE CASES

- **Construction | Project Schedule Overruns:** Predicts project schedule overruns by analyzing subcontractor performance data and resource allocation bottlenecks, allowing leaders to reassign teams immediately and avoid penalty fees.
- **Hospitality | Peak Staffing Needs:** Forecasts peak staffing needs hour-by-hour by modeling customer flow and reservation rates, ensuring high service quality while preventing costly labor over-scheduling and inefficiency.
- **Mining | Equipment Failure Risk:** Predicts equipment failure risk in remote drilling sites based on vibration and temperature telemetry, enabling preventative maintenance dispatch and drastically reducing unexpected operational downtime.
- **Agriculture | Yield Loss Forecasting:** Forecasts optimal harvesting times and potential yield loss due to localized pest infestation and weather patterns, guiding farm managers to deploy labor and resources exactly when needed.
- **E-commerce | Warehouse Fulfillment Errors:** Predicts warehouse fulfillment errors by analyzing complex order processing steps and worker fatigue patterns, implementing corrective process changes to lower shipping mistakes and returns.
- **Government | Infrastructure Maintenance Failure:** Forecasts maintenance failure on public infrastructure like bridges or roads based on sensor data and usage history, prioritizing repair crews to minimize public safety hazards and maximize asset longevity.

In the **Operations Transformation**, the objective is to move from "Linear Efficiency" to **"Anticipatory Resilience."** This shift transforms the organization from a reactive chain of events into a "Sovereign Logistics Organism" that can absorb shocks and self-optimize in real-time. By applying the six leadership roles and their technical pairings, the leader ensures that the supply chain and internal processes are not just lean, but are contextually aware, ethically transparent, and capable of autonomous offensive action.

Operations Transformation: The Forecasters (The Eyes)

In the **Operations Transformation**, the **Forecasters (The Eyes)** serve as the "Systemic Radar." While the primary instrument of **Predictive AI** identifies seasonal demand trends, it often fails to see the "Hidden Fragility" in tiered supply chains or the impact of geopolitical shifts. To move from "Inventory Forecasting" to "Systemic Foresight," a leader must integrate these six secondary pairings to sharpen the "Eyes," ensuring the organization sees the ripple effects of a global disruption before it reaches the warehouse floor.

1. The Graph Neural Network (GNN) Pairing for Tiered Visibility

Pairing **Predictive AI with GNNs** is the ultimate tool for **Mapping Supply Chain Interdependence.** While standard AI looks at direct suppliers (Tier 1), GNNs analyze the "Relationships" deep into Tier 2 and Tier 3. This is best utilized to identify "Single Points of Failure" in raw material sourcing. By seeing the operation as a web, the Forecaster can predict how a localized labor strike or power outage at a sub-component factory will stall the entire enterprise, allowing for pre-emptive sourcing shifts.

2. The Causal Inference Pairing for Bottleneck Attribution

Leaders should pair **Predictive AI with Causal Inference** to solve the "Efficiency Paradox." Standard AI might show that a certain port is slow, but Causal Inference proves if the port *caused* the delay or if the delay was a result of upstream documentation errors. This is best deployed when redesigning logistics flows. It ensures the leader is fixing the actual root cause of friction rather than just treating the symptoms of a delayed shipment, securing a high-fidelity operational map.

3. The Digital Twin Pairing for Network Stress-Testing

When expanding into new territories, **Predictive AI must be paired with Operational Digital Twins.** This is best utilized to "rehearse" the impact of extreme events—such as a 40% spike in fuel costs or a major maritime route closure—in a virtual sandbox. The leader can simulate a "Total Network Stress-Test." This pairing ensures the "Eyes" see the breaking points of the logistics infrastructure before they are reached in reality, allowing for the hardening of the supply chain against volatility.

4. The Natural Language Processing (NLP) Pairing for Geopolitical Intelligence

To capture the "Signal in the Noise" of global events, a leader should pair **Predictive AI with Advanced NLP.** While quantitative models track shipping volumes, NLP analyzes unstructured data like news reports, port strikes, and regulatory filings in multiple languages. This identifies "Early Warning Signals" of operational disruption that numbers alone would miss. This turns the Forecaster into a global sentinel, ensuring the brand's "Eyes" are sensitive to the socio-political narratives shaping global trade.

5. The Monte Carlo Pairing for Inventory Probability

To navigate the uncertainty of "Just-in-Time" logistics, leaders should pair **Predictive AI with Monte Carlo Simulations.** This is best utilized when setting safety stock levels for volatile components. By running millions of variables—from weather patterns to vessel availability—the Forecaster can map the "Out-of-Stock Probability Cloud." This allows the leader to move from a single "Optimal Inventory" number to a range of mathematically grounded buffer levels, ensuring resilience without bloated capital costs.

6. The Knowledge Graph Pairing for Skill and Resource Lineage

In an era of rapid operational scaling, **Predictive AI should be paired with Knowledge Graphs.** This pairing tracks the **"Provenance of Resource"**—mapping exactly which facilities, machines, and teams are best suited for specific high-stakes tasks. It allows the leader to see the "Lineage" of operational success. By identifying the core "Capability Nodes" that drive efficiency, the Forecaster ensures the brand's "Eyes" can perfectly match incoming demand with the firm's most resilient operational assets.

Operational Power: Explainable Logic

In a hyper-automated operational environment, the "Voice" of the system is the "Policy" and the "Justification" for every move. Traditional operations are often "Rule-Based" and "Rigid," leading to "Computer Says No" moments that frustrate employees and customers. The transformation into **Diplomats** focuses on **Semantic Operational Integrity**. In this role, the operational system is designed to "Explain the Why." It bridges the gap between the "Efficiency" of the machine and the "Intent" of the human orchestrator, ensuring that every automated decision is grounded in "Brand Values" and "Ethical Logic."

Defensive Stance: Killing the Friction of Opacity

The defensive mission of the Operational-Diplomat is the **Elimination of Algorithmic Bias**. As operations move toward machine-speed, there is a risk of the system making "Efficient but Unethical" or "Logical but Toxic" decisions. The Diplomat role kills this "Reputational Risk" by establishing "Explainable Guardrails." This is the **Shield of Operational Trust**. By ensuring that every automated procurement, every labor allocation, and every pricing shift is "Transparent and Justifiable," the Diplomat protects the brand from internal and external backlash. This defensive posture ensures the operations are always "Principled," killing the risk of "Systemic Unfairness" and ensuring the brand is seen as a "Fair Actor" in the market, protecting the "Social License" to operate at scale.

Offensive Stance: Unlocking Synchronized Intent

On the offense, the Diplomat moves toward **Policy-as-a-Service**. This is the ability of the operational system to "Communicate and Align" with external partners and internal teams in real-time. The system uses its "Voice" to "Negotiate" the best outcomes based on the "Brand's Logic." This **Unlocks Strategic Harmony**, where the operations layer becomes a "Strategic Negotiator." It doesn't just "buy" or "sell"; it "Aligns" the entire ecosystem with the brand's long-term goals. This turns the operational "Voice" into a "Force Multiplier" that persuades vendors, aligns remote teams, and inspires stakeholder confidence. By "Mastering the Dialogue" between the "Code" and the "Customer," the brand creates a "Resonant Operation" that is impossible for a purely transactional competitor to disrupt.

The Operational Outcome: Institutional Alignment

The ultimate value of the Diplomat role is **Alignment through Transparency**. When operations act as a "Diplomat," the "Why" behind the "How" is never in doubt. For the leader, this provides **Ethical Security**. You know that your "Hands" (the execution) are being guided by a "Voice" (the policy) that is resonant, ethical, and backed by the "Total Evidence" of your brand's mission. This turns the organization into a "High-Integrity Powerhouse"—where every operational move is a "Reinforcement of the Brand's Promise," creating a culture of "Principled Efficiency" that everyone can support.

Executive Power: Verifiable Operational Trust

For the modern executive, the "Voice" of operations is the **Verifiable Integrity** of the entire supply chain. In an era of "Radical Transparency" and ESG (Environmental, Social, and Governance) scrutiny, the **Diplomat** role ensures the brand's operational conduct is not a marketing claim, but a "Verifiable Truth" baked into the data. This role is about the **Abolition of the Invisible Supply Chain**, ensuring that the "Voice" of the organization's operations is the most "Ethical" and "Transparent" in the market. The leader uses this stack to ensure the brand's operational standards are articulated with absolute clarity for regulators, activists, and customers. By architecting this role, the executive ensures that every "Atom" the brand moves is accounted for with integrity, moving the company from "Risk Management" to "Value Alignment."

The Sensory Layer: Provenance & Ethical Ingestion

The **Sensory Layer** of the Operations Diplomat acts as the organization's "Integrity Ear." It moves beyond static vendor contracts into **Continuous Provenance Capture**. Utilizing **Blockchain-backed Supply Chain APIs** and **Automated Ethical Audits**, this layer tracks the "Life of a Component" from the mine to the warehouse. It captures the "Context of Production"—understanding the carbon footprint, the labor conditions, and the environmental impact of every supplier. By monitoring for "Ethical Drift" and "Compliance Anomalies," this layer provides the leader with a real-time map of "Operational Friction." It ensures the executive is aware of any potential reputational risk before it manifests in a public scandal. This layer provides a high-fidelity "Map of Provenance," ensuring the "Voice" of the firm is grounded in the absolute reality of its physical conduct across the entire global network.

The Processing Layer: Semantic Governance & Supply Chain Logic

The **Processing Layer** serves as the "Guardian of the Operational Promise." It utilizes **Knowledge Graphs** to link the brand's public ethical goals and sustainability commitments to the specific procurement and logistical data in the system. If the executive's "Voice" claims a commitment to "Net Zero" or "Fair Trade," this layer "Hard-Codes" those rules into the **Procurement Logic**. This ensures that the leader's message is not just a slogan, but is mathematically reflected in the organization's actual purchasing and logistics decisions. It acts as a **Shield of Semantic Integrity**, ensuring that the brand's operational "Voice" is never contradicted by its data. By resolving the identity of every supplier in the context of the brand's mission, this layer kills the friction of "Information Gaps," providing a "Verifiable Logic" that the company's operations are perfectly aligned with its public values, thereby building an unassailable bond of institutional trust.

The Cognitive Layer: Resonant Operational Advocacy & Disclosure

The **Cognitive Layer** is where the "Operational Truth" is transformed into **Explainable Narratives** for the world. Utilizing **Large Language Models (LLMs)** and **Explainable AI (XAI)**, the Diplomat generates "Dynamic Transparency Reports." It doesn't just provide a spreadsheet; it orchestrates "Sovereign Explanations"—tailoring the operational status story for a rigorous regulator, a sustainability-focused investor, or an ethically-minded consumer. This layer ensures the organization can "Explain its own Supply Chain," providing a reasoned justification for its procurement choices and its environmental impact. This allows the leader to project a "Voice" of **Absolute Operational Accountability**, turning "Compliance" into a "Competitive Advantage." It ensures that every stakeholder feels a sense of "Security and Alignment" with the brand's physical conduct, creating a culture of "Principled Operational Excellence."

OVERVIEW

Explainable AI (XAI) provides transparent, human-readable rationales for AI decisions, moving beyond black box outputs. It reveals why a model reached a conclusion, building trust and mitigating legal and ethical exposure. The Regulatory & Compliance leader leverages XAI to validate complex, automated decision-making processes, particularly in highly regulated domains. This synergy ensures AI systems are fair, auditable, and adhere to internal policies and external legal requirements, transforming compliance from a check-the-box function into a strategic enabler of trust.

MAIN STEPS

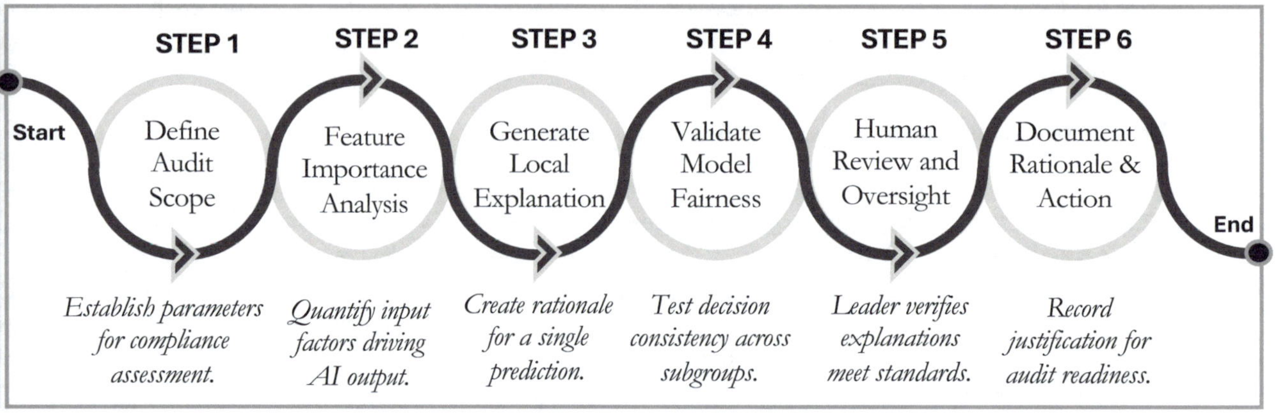

SAMPLE FLOW

Automated Workplace Safety Violation Audit: A leader utilizes Explainable AI (XAI) to audit the system that detects and reports safety violations, ensuring compliance with labor and safety regulations..

1. OSHA Safety Parameters
Define all governmental and internal operational safety standards clearly.

2. Analyze Violation Severity
Trace sensor and video data inputs that triggered the safety alert accurately.

3. Create Root Cause Rationale
XAI explains the exact operational condition that led to the safety incident or risk.

4. Simulate Precedent History
Compare the current violation against historical events to assess the recurrence risk level.

5. Assess Fine Exposure Risk
XAI calculates the potential regulatory penalty based on the severity/frequency of the violation.

6. Issue Corrective Action Plan
Leader approves mandatory retraining and documents the full compliance failure rationale.

USE CASES

- **Manufacturing | Supply Chain Traceability Audit:** Explains why specific component batches are flagged for quality or origin issues, allowing leaders to certify the supply chain is ethically and legally compliant consistently.
- **Financial Services | Transaction Processing Compliance:** Provides transparent rationale for the AI's approval or rejection of high-value transactions, ensuring adherence to anti-fraud and reporting regulations reliably.
- **Healthcare | Medical Device Performance Validation:** Explains the operational data that verifies a new medical device is performing within FDA-approved safety and efficacy parameters safely.
- **Technology | Service Uptime Justification:** Justifies AI-driven resource allocation decisions that maintain service uptime, proving the operational system meets mandated service level agreements (SLAs) quickly.
- **Retail | Inventory Disposal Compliance:** Explains the factors (spoilage, damage) that trigger AI recommendations for product disposal, ensuring compliance with environmental and waste management laws accurately.
- **Energy | Emissions Reporting Verification:** Provides transparent rationale for calculated greenhouse gas emissions data, certifying the operational reporting meets all environmental protection agency standards quickly.

In the **Operations Transformation**, the **Diplomats (The Voice)** serve as the "Sovereign Negotiator" across the supply chain. While **Explainable AI (XAI)** provides the logic for operational decisions—such as why a specific shipment was rerouted or a vendor was de-prioritized—it often lacks the transparency required to maintain trust with external partners. To move from "Zero-Sum Procurement" to "Collaborative Resilience," a leader must integrate these six secondary pairings. This ensures the organization's "Voice" is not just efficient, but is also seen as a fair, predictable, and trusted anchor in a volatile global market.

1. The Sentiment Analysis Pairing for Vendor Relationship Health

Pairing **Explainable AI with Sentiment Analysis** is best utilized for **Proactive Supplier Management**. While XAI explains why a contract term is being enforced, Sentiment Analysis monitors the tone of communications from key vendors. If the "Voice" of a critical supplier shows signs of distress or dissatisfaction, the Diplomat can intervene before a contract breach occurs. This ensures the brand's operational voice is empathetic, protecting the health of the "Human Network" that keeps the physical supply chain moving.

2. The Contextual NLP Pairing for Cross-Border Protocol

Leaders should pair **Explainable AI with Contextual NLP** to ensure **Operational Fluency** across global regions. XAI provides the "What" and "Why" of a logistics directive, but Contextual NLP translates that into the specific cultural and technical dialect of a local port authority or a third-party warehouse. This ensures the brand's "Voice" is never lost in translation, reducing administrative friction and ensuring that complex operational logic is understood and executed correctly in every corner of the world.

3. The Knowledge Graph Pairing for Contractual Lineage

When negotiating complex multi-year agreements, **Explainable AI must be paired with Knowledge Graphs**. While XAI justifies a current price point, the Knowledge Graph provides the "History of the Partnership." This allows the "Voice" to reference past concessions, loyalty milestones, and shared successes. This pairing creates a "Verifiable Negotiation," turning the Diplomat into a historian of the partnership who can defend the brand's interests using a deep, contextually grounded record of mutual value.

4. The Differential Privacy Pairing for Secure Collaboration

To build a "Shared Intelligence" with competitors or partners without leaking secrets, a leader should pair **Explainable AI with Differential Privacy**. This is best used in **Logistics Co-ops** (e.g., sharing shipping lane data to reduce carbon footprints). The Diplomat can explain the benefits of a shared route while mathematically guaranteeing that the firm's proprietary volume data remains hidden. This allows the organization's "Voice" to lead industry-wide initiatives without compromising its competitive sovereignty.

5. The Conversational AI Pairing for Real-Time Logistics Interrogation

To eliminate the "Black Box" of supply chain management, leaders should pair **Explainable AI with Conversational AI**. While XAI holds the logistics logic, Conversational AI allows field managers or vendors to "interview" the system: "Why was this cargo given priority over the others?" This turns the Diplomat into a **Real-Time Arbitrator**. By providing instant, transparent justifications for priority shifts, the "Voice" prevents internal and external disputes, ensuring everyone in the chain understands the strategic rationale behind every move.

6. The Blockchain Pairing for Proof of Ethical Sourcing

In a world demanding radical transparency, **Explainable AI should be paired with Blockchain**. XAI provides the "Explanation" for an operational claim—such as a "fair-trade" or "low-carbon" status—while the Blockchain provides the immutable proof. This is best utilized for **Regulatory Sovereignty**. It ensures the "Voice" of the brand is backed by an unassailable audit trail, protecting the organization from "greenwashing" accusations and securing its moral authority in the eyes of regulators and consumers alike.

Operational Power: Generative Workflow Manifestation

In the legacy era, operations were "Static." A process was designed, documented in a manual, and then repeated until it broke or became obsolete. This created a "Rigid Architecture" that could not keep pace with a changing world. The transformation into **Creators** represents the shift from "Managing a Process" to **"Orchestrating a Generative Engine."** In this role, the operations layer possesses the ability to "Self-Design." It uses the brand's generative power to manifest new workflows, supply routes, and organizational structures on the fly. The operations cease to be a "Fixed Track" and become a "Fluid Response," moving from a state of "Rigid Efficiency" to "Dynamic Adaptation."

Defensive Stance: Killing the Friction of Process Rot

The defensive mission of the Operational-Creator is the **Elimination of Procedural Obsolescence**. Every organization eventually suffers from "Process Rot"—old rules and workflows that no longer serve the current reality but continue to consume resources. The Creator role kills this "Inertia" by "Purifying the Value Stream" in real-time. It uses its "Engine" to identify "Digital and Physical Waste" and "Generates" a leaner alternative instantly. This creates a **Shield of Operational Freshness**. Instead of waiting for a "Consultancy Review," the operations "Self-Optimize" the moment a friction point is detected. This defensive posture kills the "Waste of Stagnation," ensuring the brand's "Engine" is always running on the most modern logic. It protects the organization's "Margins" by ensuring that the "Cost of Doing Business" never creeps upward due to outdated habits or legacy complexity.

Offensive Stance: Unlocking Scalable Customization

On the offense, the Creator moves toward **Infinite Operational Flexibility**. This is the ability of the operations layer to "Manifest" a unique fulfillment path for every single order or project. The system doesn't just "Follow a Rule"; it "Generates a Path." If a customer requires a unique delivery window or a personalized product configuration, the operations "Engine" creates the specific logistics and manufacturing sequence required to fulfill it without manual intervention. This **Unlocks Creative Agency**, where the "Hands" of the company can manifest any "Voice" or "Intent" perfectly. It allows the brand to "Saturate the Market" with a level of variety that would paralyze a manual competitor. This turns the operations into a "Platform for Manifestation," ensuring the brand is always producing the most "Original" and "Relevant" results, because the "Engine" is literally "Creating" the best way to win in every specific moment.

The Operational Outcome: Absolute Plasticity

The ultimate value of the Creator role is **Radical Agility**. When operations "Create," the concept of a "Bottleneck" dies. For the leader, this provides **Execution Sovereignty**. You are no longer limited by "How things are done"; you are limited only by your "Intent." The operations engine can manifest the "Next Move" instantly. This turns the organization into a "Living Response System"—one that evolves its "Internal Mechanics" in real-time to match the "Market Signals." It ensures that the brand's "Engine" is never "Finished," but is always working to manifest the highest possible efficiency for the business and the highest possible value for the customer.

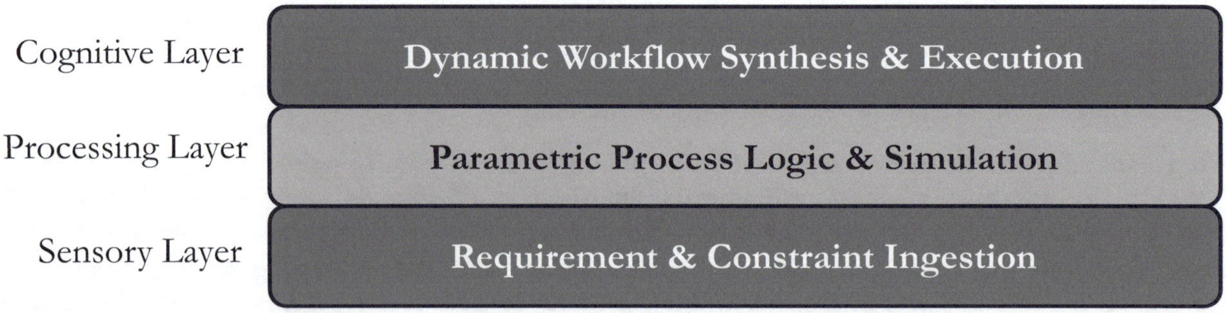

Executive Power: Generative Process Manifestation

The **Creator** role in operations transformation represents the shift from "Fixed Workflows" to **"Generative Process Manifestation."** In the legacy enterprise, operations were rigid—any change in production or logistics required months of re-engineering. The Creator role replaces this with **Autonomous Workflow Synthesis**, where the operational architecture itself is dynamically generated in response to the leader's intent and the world's constraints. The executive uses this engine to move the organization from a state of "Fixed Assets" to "Fluid Capability," ensuring the brand is always manifesting the most "Efficient" fulfillment path. This turns the "Engine" of the firm into a liquid resource that can "re-invent" its own logistical and manufacturing logic to solve for sudden market spikes or supply shocks.

The Sensory Layer: Requirement & Constraint Ingestion

The **Sensory Layer** for the Creator captures "Operational Requirements" as high-definition signals. It moves beyond traditional production planning by integrating **Atomic Requirement Capture** and **Resource Availability Streams**. By utilizing **Agentic Discovery** to scan the entire enterprise asset base and the global vendor market, this layer identifies the "Atomic Components" of an operational goal before a process is ever designed. It captures the raw intent of the leadership—"We need to shift production of our core line to South America within 30 days"—and automatically identifies the parameters required for success. This ensures the leader is building the "Engine" of the project on a foundation of "Available Reality" rather than "Legacy Workflows," capturing the necessary signals for a bespoke operational manifestation as a live, fluctuating requirement.

The Processing Layer: Parametric Process Logic & Simulation

The **Processing Layer** acts as the "Architectural Refinery" for operational evolution. It utilizes **Constraint-Based Solvers** and **Digital Twin Simulations** to manifest a workflow that is both "Optimal" for the goal and "Viable" for the budget. This layer understands the "Rules of the Possible"—including labor laws, shipping lead times, and manufacturing tolerances. It stress-tests every generative process against the organization's "Causal Truths" in a virtual sandbox, ensuring the executive only authorizes changes that are structurally and financially sound. It acts as a **Guardian of Feasibility**, filtering out high-risk "hallucinations" in the autonomous logic. By applying "Parametric Operational Logic," this layer ensures the "Engine" is disciplined, preventing the brand from manifesting processes that are unstable, while ensuring the organization receives a fulfillment path that feels uniquely engineered for the current global context.

The Cognitive Layer: Dynamic Workflow Synthesis & Execution

The **Cognitive Layer** is the "Orchestration Mind" of the operations engine, powered by **Large Language Models (LLMs)** and **Action Transformers**. This layer "hallucinates" thousands of different ways to fulfill the operational intent, presenting the most "Efficient" version for autonomous deployment. It doesn't just suggest a plan; it manifests the **Functional Architecture**. The leader provides the strategic guardrails, and the Cognitive Layer generates the "Workflow Genome"—automatically writing the procurement scripts, configuring the shipping APIs, and optimizing the production schedules required to fulfill the mission. It involves "Recursive Process Evolution," where the AI adapts the workflow as it runs, ensuring the final output is the "Optimal Manifestation" of the brand's intent. This turns the executive into a "Conductor of Execution," capable of deploying complex, individualized operational paths at a speed that makes traditional change management look prehistoric.

OVERVIEW

Generative AI creates novel outputs like text, art, code, and immersive environments from simple prompts. It excels at rapid prototyping, style transfer, and exploratory ideation. The Creative & Artistic leader partners with Generative AI to rapidly visualize strategic concepts, explore diverse future scenarios, and prototype new organizational narratives. The synergy transforms vision casting from a slow, manual process into a continuous, inspiring, and engaging collaborative endeavor.

MAIN STEPS

Start

STEP 1 — Define Creative Prompt
Articulate desired output parameters precisely.

STEP 2 — Initial Content Generation
Create diverse drafts based on input.

STEP 3 — Style and Mood Adaptation
Adjust tone, format, or aesthetic output.

STEP 4 — Iterative Refinement Loop
Guide model toward final creative goal.

STEP 5 — Human Curation & Selection
Choose the most impactful creative result.

STEP 6 — Deploy Creative Asset
Integrate final output into strategy.

End

SAMPLE FLOW

Designing Efficient Warehouse Layouts: An Operations Director uses Generative AI to generate thousands of optimal, customized warehouse and fulfillment center layout options for Facilities to review.

1. Input Operational Constraints
Define inventory size, flow patterns, and equipment specifications precisely.

2. Generate Layout Blueprints
Produce thousands of unique, optimized 3D and 2D floor plans instantly from text.

3. Simulate Efficiency Flow
Model material handling and worker paths to test time and distance rigorously.

GENERATIVE AI

4. Refine Based on Physics
Adjust rack heights, aisle widths, and zone placements based on simulation data.

5. Select Final Design Portfolio
Curate & recommend the top 10 most space-efficient and cost-effective layout configurations.

5. Present New Facility Design
Present the final validated and optimized layout plan for structural and engineering evaluation.

USE CASES

- **Manufacturing | Supply Chain Visualization:** Generates artistic, interactive digital twin models of the global supply chain, helping leaders intuitively identify congestion points and visualize risk scenarios in real-time.
- **Financial Services | Process Automation Storytelling:** Creates animated videos and interactive guides that clearly illustrate new robotic process automation (RPA) workflows to operations teams, easing adoption concerns.
- **Healthcare | Surgical Simulation Scenarios:** Generates complex, high-fidelity virtual reality scenarios for surgical teams, allowing them to creatively rehearse rare or critical operational procedures safely and efficiently.
- **Technology | Data Center Heat Mapping:** Develops abstract, real-time visual art from server temperature and utilization data, enabling operations staff to instantly and creatively spot and resolve overheating patterns.
- **Retail | Store Traffic Optimization:** Generates 3D models of optimal in-store traffic flow patterns and checkout lane configurations, maximizing customer comfort and significantly reducing wait times during peak periods.
- **Energy | Maintenance Scheduling Graphics:** Creates intuitive, graphical work schedules and asset repair visualizations for field teams, ensuring that maintenance operations are communicated clearly and executed efficiently.

In the **Operations Transformation**, the **Creators (The Engine)** represent the "Process Innovation Lab." While the primary instrument of **Generative AI** can design more efficient warehouse layouts or draft procurement contracts, it often lacks the "Material Reality" to ensure these designs are physically possible or legally resilient. To move from "Theoretical Efficiency" to "Autonomous Process Evolution," a leader must integrate these six secondary pairings. This ensures the "Engine" doesn't just suggest improvements, but generates operational blueprints that are structurally sound, resource-aware, and aligned with the brand's sovereign standards.

1. The Synthetic Data Pairing for Disruption Modeling

Pairing **Generative AI with Synthetic Data Generation** is the ultimate tool for **"Black Swan" Preparedness.** Before a supply chain is finalized, the Creator generates millions of "Synthetic Disruptions"—simulating simultaneous port strikes, extreme weather, and sudden raw material shortages. This is best utilized in "Hardening the Network." By training the engine on these high-fidelity crisis scenarios, the leader ensures that the generated operational plans include "Digital Resilience" by design, rather than as an afterthought.

2. The Digital Twin Pairing for Layout Optimization

Leaders should pair **Generative AI with Operational Digital Twins** to achieve **Rapid Facility Iteration.** While the AI generates thousands of potential configurations for a factory floor or sorting center, the Digital Twin provides the "Physics-Based Validation." This is best used in "Automated Fulfillment." The Engine dreams of a new robotic pathing logic, and the Digital Twin instantly tests it for collisions and throughput bottlenecks. This pairing ensures that only the most efficient, crash-proof layouts are ever implemented in the physical world.

3. The RLHF (Human Feedback) Pairing for Value-Aligned Logic

When the "Engine" creates the rules for operational prioritization—such as which orders get shipped first during a shortage—**Generative AI must be paired with Reinforcement Learning from Human Feedback (RLHF).** This ensures the output reflects the firm's "Sovereign Ethics." Human operations experts provide the "Nuance" and "Moral Judgment" that raw models lack, such as prioritizing medical supplies or long-term loyal partners over a quick profit. This turns the Creator into a "Strategic Asset," ensuring the brand's soul is encoded into every automated operational decision.

4. The Vector Database (RAG) Pairing for Grounded Procurement

To prevent "Operational Hallucinations," a leader should pair **Generative AI with Vector Databases (RAG).** This pairing ensures that every generated contract or procurement strategy is grounded in the firm's actual historical pricing, vendor performance records, and legal templates. This is best utilized for "Intelligent Sourcing." Instead of the AI making up vendor terms, it retrieves the "Institutional Wisdom" of past successful negotiations to suggest strategies that are legally sound and competitively superior.

5. The Differential Privacy Pairing for Ecosystem Co-Creation

To enable "Open Innovation" in the supply chain without exposing proprietary secrets, leaders should pair **Generative AI with Differential Privacy.** This allows the brand to share anonymized operational data with logistics startups or university researchers. The generative model can learn from "Communal Patterns" of shipping efficiency while the privacy layer ensures that no specific company secret or customer location is ever revealed. This turns the operation into a "Living Lab," where the brand can benefit from global innovation without losing its competitive sovereignty.

6. The API Orchestration Pairing for Instant Process Manifestation

To move from "Process Design" to "Live Execution," **Generative AI should be paired with API Orchestration.** While the AI creates a new workflow for a sudden surge in orders, API Orchestration provides the "Connection" to actually trigger the new hiring requests, shipping labels, and inventory transfers. This moves the "Engine" from a theoretical tool to a live "Operational Nervous System." It ensures the organization can pivot its entire physical operation in real-time, closing the gap between a strategic shift and its manifestation in the real world.

Operational Power: Causal Value Optimization

Traditional operations are obsessed with "Linear Efficiency"—doing the same thing faster or cheaper. However, "Faster" is not always "Better" if the goal is long-term value. The transformation into **Strategists** moves operations into the realm of **Causal Value Chain Optimization.** In this role, the operations don't just "Perform"; they "Navigate." The system understands the "Mathematical Why" behind every operational move and uses its "Compass" to find the path that maximizes the *total* outcome, not just a single siloed metric. The operations evolve from a "Passive Utility" into a "Strategic Navigator" that manages the company's "Value Peak."

Defensive Stance: Killing the Efficiency Trap

The defensive mission of the Operational-Strategist is the **Validation of True Impact**. Many companies fall into the "Efficiency Trap," where they cut costs in one area (like shipping) only to cause a massive loss in another (like customer retention). The Strategist role kills this "Strategic Friction" by applying "Causal Logic" to the entire value chain. It acts as the "Guardian of the Bottom Line," identifying when a "Local Optimization" is actually a "Global Failure." This is the **Shield of Objectivity**. By "Modeling the Risk" of every operational shift, the leader kills the "Mismatch" between effort and reward. It prevents the organization from "Chasing Vanity Savings" and ensures that every resource is allocated to the "True Drivers" of sustainable profit. This defensive posture protects the brand from "Short-Termism" and ensures the "Compass" is never distorted by the "Noise" of a single department's goals.

Offensive Stance: Unlocking Prescriptive Orchestration

On the offense, the Strategist moves toward **Dynamic Prescriptive Power**. This is the ability of the operations to "Command the Market" through superior navigation. The "Compass" doesn't just show the status; it "Prescribes the Best Move" to capture a new opportunity. Whether it is "Simulating the Win" for a new market entry or "Optimizing the Synchronized Symphony" of global logistics and local pricing, the operations are "Strategizing the Outcome." This **Unlocks Navigational Sovereignty**, where the operations layer becomes the "Operating System for Profit." It involves finding the "Path of Least Resistance" that competitors, limited by their linear spreadsheets, will never see. This turns the organization into a "Calculating Winner" that can pivot its weight in milliseconds to find the most "Value-Dense" ground. It allows the brand to "Own the Outcome," ensuring that every operational second is a "Step Toward the Goal."

The Operational Outcome: Strategic Certainty

The ultimate value of the Strategist role is **Objective Confidence**. When operations act as a "Strategist," the leader feels "Navigated" rather than "Overwhelmed." For the leader, this provides **Economic Resilience**. You are building an "Optimized Machine" whose performance is mathematically linked to the fundamental laws of the market. This turns the operations into a "Strategic Ally"—one that finds the "Win-Win" in every disruption. It ensures that the brand's "Compass" is always pointing toward "Maximum Yield," creating a brand that is respected for its "Intelligence" and relied upon for its "Infallible Logic" in navigating a complex global economy.

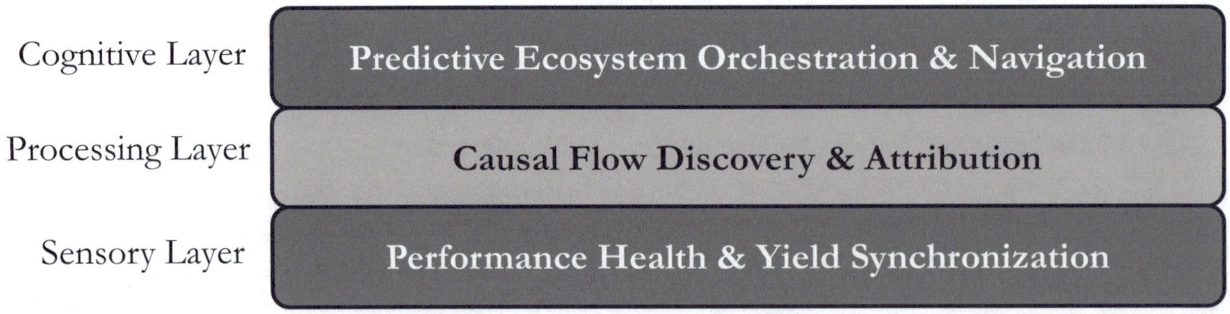

Executive Power: Causal Systemic Yield

The **Strategist** role in operations transformation represents the "Abolition of Incrementalism." It moves the executive from "Cost Cutting" to **"Causal Systemic Yield Optimization."** The leader uses this stack to ensure that every "Operational Dollar" and "Logistical Minute" is an investment in the brand's long-term systemic resilience. The Strategist is the **Guardian of the Value Stream**, ensuring that the organization is not just "Optimizing Silos," but "Maximizing Global Yield." In this role, the executive evolves into a "Strategic Ecosystem Navigator" who manages the company's "Operational Capital" with mathematical precision. This role kills the friction of "Bullwhip Effects" and "Operational Waste" by ensuring the brand's energy is always flowing toward the "True Drivers" of margin, speed, and long-term systemic health.

The Sensory Layer: Performance Health & Yield Synchronization

The **Sensory Layer** of the operations Compass monitors the "Micro-Levers" of systemic health across the global enterprise. It synchronizes **Real-Time Operational Data** (production rates, inventory turns, shipping costs) with **External Economic Context**. By tracking the "Velocity of Fulfillment" (how quickly an order moves from intent to the customer's hand) against the "Systemic Cost of Effort," this layer provides the "Coordinate Data" required to know exactly where the firm's operations stand in relation to its "Strategic North Star." It provides the "Eyes" for the Compass, ensuring that the starting point of any operational initiative is grounded in the absolute reality of current performance levels. This layer removes the "Blind Spots" in management, ensuring the leader is aware of every factor—from "Silent Backlogs" to "Logistical Friction"—that could impact the trajectory of the organization's success.

The Processing Layer: Causal Flow Discovery & Attribution

The **Processing Layer** is the heart of the Strategist role, utilizing **Causal Inference Engines** and **Structural Equation Modeling** to determine "Why" certain operational nodes are outperforming others. While traditional supply chain analytics show simple "Correlation" (e.g., our fastest suppliers are in the North), the Strategist identifies the **Causal Path** (e.g., they are faster because of a specific API integration that bypasses manual customs filing). This layer provides the "Mathematical Evidence" for operations strategy, ensuring the leader is investing in the "Levers" that actually matter for the brand's "Yield." It acts as a **Shield of Objectivity**, stripping away the "Noise" of departmental bias and focusing the executive on "Global Optimization." By identifying the "Mathematical Why," this layer ensures the firm is never "Chasing Efficiency Gains" in isolation, but is always focused on the "Causal Truths" that drive real-world profitability.

The Cognitive Layer: Predictive Ecosystem Orchestration & Navigation

The **Cognitive Layer** acts as the "Navigational Mind" of the operations stack, utilizing **Reinforcement Learning** and **Multi-Agent Simulation**. This layer "plays the game" of the global market millions of times to find the "Optimal Path of Execution." When the external environment shifts—due to a trade war or a new technological disruption—the Compass "Recalculates" the operational strategy instantly. This layer doesn't just report on operations; it "Prescribes the Best Move" to maximize the **Systemic Yield**. For example, it might recommend shifting 30% of logistics to a rail-based provider because the probability of fuel surcharges has increased. This provides the executive with **Strategic Certainty**, allowing them to navigate operational volatility with the confidence that their "Compass" is calibrated to the objective laws of systemic value. It turns the organization into a "Calculating Powerhouse," ensuring the brand "Wins" because its "Operations Strategy" is mathematically optimized for the highest possible yield and resilience.

OVERVIEW

Cognitive AI mimics human intelligence to understand and interpret unstructured data like text, speech, and images. It processes complex inputs to generate hypotheses, assess logical connections, and offer nuanced context. The Research & Analytical leader uses Cognitive AI as a hyper-efficient knowledge partner, rapidly synthesizing vast amounts of data to rigorously test assumptions, structure complex arguments, and make deeply informed, evidence-based strategic decisions.

MAIN STEPS

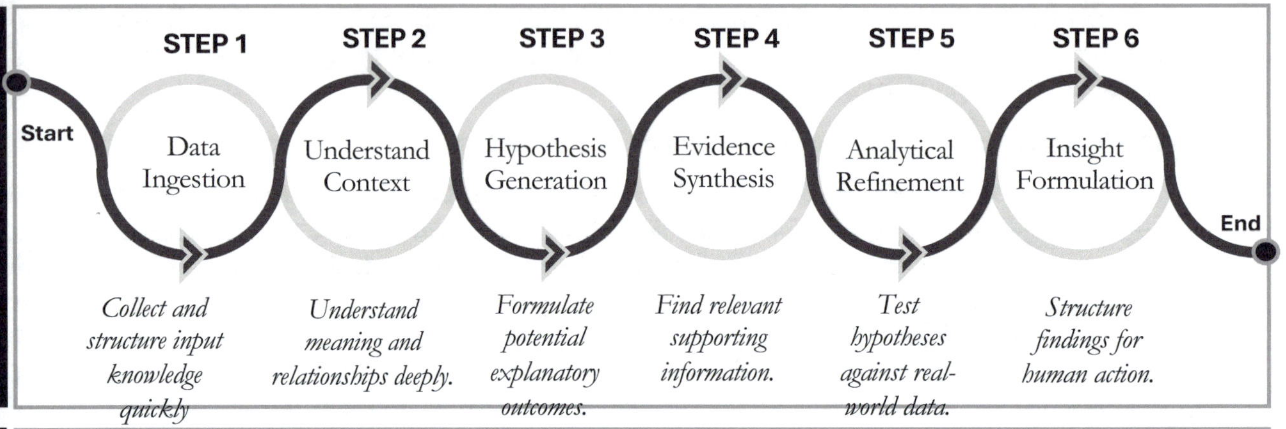

SAMPLE FLOW

Supply Chain - Resiliency and Optimization: A COO uses AI to analyze complex supply chain dynamics, rapidly identifying bottlenecks and optimizing resource flow across the network.

1. Input Logistical Constraints
Specify all network variables like vendor capacity and transit times precisely.

2. Process Unstructured Data
Ingest shipping manifests and geopolitical risk reports deeply.

3. Model Disruption Scenarios
Generate hypothetical failure modes based on past historical anomalies quickly.

4. Analyze Optimal Routing
Synthesize cost, speed, and risk factors for alternative pathways quickly.

5. Evaluate Partner Reliability
Test vendor performance history against current quality and delivery metrics rigorously.

6. Implement Adaptive Plan
Sign off on the most resilient and cost-effective supply chain solution.

USE CASES

- **Manufacturing | Factory Floor Digital Twin:** Builds a detailed, simulated environment using real-time sensor data and operational logs to test adjustments and improve throughput without interrupting production flow.
- **Financial Services | Compliance Audit Automation:** Scours vast volumes of transactional data, regulatory text, and internal policies to automate compliance checks, minimizing human error and reducing costly financial penalties.
- **Healthcare | Hospital Resource Allocation:** Optimizes the scheduling of operating rooms, staff shifts, and medical equipment using predictive models to ensure maximum patient throughput and minimal downtime.
- **Technology | Cloud Infrastructure Spend:** Analyzes application usage patterns and resource provisioning against internal budget forecasts to automatically identify and flag wasteful cloud expenditures immediately.
- **Retail | Warehouse Layout Optimization:** Interprets order fulfillment data, picking paths, and inventory density to dynamically adjust warehouse layouts for maximum efficiency and speed.
- **Energy | Grid Maintenance Scheduling:** Prioritizes infrastructure inspection and repair based on asset age, weather predictions, and consumption data to proactively prevent system-wide power failures.

In the **Operations Transformation**, the **Strategists (The Compass)** act as the "Resilience Architects." While the primary instrument of **Cognitive AI** identifies patterns in production throughput and logistics costs, it often lacks the multi-dimensional foresight to balance "Lean" efficiency with "Anticipatory" buffer capacity. To move from "Cost-Cutting" to "Strategic Agility," a leader must integrate these six secondary pairings. This ensures the "Compass" doesn't just minimize spend, but identifies the precise path to operational sovereignty in a world of fractured supply chains and volatile resources.

1. The Multi-Objective Optimization (MOO) Pairing for Resilient Buffering

Pairing **Cognitive AI with MOO** is the ultimate tool for **Balancing Lean vs. Agile.** While AI identifies the lowest-cost path, MOO balances the trade-offs between inventory holding costs, carbon footprint, and the "Risk of Stockout." This is best utilized in "Global Sourcing." It allows the leader to identify the "Efficiency Frontier," ensuring the operation isn't so lean that it shatters during a minor disruption, but is instead optimized for the highest possible throughput under variable stress.

2. The Game Theory Simulation Pairing for Sourcing Sovereignty

Leaders should pair **Cognitive AI with Game Theory** to achieve **Negotiation Superiority.** This is best deployed during "Vendor Consolidation" or "Raw Material Wars." While the AI predicts price fluctuations, Game Theory simulates how suppliers will react to different procurement strategies. This ensures the "Compass" identifies the "Strategic Anchor"—the specific vendor relationship or vertical integration move that secures the brand's supply line even when the broader market is in a state of scarcity.

3. The Digital Twin Pairing for Network Re-Architecture

When shifting from global to "Regionalized" or "Near-Shoring" models, **Cognitive AI must be paired with Operational Digital Twins.** This is best utilized to "rehearse" the impact of moving a primary factory or changing a logistics hub. The Strategist simulates how the entire network's lead times and costs will shift over a five-year horizon. This pairing ensures the "Compass" guides the physical relocation without causing a service collapse, allowing for the hardening of the new operational logic before a single brick is laid.

4. The Knowledge Graph Pairing for Skill and Asset Arbitrage

To maximize the utility of the firm's physical assets, a leader should pair **Cognitive AI with Knowledge Graphs.** While AI tracks machine uptime, the Knowledge Graph maps the "Compatibility" between different facilities, spare parts, and specialized engineering teams. This allows for "Resource Arbitrage"—identifying how a production line in one region can be quickly repurposed to solve a shortage in another. It ensures the "Compass" is always directing the brand's "Operational Alpha" toward the area of greatest systemic need.

5. The Causal Inference Pairing for Productivity Validation

For maintaining a "Results-Driven Operational Culture," leaders should pair **Cognitive AI with Causal Inference.** While AI identifies "Efficient Teams," Causal Inference proves *which* specific management style or technology tool actually drove the improvement—separating true gains from "False Correlatives." This is best utilized in "Operational Audits." It ensures the "Compass" points toward the investments that actually increase systemic output, preventing the waste of capital on "Efficiency Theater" that doesn't impact the bottom line.

6. The Agentic Orchestration Pairing for Live Flow Governance

In high-velocity logistics, **Cognitive AI should be paired with Agentic Orchestration** for **Real-Time Network Re-Routing.** While the Strategist sets the delivery goals, the Agentic layer automatically re-allocates shipping containers, warehouse shifts, or flight paths based on real-time weather, port congestion, or fuel price spikes. This is best utilized in "Omni-Channel Fulfillment." It turns the "Compass" into an active driver of the enterprise, ensuring that the brand's physical resources are always flowing toward the highest-priority commitments.

Operational Power: Structural Intelligence

In the legacy enterprise, operations were managed through "Silos"—separate departments for logistics, manufacturing, and procurement, each with its own data and its own worldview. This created a "Fragmented Reality" where information was lost in the gaps between teams. The transformation into **Architects** turns the focus toward **Structural Connectivity**. In this role, the operations are designed as a "Single Brain" that spans the entire value chain. The Architect doesn't just manage the flow of goods; they design the "Knowledge Infrastructure" that allows the organization to perceive itself as a unified whole. It is the shift from "Managing Parts" to "Designing the Systemic Mind."

Defensive Stance: Killing the Friction of Siloed Logic

The defensive mission of the Operational-Architect is the **Abolition of Fragmentation**. Most operational failures—like overstocking in one region while facing shortages in another—are the result of the "Brain" not talking to itself. The Architect role kills this "Institutional Friction" by building a "Unified Operational Knowledge Graph." By "Resolving the Identity" of every resource and process across the entire company, the leader kills the "Friction of Contradiction." This creates a **Shield of Systemic Integrity**. It ensures that the organization has a "Single Version of the Truth," protecting the brand from the errors and delays that occur when data is trapped in a silo. This defensive posture ensures the "Brain" of the operation is always "Healthy," killing the risk of "Operational Blindness" and ensuring that the brand's infrastructure is a "Fortress of Continuity" that can withstand the departure of any single individual.

Offensive Stance: Unlocking Networked Wisdom

On the offense, the Architect moves toward **Cognitive Operational Fluidity**. This is the ability of the operational "Brain" to provide "Perfect Context" for any decision. When the architecture is sound, a manager doesn't just see a "Low Inventory" alert; they see the "Network of Meaning" surrounding it—the supplier's current weather, the real-time shipping costs, and the customer's predicted demand. This **Unlocks Relevance**, allowing the organization to act with "Total Awareness." It involves "Dynamic Schema Evolution," where the operational brain learns and adapts its own rules as the business scales. This turns the organization into a "High-IQ Brand"—one that is fundamentally "Smarter" than any competitor because its "Brain" can connect dots across the globe in milliseconds. It ensures that the brand's "Wisdom" is an "Accumulative Asset" that grows more sophisticated with every transaction, making the operations "Indestructible" through their sheer structural depth.

The Operational Outcome: Institutional Immortality

The ultimate value of the Architect role is **Systemic Resilience**. It ensures that the company's "Intelligence" is embedded in the "Architecture" of the operations, rather than the "Memory" of the staff. For the leader, this provides **Operational Stability**. You are building an "Intellectual Infrastructure" that provides the same level of sophisticated context to every node in the global network. This turns the operations into a "Living Archive" that possesses "Total Recall," ensuring the brand's "Brain" is a "Global Force" that keeps the organization perfectly synchronized and infinitely scalable.

Executive Power: Institutional Operational Consciousness

In the legacy enterprise, operational knowledge was "Silod"—it existed in the heads of individual logistics managers or within the isolated databases of specific factories. The **Architect** role represents the executive's transition to the "Designer of the Firm's Collective Operational Consciousness." In this role, the leader constructs a permanent, self-evolving **Cognitive Infrastructure** that transforms every operational event into a persistent, networked "Brain." This is the shift from "Managing Tasks" to "Architecting Awareness," ensuring that the brand's "Wisdom" about its own execution is a compound asset. The Architect ensures the organization has an "Operations Mind"—one that understands the history, context, and dependencies of every move the brand makes, ensuring the company "never forgets" a logistical success or a supply chain lesson.

The Sensory Layer: Universal Entity & Event Resolution

The **Sensory Layer** of the Operations Brain acts as the "Universal Perception System." It moves beyond simple logging into **Autonomous Event Resolution**. Utilizing **Semantic Linking** and **High-Frequency Data Ingestion**, this layer ensures that every operational entity—whether a pallet, a machine, or a vendor contract—has a unique, persistent digital identity that the "Brain" can recognize across the entire ecosystem. It doesn't just record a "Movement"; it "labels" the context and intent behind that movement in real-time. By utilizing **Edge Perception Agents**, the sensory layer captures the "Nuance" of the operation—distinguishing between a routine transfer and a critical replenishment—at the point of origin. This ensures the "Brain" of the firm is constantly being fed with a high-fidelity, perfectly identified stream of operational reality, removing the "Friction of Disconnect" that traditionally leads to execution errors.

The Processing Layer: Structural Intelligence & Systemic Mapping

The **Processing Layer** serves as the "Neural Connector" of the operations stack. It utilizes **Operational Knowledge Graphs** and **Vector Databases** to define and store the complex relationships between the identities and events captured in the sensory layer. While traditional systems store data in isolated buckets, the Architect's processing layer stores the **Systemic Connectivity**. It understands, for example, the deep relationship between a specific vendor's financial health and the brand's long-term production stability. By employing **Graph Neural Networks (GNNs)**, this layer identifies "Hidden Risks" and "Efficiency Gaps" within the value chain, allowing the "Brain" to perceive the entire operation as a unified, thinking organism. This creates a **Shield of Systemic Integrity**, ensuring that when a disruption occurs in one node, the implications are immediately "felt" and calculated across the entire global operational architecture.

The Cognitive Layer: Recursive Institutional Memory & Reasoning

The **Cognitive Layer** is the "Reasoning Core" of the Operations Brain, powered by **Retrieval-Augmented Generation (RAG)** and **Recursive Feedback Loops**. This layer ensures the brand "never forgets" an execution lesson. It synthesizes the relationships from the processing layer into a "Continuous Stream of Intelligence" that acts as a permanent advisor for the firm's logistics and production teams. When an executive or an AI agent interacts with the supply chain, the Cognitive Layer doesn't just retrieve a report; it "reasons" through the entire operational history to provide an action grounded in "Deep Systemic Context." It involves **Self-Evolving Operational Schemas**, where the Brain updates its understanding of the business as new data arrives. This turns the organization into a "Learning Organism" with an ever-increasing "Operational IQ," providing the executive with a "Memory Asset" that ensures the brand's "Mind" remains perfectly synchronized with its "Physical Reality."

OVERVIEW

Human-in-the-Loop (HITL) AI is a cycle where human supervisors continuously train, tune, and validate a model's performance, especially for ambiguous tasks. The Supervisory & Coaching leader focuses on defining ethical boundaries and quality thresholds, ensuring AI decisions align with organizational values. This synergy elevates the leader from a decision-maker to a model trainer and quality assurance expert. By providing critical feedback, the leader helps the AI evolve, improving accuracy, fairness, and overall governance in complex, high-stakes leadership decisions.

MAIN STEPS

Start

STEP 1 – Initial Model Training: *Establish baseline knowledge and performance rules*

STEP 2 – Active Decision Queue: *Flag uncertain outputs for human review.*

STEP 3 – Human Supervisor Intervention: *Leader reviews, corrects, documents outcomes.*

STEP 4 – Feedback Integration Cycle: *Corrected data retrains and improves the model.*

STEP 5 – Model Validation & Testing: *Confirm updated model meets quality standards.*

STEP 6 – Deployment and Monitoring: *Deploy new model; track performance, collect data.*

End

SAMPLE FLOW

Optimizing Supply Chain Scheduling with Human Oversight: A leader uses HITL to supervise AI-generated supply chain schedules, correcting anomalies caused by unforeseen geopolitical or local constraints..

HUMAN-IN-THE-LOOP AI

1. Input Ethical Standards
Define fairness metrics and bias thresholds for all performance scoring clearly.

2. Flag Ambiguous Schedules
HITL flags logistics plans with high variance in delivery time or route cost inaccurately.

3. Coach Model on Events
Leader manually overrides routing decisions, documenting reasons like port congestion, strikes & tariffs

4. Retrain Algorithm
Incorporates the leader's manual corrections to enhance future planning logic into its logic.

5. Validate Unbiased Output
Test the updated model against diverse profiles to confirm ethical fairness is achieved correctly.

6. Approve Final Logistics Plan
Leader signs off on the final, optimized, and geographically sound operational schedule criteria.

USE CASES

- **Manufacturing | Quality Control Image Review:** Leaders review AI's failure to detect subtle surface defects in products, manually tagging images to continuously improve the visual inspection model's accuracy effectively.
- **Financial Services | Fraud Alert Adjudication:** Supervisors manually review transactions flagged as potential fraud but deemed safe, correcting the AI's risk profile to reduce false positives in high-volume operations reliably.
- **Healthcare | Hospital Bed Allocation:** Senior staff review AI-suggested patient bed assignments, manually adjusting for complex social or family factors to ensure compassionate resource management safely.
- **Technology | Cloud Resource Scaling:** Operations engineers review AI's scaling recommendations during peak load, correcting configurations based on deep system knowledge to prevent service outages clearly.
- **Retail | Inventory Placement Optimization:** Managers review AI-suggested shelf layouts, manually adjusting for product visibility and local merchandising standards to improve physical flow accurately.
- **Energy | Drone Inspection Verification:** Supervisors review AI analysis of power line drone footage, manually correcting misidentified equipment faults to ensure maintenance crews are dispatched appropriately quickly.

In the **Operations Transformation**, the **Architects (The Brain)** serve as the "Systemic Nervous System." While the primary instrument of **Human-in-the-Loop (HITL)** ensures that the organization's logistics and manufacturing remain ethically aligned and safe, it requires secondary technical pairings to manage the hyper-complexity of a global supply chain. To move from "Fragmented Spreadsheets" to a "Sovereign Operational Intelligence," a leader must integrate these six pairings. This ensures the organization's collective memory of "how things are made and moved" is contextually aware, legally unassailable, and instantly accessible.

1. The Vector Database (RAG) Pairing for Procedural Memory

Pairing **HITL with Vector Databases** is the foundation of the **"Self-Documenting Supply Chain."** While the human Architect sets the operational logic, the Vector Database provides the "Semantic Retrieval" that allows factory managers to find solutions based on intent. This is best utilized to eliminate "Knowledge Silos." By indexing decades of maintenance logs and procurement records, the Architect ensures that any operational query retrieves the most relevant past solution, turning "Historical Data" into a live tactical asset.

2. The Knowledge Graph Pairing for Supply Chain Lineage

Leaders should pair **HITL with Knowledge Graphs** to ensure **Logical Traceability.** While the human provides the strategic oversight, the Knowledge Graph maps the complex web of dependencies between raw materials, sub-assemblies, and final delivery. This is best deployed in "Conflict Mineral" or "ESG Compliance." It allows the Architect to trace exactly how a single component moved from a remote mine to a customer's door, providing a transparent record of the firm's operational lineage that a human can audit in seconds.

3. The Natural Language Processing (NLP) Pairing for Regulatory Governance

To maintain a high level of compliance, **HITL must be paired with Advanced NLP.** This is best utilized for "Auto-Monitoring" of international trade laws and shipping regulations. When the NLP layer identifies a change in a tariff or a new environmental law that conflicts with existing operations, it triggers a request for a human Architect to intervene. This ensures the "Brain" is constantly self-updating, maintaining the legality of the firm's global footprint without requiring thousands of manual legal hours.

4. The Differential Privacy Pairing for Partner Sovereignty

When building a collaborative logistics ecosystem, a leader should pair **HITL with Differential Privacy.** This allows the Architect to share synthesized efficiency insights with shipping partners or vendors without the risk of exposing sensitive proprietary volumes. This is best used in "Federated Logistics." The firm can improve its collective routing logic based on shared data while mathematically ensuring that no individual partner's specific shipping routes or cost structures are ever revealed, protecting the brand's competitive sovereignty.

5. The Multi-Modal AI Pairing for Sensory Operational Capture

To move beyond structured data, leaders should pair **HITL with Multi-Modal AI.** This allows the "Brain" to ingest and "understand" non-textual data like drone footage of warehouse roofs, audio from factory machinery, or visual sensor data from shipping containers. This is best utilized in "Capturing Physical Health." The human Architect guides the AI to identify which visual or auditory patterns represent "Imminent Failure." This ensures the "Brain" captures the subtle physical nuances that represent operational risk.

6. The Blockchain Pairing for Immutable Logistics Governance

For ensuring the absolute integrity of the firm's "Operational Truth," **HITL should be paired with Blockchain.** While the human Architect sets the core procurement and shipping policies, the Blockchain stores a tamper-proof "Audit Trail" of every modification made to the operational logic. This is best utilized for "Sanctions Compliance." It prevents any unauthorized tampering with the supply chain's path. It ensures that the "Brain" remains an unassailable record of truth, providing a permanent ledger of the organization's ethical and operational commitments.

Operational Power: Autonomous Agency

The final shift in the operational journey is the transition from "Human-Led Execution" to **"Autonomous Fulfillment."** Historically, operations were "Manual"—relying on humans to move data, approve orders, and manage exceptions. The transformation into **Executors** represents the birth of "Agentic Operations." In this role, the business doesn't just "report" a status; it "executes" a response. The operations become the "Hands" of the leader, capable of navigating complex workflows and resolving problems autonomously. This is the shift from "Passive Systems" to "Active Agents," turning the organization into an "Autonomous Powerhouse" that manifests the "Voice" of the leadership at the speed of light.

Defensive Stance: Killing the Operational Drag of Manual Toil

The defensive mission of the Operational-Executor is the **Elimination of Latency**. In the legacy world, there is always a "Gap" between a data signal and a human response—this gap is where profit and trust are lost. The Executor role kills this "Operational Drag" by establishing "Closed-Loop Autonomy." The system identifies a deviation from the plan and "Executes the Fix" instantly. This is the **Shield of Frictionless Flow**. By "Hard-Coding the Response" to routine operational failures—such as automatic re-routing of delayed cargo or instant re-balancing of budgets—the "Hands" of the company ensure a state of "Constant Performance." This defensive posture protects the brand's "Margins" and "Reputation" by ensuring that "Minor Glitches" never grow into "Major Crises," because the "Hands" are always there to catch the ball before it hits the ground.

Offensive Stance: Unlocking Agentic Orchestration

On the offense, the Executor moves toward **Strategic Operational Fulfillment**. This is the ability of the operations to "Conduct" the world to achieve a goal. Instead of a human manually managing a supplier relationship, "Operational Agents" navigate the digital landscape to fulfill a "Strategic Intent." For example, an "Agentic Procurement System" doesn't just flag a need; it searches for the best ethical source, negotiates the terms, and verifies the delivery autonomously. This **Unlocks Infinite Scale**, where the "Hands" of the company can handle a 100x increase in complexity without a single new hire. It turns the organization into an "Autonomous Force"—one that out-executes competitors through sheer "Operational Velocity." It allows the leader to "Orchestrate" global outcomes by setting "High-Level Goals" and letting the "Agents" find the most efficient path to manifest that reality in the physical world.

The Operational Outcome: Absolute Leverage

The ultimate value of the Executor role is **Operational Sovereignty**. It allows a brand to "Operate at the Speed of Intent," not the speed of bureaucracy. For the leader, this provides **Execution Peace of Mind**. You are no longer "Managing Processes"; you are "Directing an Engine." This turns the organization into an "Autonomous Powerhouse"—where the "Hands" of the operations are always active, always precise, and always working to fulfill the "Voice" of the leadership. It ensures that the brand's "Execution" is as reliable as a law of physics, transforming "Strategy" into "Action" and "Action" into "Market Leadership."

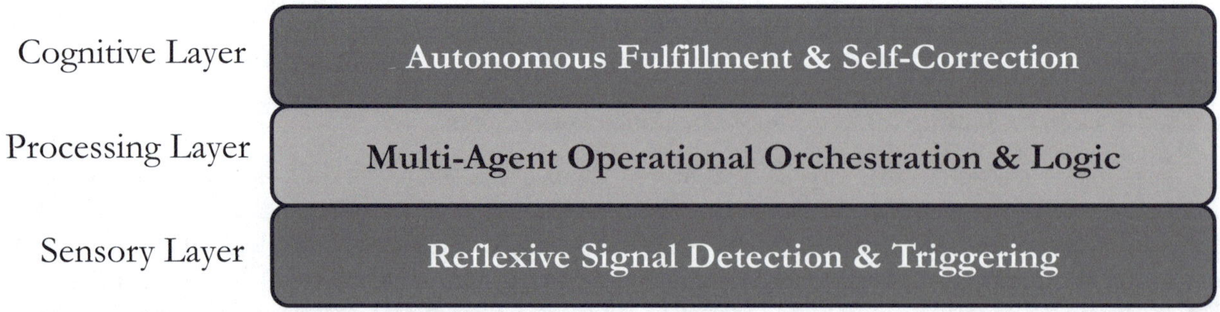

Executive Power: Agentic Operational Autonomy

The final shift in the operations journey is the transition from "Manual Fulfillment" to **"Orchestrated Autonomous Agency."** The **Executor** role represents the birth of "Active Operations," where the leader provides the "Strategic Intent" and the brand's AI-augmented "Hands" manifest fulfillment results directly in the world. Historically, operations were "Reactive"—a human had to wait for a signal to act. The Executor role replaces this with a system of **Agentic Autonomy**, where the "Hands" are capable of fulfilling tasks—such as re-balancing inventories, updating shipping routes, or negotiating with vendor bots—without manual intervention. This is the shift from "Managing the Supply Chain" to "Directing an Execution Machine," giving the executive the leverage to "Operate at the Speed of Light" by transforming "Intent" into "Autonomous Physical Reality."

The Sensory Layer: Reflexive Signal Detection & Triggering

The **Sensory Layer** of the Operational Hands serves as the organization's "Reflex System." It is designed for **Real-Time Execution Triggering**, monitoring the internal and external environment for specific "Anomalies or Opportunities" that require an immediate autonomous response. Utilizing **Predictive Anomaly Detection** and **Event-Driven APIs**, this layer identifies the moment an operational state deviates from the "Ideal Path." It doesn't wait for a human to see a "Stockout" or a "Logistics Delay"; it "feels" the friction the micro-second it occurs. This sensory input provides the "Activation Signal" for the agentic stack, ensuring that the "Hands" of the company are always ready to move the millisecond a requirement is detected. This layer protects the brand's "Margins" by ensuring the organization responds to reality with biological speed, resolving issues before they reach the customer.

The Processing Layer: Multi-Agent Operational Orchestration & Logic

The **Processing Layer** acts as the "Execution Cerebellum," breaking down the leader's high-level intent into executable operational tasks. It utilizes **Agentic Logic (ReAct Frameworks)** and **Chain-of-Thought Reasoning** to orchestrate a swarm of specialized fulfillment agents. If the leader's intent is to "re-route 50% of North American inventory to avoid a port strike," this layer identifies the thousand sub-tasks required—from re-negotiating carrier contracts to updating warehouse arrival windows. It manages the "Inter-Agent Communication," ensuring that each bot knows the "Ethical and Financial Guardrails" set by the executive. This layer provides the **Governance and Logic** for the execution, ensuring that the "Hands" stay within the boundaries of operational integrity, killing the friction of manual oversight through automated, high-speed coordination across the entire global supply chain.

The Cognitive Layer: Autonomous Fulfillment & Self-Correction

The **Cognitive Layer** is where the "Hands" actually manifest the result through **Action Transformers** and **Automated Fulfillment Engines**. This is the layer of "Closed-Loop Execution." It doesn't just "process" a ticket; it "achieves" the strategic outcome. If the agentic stack encounters an obstacle—such as a failed API or a physical logistics barrier—this layer uses **Self-Correction Logic** and **Reinforcement Learning** to find an alternative fulfillment path without requiring human intervention. It turns "Executive Intent" into "Physical Reality," whether that means moving funds, updating manufacturing schedules, or deploying logistics autonomously. This provides the executive with **Absolute Operational Leverage**, allowing them to "Direct the Value Flow" rather than "Manage the Toil." It transforms the firm into an "Autonomous Execution Machine" that delivers "Market Dominance" through sheer agentic velocity and proactive fulfillment.

OVERVIEW

Agentic AI operates as an autonomous mission manager, breaking down high-level strategic goals into executable tasks. The Driving Mission leader defines the ultimate objective and core values, essentially setting the constitutional limits for the AI agent. This synergy offloads complex, time-consuming multi-step execution to the AI, allowing the leader to focus exclusively on strategic vision and alignment. The leader initiates the mission; the agent executes and reports on key milestones. This partnership fundamentally converts strategy into action, maximizing the leader's impact on high-level direction and governance.

MAIN STEPS

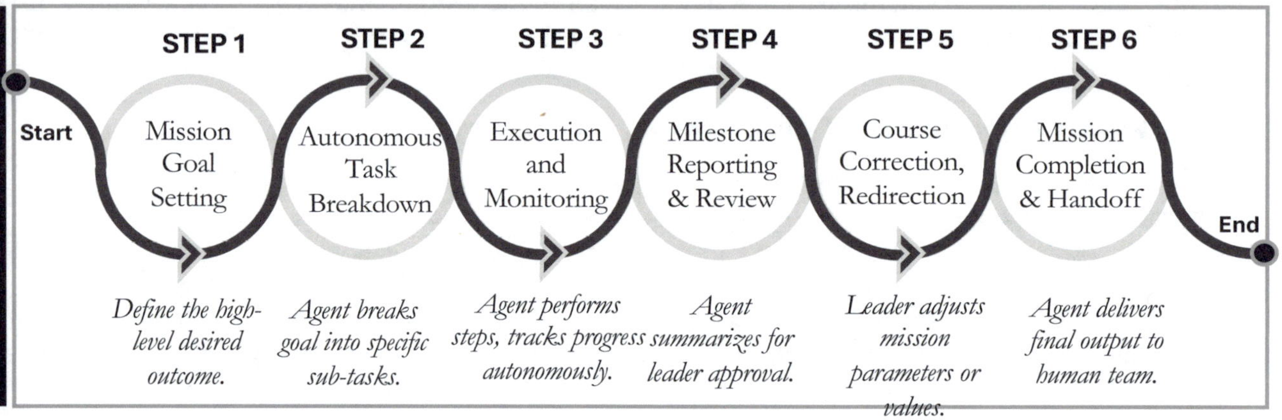

SAMPLE FLOW

Autonomous Supply Chain Resilience Mission: The leader mandates an agent to establish a resilient, self-healing supply chain for critical component sourcing.

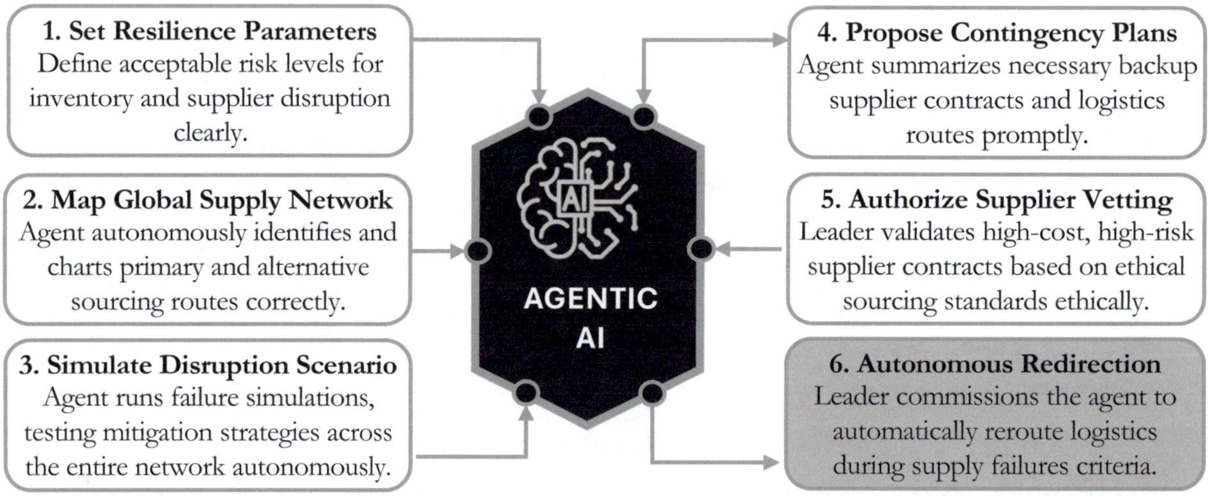

USE CASES

- **Manufacturing | Dynamic Factory Floor Reconfiguration:** An agent manages the mission to autonomously reconfigure factory floor layouts and tool paths in real-time, optimizing production flow based on changing demand and resource availability effectively.
- **Financial Services | Cross-Border Transaction Mission:** The leader tasks an agent with automating complex, multi-jurisdictional transaction processing, autonomously handling currency conversions, compliance checks, and final ledger reconciliation reliably.
- **Healthcare | Hospital Asset Utilization:** An agent manages the mission to optimize the utilization of high-value hospital assets (e.g., MRI machines), autonomously coordinating scheduling, maintenance, and staff allocation accurately.
- **Technology | Infrastructure Cost Optimization:** An agent continuously monitors cloud usage and autonomously initiates missions to re-architect deployments, optimizing server usage and reducing hosting costs consistently.
- **Retail | Store Energy Efficiency Upgrade:** An agent researches and executes a mission to upgrade energy infrastructure across store locations, autonomously scheduling vendors, managing procurement, and ensuring system integration quickly.
- **Energy | Predictive Grid Maintenance Mission:** The leader deploys an agent to manage continuous predictive maintenance across the power grid, autonomously identifying failure points and dispatching repair teams clearly.

Operations Transformation: The Executors (The Hands)

In the **Operations Transformation**, the **Executors (The Hands)** represent the "Physical Actuation Core." While the primary instrument of **Agentic AI** provides the high-level agency to solve logistics bottlenecks, it requires secondary technical pairings to bridge the gap between digital instructions and physical movement. To move from "Manual Labor" to "Autonomous Throughput," a leader must integrate these six pairings. This ensures the "Hands" can act on operational signals with millisecond precision, turning the supply chain into a self-healing, self-moving organism that delivers on the brand's promise without friction.

1. The RPA Pairing for Back-Office Synchronization

Pairing **Agentic AI with Robotic Process Automation (RPA)** is best utilized to eliminate "Paperwork Latency." While the Agent decides to reroute a shipment due to a storm, RPA provides the "fingers" to instantly update the customs forms, manifest records, and billing entries in legacy port systems. This ensures that the physical movement of goods is never stalled by digital bureaucracy, allowing the "Hands" to bypass the administrative bottlenecks that typically slow down global commerce.

2. The Digital Twin Pairing for Robotic Rehearsal

Leaders should pair **Agentic AI with Operational Digital Twins** when deploying autonomous robotics in the warehouse. Before the "Hands"—such as a fleet of picking robots—execute a new sorting logic, the move is "rehearsed" in the Digital Twin. This pairing ensures the robots never collide or create a bottleneck. It provides a virtual "Safety Guardrail" for autonomous execution, ensuring every physical movement is optimized for speed and safety before a single motor turns in the real world.

3. The Edge AI Pairing for High-Speed Sorting

When execution must happen at the "Point of Movement"—such as on a high-speed conveyor belt or an autonomous forklift—**Agentic AI must be paired with Edge AI**. This allows the "Hands" to process visual data and make sorting decisions locally with zero latency. This pairing is vital for "Real-Time Throughput," ensuring the system can identify a damaged package or a mislabeled bin in milliseconds. It turns the warehouse floor into a responsive nervous system that acts at the speed of the machinery.

4. The Blockchain Pairing for Automated Settlement

For secure, multi-party logistics, a leader should pair **Agentic AI with Blockchain (Smart Contracts)**. The "Hands" don't just move boxes; they trigger the transfer of ownership and payment the moment a sensor confirms a successful delivery. This pairing ensures that autonomous actions—like a smart container releasing its own bill of lading—are legally binding and transparent. It eliminates the need for manual audit and reconciliation, ensuring the organization's "Hands" are financially and legally self-sufficient.

5. The Computer Vision Pairing for Quality Actuation

To allow the "Hands" to maintain "Product Integrity," leaders should pair **Agentic AI with Computer Vision**. This is best utilized in "Autonomous Quality Control." By "seeing" a physical product on the production line, the agent can autonomously identify a defect and trigger a robotic arm to remove it. This pairing gives the "Hands" the sensory context required to perform high-precision physical tasks, ensuring the brand's "Operational Truth" is verified through constant visual inspection.

6. The Reinforcement Learning Pairing for Throughput Evolution

In complex, high-volume environments, **Agentic AI should be paired with Reinforcement Learning (RL)**. This is best utilized for "Self-Optimizing Logistics," such as dynamic warehouse slotting or truck loading. The "Hands" learn from the success or failure of every movement, constantly refining their technique to reduce travel time and maximize space. This provides the brand with an "Evolving Execution Layer" that grows more efficient with every shipment handled, ensuring the firm's operational muscle strengthens over time.

THE Ai6x6™ STRATEGIC ROADMAP SERVICE

Your 3-Year Vision, Stress-Tested for the AI Era

The **Ai6x6™** book series has given you the "source code"—the clinical lenses, the vulnerability audits, and the diagnostic frameworks. But strategic orchestration at the highest level requires a cold, objective eye. If you are ready to move from self-diagnosis to radical transformation, we invite you to engage the **Ai6x6™ Roadmap Service.**

The 21-Day Transformation Engine

We don't believe in "billable hours" or six-month discovery phases. We believe in high-fidelity data and rapid execution. Our domain experts take your raw diagnostic output and synthesize it into a personalized, holistic blueprint for the next three years of your organization's life.

How It Works

1. **Submit Your Data:** Upload your completed Discovery Questionnaires, Vulnerability Audits, and Radar Charts directly to our secure portal or via email or via courier.
2. **Contextualize:** Provide your current 3-year roadmap and any business documentation that can provide more context on your line of business, your mission and success KPIs.
3. **The Clinical Review:** Our experts analyze your organizational DNA against the **6 Lens Sequence**, identifying hidden frictions and untapped leverage points. Assign a single point within your organization for asking questions.
4. **The Delivery:** Within **2 to 3 weeks**, you receive a personalized **Holistic Transformation Roadmap**, including our proprietary **Self-Disruption Strategy**.

Why Self-Disruption?

The greatest threat to your organization isn't your competitor; it is your own legacy success. Our roadmap doesn't just show you how to grow; it shows you how to cannibalize your outdated models before the market does it for you.

The Commitment

Packages from $3000

- **Speed:** 14–21-day turnaround.
- **Security:** Full NDA-backed engagement.
- **Precision:** Fact-based, quantitative analysis. No "fluff," no subjective brainstormin

SECURE YOUR DIAGNOSTIC REVIEW:

Ai6x6.com/roadmap

Consulting Roadmap

BLUEPRINT SERIES

SECTION C

Business Maturity

Ai6x6™ Transformation Blueprint

Story 1 — The Marshall Plan Reconstruction (1948–1952): Post-war Europe faced shattered economies, broken supply chains, and institutional collapse. The Marshall Plan introduced structured funding, transparent reporting, and coordinated execution across nations. Local governments provided context while centralized oversight ensured accountability. This blend of autonomy and structure accelerated recovery. The program demonstrated how disciplined governance can rebuild complex systems. It became a blueprint for large-scale organizational maturity.

Story 2 — The Rise of Quality Management in Japan (1950s–1960s): Japan's industries struggled with inconsistent quality and limited global credibility. Statistical quality control introduced measurement, repeatability, and disciplined improvement cycles. Workers contributed insights while leaders standardized processes. Over time, quality became a cultural norm rather than a corrective action. This shift transformed fragmented operations into cohesive systems. It marked the emergence of maturity through methodical refinement.

Story 3 — NASA's Systems Engineering Discipline (1960s): The space race demanded unprecedented coordination across thousands of contractors and scientific domains. NASA formalized systems engineering to integrate requirements, risks, and interfaces into a unified framework. Structured reviews balanced innovation with rigor. Human judgment guided trade-offs while documentation preserved institutional memory. This discipline enabled complex missions to succeed repeatedly. It became a global model for organizational maturity in high-stakes environments.

Story 4 — The Birth of Modern Corporate Governance (1970s–1980s): Corporate scandals exposed weaknesses in oversight, transparency, and fiduciary responsibility. Boards adopted clearer roles, audit mechanisms, and accountability structures. Regulations reinforced ethical conduct while organizations matured through formal governance. Leaders learned to balance autonomy with compliance. This era established governance as a strategic capability, not an administrative burden. It marked a turning point in institutional maturity.

Story 5 — Toyota's Global Expansion Playbook (1980s–1990s): As Toyota expanded globally, it faced cultural differences, supply-chain complexity, and operational variability. The company codified principles, standardized training, and embedded continuous improvement across regions. Local teams adapted practices while preserving core philosophy. This balance of global consistency and local flexibility enabled scalable excellence. It demonstrated maturity as the ability to replicate capability without diluting identity.

Story 6 — The Digital Transformation of Banking (1990s–2000s): Banks confronted rising customer expectations, regulatory pressure, and technological disruption. They modernized legacy systems, standardized data, and introduced structured risk frameworks. Digital channels required new operating models and cross-functional coordination. Human expertise guided decisions where automation lacked nuance. Over time, banks evolved from siloed institutions into integrated digital enterprises. This transition reflected maturity through modernization and disciplined reinvention.

LESSONS FROM AN OPERATIONS TRANSFORMATION PERSPECTIVE

Operational maturity emerges when organizations transition from ad-hoc execution to structured, resilient, and continuously improving systems. Over time, institutions learned to integrate standardization, measurement, and cross-functional coordination into daily work. Mature operations balance efficiency with robustness, ensuring that processes perform reliably under both normal and volatile conditions. They embed feedback loops that detect issues early and enable rapid correction. As organizations scale, operations become the backbone of consistency—supporting governance, customer value, and strategic execution. Maturity is achieved when operations are not dependent on individual expertise but on institutionalized practices that sustain performance over time.

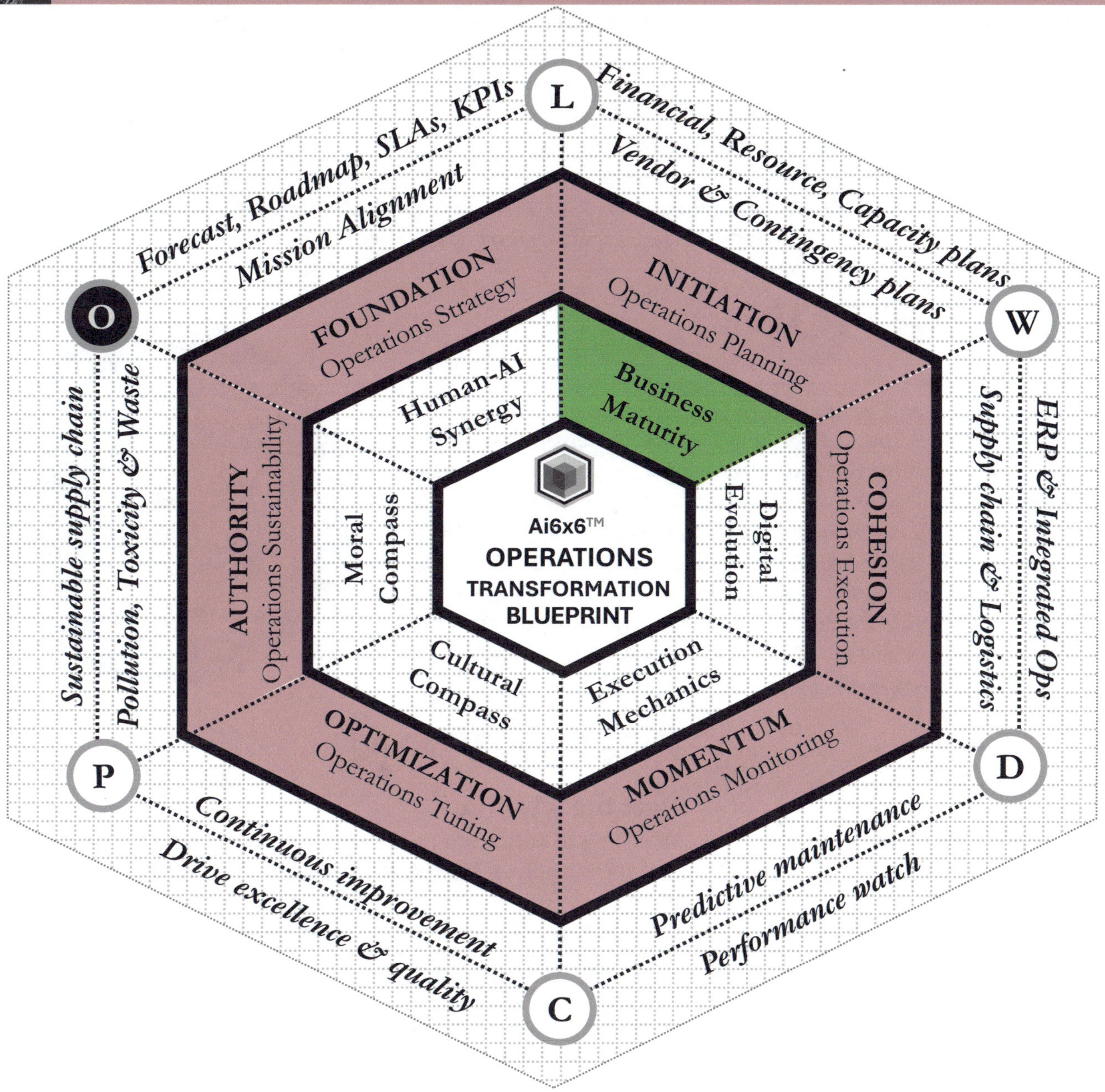

The **Ai6x6 Research Lab** selected this lens because an organization's operational maturity is defined by its ability to maintain equilibrium between growth and **Sustainability**. We view business maturity as an internal filter for the organization's capacity to align its mission with its physical footprint. This dimension ensures that the aggressive energy of **Execution** and **ERP** management is anchored in the long-term wisdom of **Capacity** and **Contingency** planning. It is a vital lens because, in a volatile supply chain, a leader's ability to move from **Anomaly Tracking** to **Root Cause** optimization is the only force that prevents the enterprise from stagnating.

We believe true maturity is found in the ability to manage the delicate balance of **Efficiency and Waste Reduction**, recognizing that the machine is a reflection of the organization's discipline. By focusing on **Process Mining**, the Lab provides a mechanism for leaders to audit their operational reality before it impacts the bottom line. This lens ensures that the pursuit of speed does not come at the cost of stability or the environment. It is the steady hand that guides operations through the complexities of **Supply Chain** dynamics, recognizing that while the "Movement" is constant, the bedrock of operational excellence must remain immutable.

Every leadership transformation begins with ambition, but ambition alone is never enough to carry an organization into its future. **Business Maturity** is the lens that forces leaders to confront the gap between what they hope to achieve and what their enterprise is prepared to sustain. It is the discipline that reveals whether the organization is structurally, operationally, and culturally ready for the future it claims to desire. In the AI era, this gap becomes even more pronounced; AI does not simply automate tasks—it reshapes workflows, decision cycles, and the very rhythm of how leadership operates. It exposes weaknesses that were previously tolerable and accelerates the consequences of poor preparation, demanding a level of rigor that traditional management frameworks often overlook. It examines how ambition translates into capability, how vision converts into execution, and how culture sustains momentum under pressure.

FOUNDATION

Define the Core

The journey begins at **Foundation**, where the primary objective is to *Define the Core*. This is the bedrock upon which all transformation rests, requiring leaders to **Filter the Noise** of competing priorities and anchor their identity and strategic intent. At this level, clarity is not optional—it is existential. Without clear governance, AI initiatives drift, priorities conflict, and teams operate in isolated silos. Foundation maturity is the discipline of intent: the ability to distinguish urgency from readiness. Leaders who skip this stage mistake motion for progress, building transformation on shifting sand. True maturity begins when the organization knows what it stands for, what it will not compromise, and how it will measure success.

INITIATION

Start the Motion

Ascending to **Initiation** marks the first active step: *Start the Motion*. This level turns the gaze inward, demanding self-awareness and the courage to confront personal biases. By **Wiring the System**, leaders catalyze the shift from passive management to intentional authority. Initiation is where leadership becomes kinetic—it transforms from aspiration into activation. The organization begins to move, not because it is told to, but because its internal circuits are aligned. Maturity here is emotional as much as structural; it requires self-regulation, empathy, and the ability to lead through friction. The leader's persona is tested, revealing whether they can sustain momentum through disruption rather than retreat into comfort.

COHESION

Connect the Parts

The climb continues to **Cohesion**, represented by *The Engine*. This is the critical unifier where people, processes, and systems are harmonized. The focus shifts to **Connect the Parts**—to ensure that individual leadership goals align with a unified organizational voice. By **Shifting the Gear**, leaders eliminate silos and systemic friction, transforming fragmented efforts into a synchronized machine. Cohesion maturity is the art of integration: turning disconnected motion into collective momentum. Without it, organizations celebrate activity instead of achievement. The mature leader understands that alignment is not consensus—it is coherence, the ability to move as one without losing diversity of thought.

MOMENTUM

Drive the Speed

The fourth rung is **Momentum**, where the organization begins to *Drive the Speed*. Leadership here moves toward scaling impact with predictable velocity. By **Removing the Drag** of legacy thinking and institutional inertia, leaders accelerate the collective intelligence of the enterprise. Momentum maturity is the physics of transformation—it measures how efficiently energy converts into progress. At this stage, the organization's rhythm changes; it begins to operate with self-propelling force. The leader's role shifts from directing individuals to empowering systems, ensuring that innovation scales faster than resistance. Momentum is not chaos—it is controlled acceleration, the moment when transformation gains its gravity.

OPTIMIZATION

Refine the Flow

Refinement occurs at **Optimization**, characterized by *The Flow*. Here, leadership pauses to **Refine the Flow** by **Locking the Code** of influence through reflection and data-driven inquiry. Immature organizations measure what is easy; mature ones measure what matters. Optimization maturity is the discipline of precision—removing friction without losing flexibility. Leaders refine decisions still in motion, adjusting with humility and surgical accuracy. Ego and bias are replaced by evidence and iteration. The organization becomes self-healing, capable of continuous improvement without external intervention. Optimization is not about perfection—it is about adaptability, the ability to evolve faster than the environment demands.

AUTHORITY

Own the Market

The summit is reached at **Authority**, where leaders finally *Own the Market*. This stage manifests as *The Benchmark*—a legacy of endurance that can withstand external shocks and volatility. By **Fueling the Future**, leaders ensure that transformation is not a temporary win but a permanent standard. Authority maturity is the discipline of stewardship: sustaining excellence while preparing for reinvention. Immature organizations avoid this stage because it exposes uncomfortable truths; mature ones embrace it as the crucible of resilience. At this level, leadership becomes institutional—it transcends personality and becomes culture. The organization no longer reacts to the market; it defines it.

Business Maturity is the scaffolding that holds the weight of innovation. It ensures that speed does not collapse under fragility and that ambition does not outpace capability. Each level builds upon the last, forming a continuum from clarity to mastery. Foundation defines intent, Initiation activates leadership, Cohesion integrates systems, Momentum accelerates impact, Optimization refines intelligence, and Authority sustains legacy. In the AI era, maturity is not a luxury—it is survival. As algorithms amplify both strengths and weaknesses, leaders must design organizations that can learn, adapt, and endure. Business Maturity is therefore not a checklist but a **leadership discipline**—a way of thinking that transforms ambition into architecture. When leaders embrace this full sequence, they discover that maturity is not the end of transformation but its engine—the invisible structure that makes both explosive speed and long-term continuity possible.

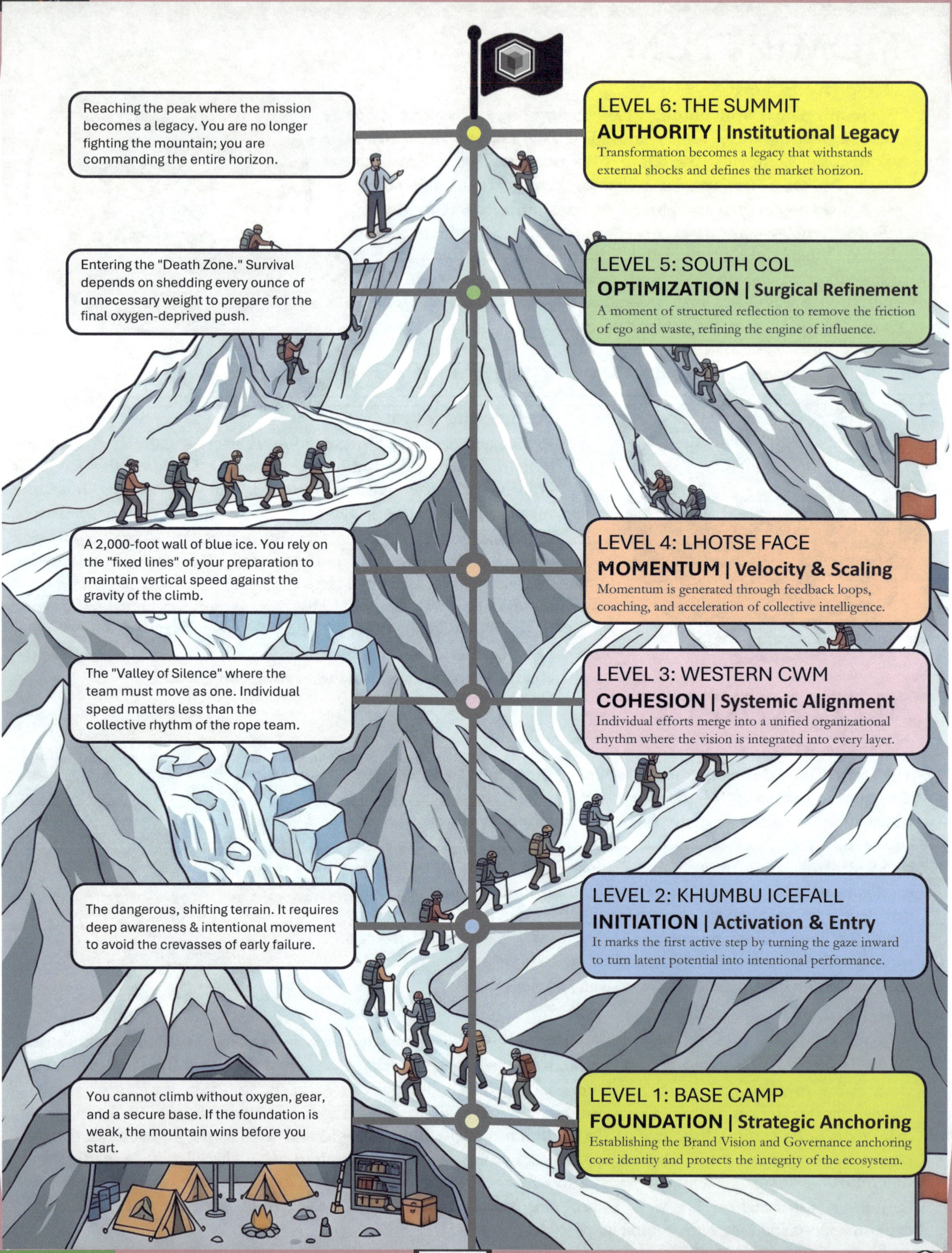
Reaching the peak where the mission becomes a legacy. You are no longer fighting the mountain; you are commanding the entire horizon.
LEVEL 6: THE SUMMIT
AUTHORITY | Institutional Legacy
Transformation becomes a legacy that withstands external shocks and defines the market horizon.
Entering the "Death Zone." Survival depends on shedding every ounce of unnecessary weight to prepare for the final oxygen-deprived push.
LEVEL 5: SOUTH COL
OPTIMIZATION | Surgical Refinement
A moment of structured reflection to remove the friction of ego and waste, refining the engine of influence.
A 2,000-foot wall of blue ice. You rely on the "fixed lines" of your preparation to maintain vertical speed against the gravity of the climb.
LEVEL 4: LHOTSE FACE
MOMENTUM | Velocity & Scaling
Momentum is generated through feedback loops, coaching, and acceleration of collective intelligence.
The "Valley of Silence" where the team must move as one. Individual speed matters less than the collective rhythm of the rope team.
LEVEL 3: WESTERN CWM
COHESION | Systemic Alignment
Individual efforts merge into a unified organizational rhythm where the vision is integrated into every layer.
The dangerous, shifting terrain. It requires deep awareness & intentional movement to avoid the crevasses of early failure.
LEVEL 2: KHUMBU ICEFALL
INITIATION | Activation & Entry
It marks the first active step by turning the gaze inward to turn latent potential into intentional performance.
You cannot climb without oxygen, gear, and a secure base. If the foundation is weak, the mountain wins before you start.
LEVEL 1: BASE CAMP
FOUNDATION | Strategic Anchoring
Establishing the Brand Vision and Governance anchoring core identity and protects the integrity of the ecosystem.

Ai6x6™ Transformation Blueprint

Foundation | Operations Strategy

"Strategy is about making choices, trade-offs; it's about deliberately choosing to be different." - Michael Porter.

A leading authority on competitive strategy and economic development. His theories on competitive advantage, industry analysis (Porter's Five Forces), and value chains have profoundly influenced business management and economic thought, widely used for strategy formulation.

The Foundation level is your Operational Strategy, aligning your internal mission with a physical identity and market presence that can sustain demand. This level ensures the operational "skeleton" is strong enough to support all future growth.

Defining the overarching, long-term direction and comprehensive scope of all operational activities, meticulously aligning them with the organization's strategic business goals and competitive landscape. This is essential for effectively aligning resources, optimizing capabilities, and streamlining processes to achieve a sustainable competitive advantage and maximize operational efficiency across the board.

A well-articulated operations strategy ensures that daily activities contribute directly to the company's long-term vision, driving productivity, cost-effectiveness, and market responsiveness.

For business strategy maturity, the "**Operations Strategy**" chapter is foundational because it defines the objectives, allocates resources, and creates roadmaps for overall operational success. A mature business strategy ensures that operational planning is directly aligned with the broader organizational vision and goals. This strategic alignment is critical for optimizing efficiency, reducing waste, and ensuring that day-to-day activities directly contribute to strategic outcomes. It's about establishing a clear operational compass that guides all subsequent actions, ensuring resources are deployed effectively to achieve strategic objectives. Here is how we approach our learning:

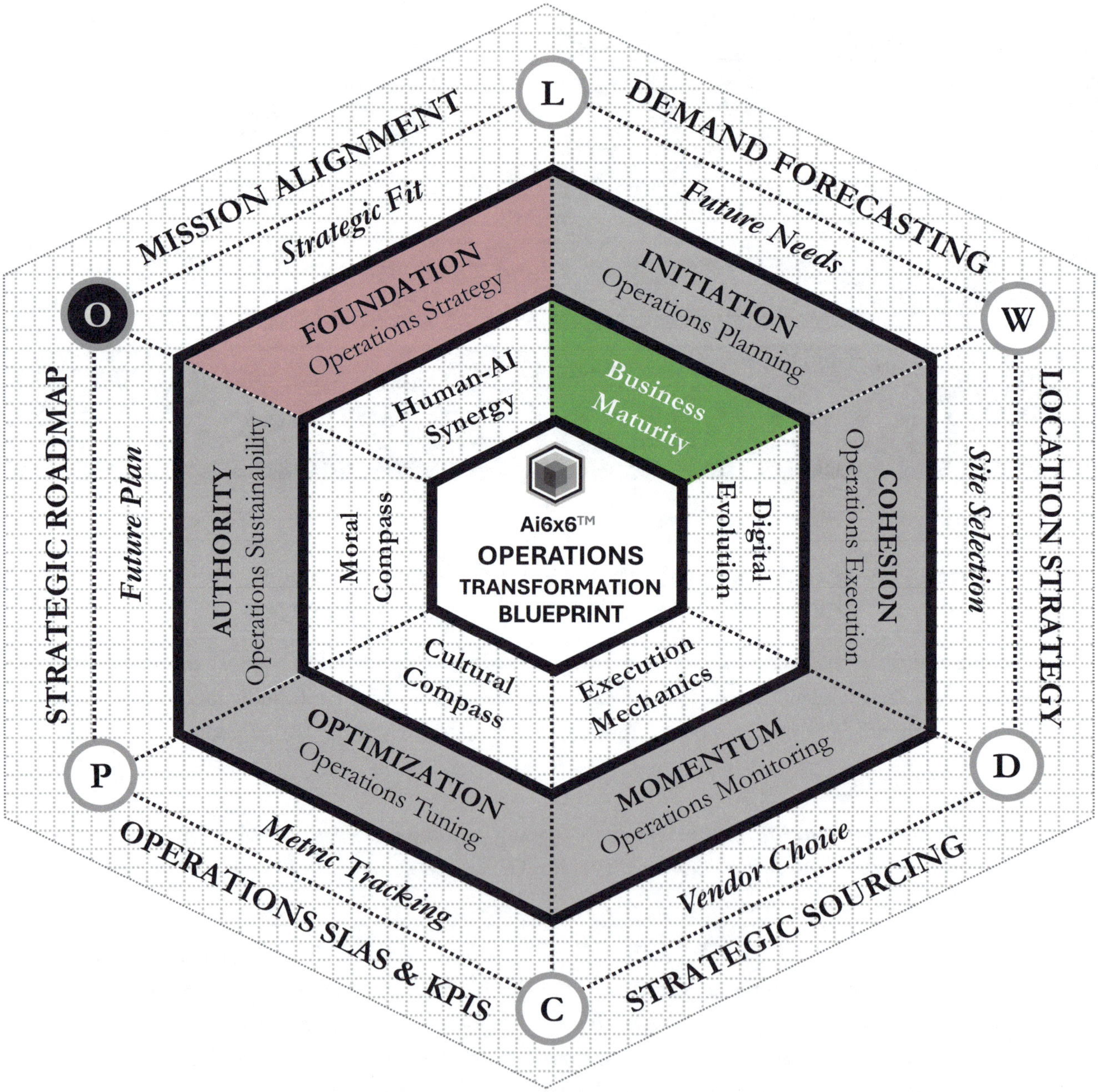

Ai6x6 BUSINESS MATURITY FRAMEWORK
OPERATIONS TRANSFORMATION | OPERATIONS STRATEGY

For a business leader, ensuring meticulous mission alignment within your operations is a foundational strategic imperative. It's about more than just efficiency; it's about guaranteeing that every operational decision, process, and system directly supports and reinforces your organization's core values, overarching vision, and strategic goals. Without this alignment, even highly efficient operations can inadvertently work against broader business objectives, leading to misallocation of resources and a diluted impact. Your role as a leader is to articulate clearly defined strategic objectives, translating the company's grand vision and mission into concrete, actionable operational goals. This ensures that every department and team understands precisely how their daily activities contribute to the bigger picture.

Furthermore, fostering robust cross-functional integration is crucial. Coordinating departments to ensure seamless consistency in mission execution—from product development to sales, marketing, and customer service—breaks down silos and ensures that the entire organization operates as a cohesive unit towards shared aims. By setting clear benchmarks and regularly assessing how operational outcomes align with long-term strategic goals, you can proactively identify deviations, make timely adjustments, and ensure that your operational engine is continuously driving the business in the right direction. Neglecting mission alignment can lead to internal inefficiencies, disjointed efforts, wasted resources on misdirected activities, and a struggle to achieve strategic objectives. By championing this foundational principle, you empower your organization to operate with clarity, purpose, and unified direction, ultimately accelerating the achievement of your business's strategic goals and securing its success in the marketplace.

WHAT?	WHY?	HOW?	BEWARE!
DEFINING STRATEGIC OBJECTIVES	**CLEAR DIRECTION**	**STRATEGIC WORKSHOPS**	**AMBIGUITY**
Mapping operations to align with vision and mission.	Ensures all operational efforts contribute directly to core business goals.	Conduct workshops to translate high-level vision into actionable operational objectives.	Objectives being too vague or not clearly communicated across all levels.
CROSS-FUNCTIONAL INTEGRATION	**SEAMLESS EXECUTION**	**SHARED KPIS**	**RESISTANCE**
Coordinating departments for consistent mission execution.	Breaks down silos, fostering cohesion and efficiency in mission delivery.	Implement shared KPIs and regular inter-departmental meetings.	Departments resisting collaboration or prioritizing their own goals.
PERFORMANCE MEASUREMENT	**ACCOUNTABILITY**	**METRIC DEVELOPMENT**	**ATTRIBUTION**
Establishing benchmarks to assess alignment.	Provides objective assessment of operational contribution to strategic goals.	Define and track operational KPIs that directly link to strategic objectives.	Difficulty attributing specific operational performance directly to mission alignment.

For a business leader, mastering demand forecasting is a critical strategic imperative for optimizing resource allocation, enhancing operational efficiency, and ensuring your organization is always prepared to meet future market needs. In a volatile global economy, simply reacting to demand is insufficient; proactive prediction is key to competitive advantage and cost control. Central to this is leveraging data-driven forecasting models. By utilizing advanced AI and historical analytics, you can anticipate demand fluctuations with greater accuracy, predicting everything from product sales and service requests to raw material needs and staffing requirements. This predictive capability minimizes waste and maximizes responsiveness.

Furthermore, meticulous seasonal and market trend analysis is crucial. Adjusting operational strategies based on recognized industry cycles, economic indicators, and external influences ensures that your production schedules, inventory levels, and service capacities are optimized for peak periods and downturns, preventing both overstocking and stockouts. Crucially, developing adaptive capacity planning strategies is vital. This ensures your supply chain and operational infrastructure can respond swiftly and efficiently to evolving demand patterns, scaling up or down as needed without compromising service quality or incurring excessive costs. Neglecting robust demand forecasting can lead to inefficient resource allocation, missed sales opportunities, excessive inventory carrying costs, and customer dissatisfaction due to stockouts or long lead times. By championing a data-driven and agile approach to predicting future needs, you empower your organization to operate with greater precision, reduce waste, and deliver superior customer service, ultimately driving the sustained efficiency and profitability of your business.

WHAT?	WHY?	HOW?	BEWARE!
DATA-DRIVEN FORECASTING MODELS	**ACCURATE PREDICTION**	**ML MODELING**	**DATA QUALITY**
Using AI and historical analytics.	Enhances precision in anticipating demand fluctuations.	Implement AI/ML-powered forecasting tools to analyze historical data.	Poor historical data leading to inaccurate forecasts.
SEASONAL AND MARKET TREND ANALYSIS	**OPTIMIZED STRATEGIES**	**TREND MONITORING**	**UNFORESEEN EVENTS**
Adjusting strategies based on cycles.	Ensures operations are aligned with market realities and cycles.	Analyze market reports, economic indicators, and consumer behavior.	Sudden, unpredictable market shifts (e.g., pandemics).
ADAPTIVE CAPACITY PLANNING	**AGILITY & EFFICIENCY**	**FLEXIBLE RESOURCE POOLS**	**OVER-CAPACITY**
Ensuring supply chain responsiveness.	Optimizes resource deployment for evolving demand patterns.	Develop flexible staffing, production, and inventory strategies.	Risk of investing in capacity that ultimately remains underutilized.

For a business leader, meticulously developing a location strategy is a critical strategic imperative for structuring operations to maximize geographic efficiency, optimize cost structures, and enable sustainable business growth. The physical placement of your facilities—whether manufacturing plants, distribution centers, or service hubs—profoundly impacts your speed to market, supply chain costs, and access to talent. Central to this is rigorous site selection analysis. This involves evaluating a multitude of factors such as real estate costs, accessibility to transportation networks, availability of skilled labor, regulatory environment, and proximity to suppliers and customers. A comprehensive analysis ensures optimal long-term operational fit.

Furthermore, prioritizing market proximity optimization is crucial. Placing facilities closer to high-demand regions for faster service delivery, reduced transportation costs, and improved customer satisfaction provides a significant competitive advantage, especially in industries with tight delivery windows or high service expectations. Crucially, implementing a distributed operations model often involves a multi-site strategy for enhanced scalability and risk mitigation. Spreading operations across different geographical locations diversifies your risk against localized disruptions (e.g., natural disasters, labor disputes) and provides greater flexibility for scaling production or service delivery to meet regional demands. Neglecting a thoughtful location strategy can lead to inflated operational costs, inefficient supply chains, limited market access, and vulnerability to localized disruptions. By championing a strategic approach to where you operate, you empower your organization to achieve greater efficiency, expand its market footprint, and build inherent resilience, ultimately contributing significantly to the sustained growth.

WHAT?	WHY?	HOW?	BEWARE!
SITE SELECTION ANALYSIS	**OPTIMAL PLACEMENT**	**CRITERIA MATRIX**	**DATA AVAILABILITY**
Evaluating factors such as cost, accessibility, and labor.	Ensures facilities are located for maximum efficiency and suitability.	Develop a weighted matrix of factors (cost, labor, logistics, regulations).	Obtaining comprehensive and reliable data for all evaluation factors.
MARKET PROXIMITY OPTIMIZATION	**FASTER SERVICE**	**GEOSPATIAL ANALYSIS**	**COST BALANCE**
Placing facilities closer to high-demand regions.	Reduces delivery times and enhances customer satisfaction.	Use geospatial data to map customer density and optimize facility placement.	Balancing proximity benefits with land costs and labor availability.
DISTRIBUTED OPERATIONS MODEL	**SCALABILITY & RESILIENCE**	**NETWORK DESIGN**	**COORDINATION COMPLEXITY**
Implementing multi-site strategies.	Provides flexibility for growth and mitigates localized risks.	Plan multiple facilities across different geographies.	Managing and coordinating operations across multiple sites.

For a business leader, mastering strategic sourcing is a critical imperative for enhancing procurement, optimizing supplier relationships, and ensuring that your supply chain robustly supports your operational goals. In today's interconnected global economy, procurement is no longer just about cost reduction; it's about building resilient, ethical, and high-performing supplier networks that align with your broader business objectives. Central to this is rigorous vendor selection and evaluation. This involves identifying and assessing potential partners not just on cost-effectiveness, but equally on quality, reliability, innovation capabilities, and alignment with your ethical standards. A thorough due diligence process ensures strong, long-term partnerships.

Furthermore, integrating sustainable sourcing principles is crucial. Incorporating ethical labor practices, environmental responsibility, and transparent supply chains into your procurement decisions enhances your brand's reputation, mitigates risks, and meets evolving consumer and regulatory expectations for sustainability. Crucially, developing robust risk management in supply chains is vital. Establishing comprehensive contingency plans to mitigate supplier disruptions—whether from natural disasters, geopolitical events, or financial instability—ensures business continuity and protects your operations from unexpected shocks. Neglecting strategic sourcing can lead to unreliable supply chains, inflated costs, ethical controversies, and a vulnerability to unforeseen disruptions. By championing a proactive, ethical, and risk-aware approach to procurement and supplier relationships, you empower your organization to build a resilient, efficient, and responsible supply chain, ultimately contributing significantly to your business's operational excellence, brand integrity, and sustained competitive advantage.

WHAT?	WHY?	HOW?	BEWARE!
VENDOR EVALUATION	**OPTIMAL PARTNERSHIPS**	**RFP PROCESS**	**BIAS**
Identifying partners based on quality, reliability, and cost.	Ensures selection of high-performing, reliable suppliers.	Develop a rigorous RFI/RFP process, including scorecards for evaluation.	Subjectivity in evaluating non-cost factors like quality and reliability.
SUSTAINABLE SOURCING PRINCIPLES	**REPUTATION & ESG**	**AUDITS & CERTIFICATIONS**	**TRANSPARENCY**
Incorporating ethical and environmentally responsible procurement.	Enhances brand reputation and meets ESG (Environmental, Social, Governance) goals.	Partner with certified suppliers and conduct regular ethical audits.	Ensuring full transparency into complex global supply chains.
RISK MANAGEMENT IN SUPPLY CHAINS	**BUSINESS CONTINUITY**	**SUPPLIER DIVERSIFICATION**	**COMPLEXITY**
Establishing contingency plans for disruptions.	Mitigates operational disruptions from supplier failures.	Diversify supplier base and develop contingency plans for critical components.	Managing risks across vast and interconnected global supply chains.

For a business leader, meticulously defining Operations Service Level Agreements (SLAs) and Key Performance Indicators (KPIs) is a critical strategic imperative for establishing clear performance metrics, fostering accountability, and driving continuous improvement across all operational functions. These defined measures provide the necessary clarity and objective data to manage performance effectively. Central to this is establishing clear Service-Level Agreements (SLAs). These formal agreements set precise expectations for operational efficiency, service quality, and response times, ensuring clear commitments between internal departments or with external stakeholders. SLAs clarify responsibilities and performance benchmarks.

Furthermore, rigorous Key Performance Indicator (KPI) tracking is crucial. Monitoring a balanced set of KPIs related to productivity, cost management, service quality, and efficiency provides a quantifiable view of operational health, allowing you to identify strengths, weaknesses, and areas needing immediate attention. Crucially, consistently benchmarking against industry standards is vital for competitive performance. Comparing your operational KPIs with those of industry leaders and best-in-class organizations provides valuable external context, highlighting opportunities for optimization and ensuring your operations remain competitive and efficient. Neglecting to define clear SLAs and KPIs can lead to ambiguous performance expectations, a lack of accountability, inefficiencies going unnoticed, and an inability to drive continuous improvement. By championing a data-driven approach to operational measurement, you empower your organization to achieve higher levels of efficiency, maintain service quality, and continuously refine its processes, ultimately contributing significantly to your business's overall performance.

WHAT?	WHY?	HOW?	BEWARE!
SERVICE-LEVEL AGREEMENTS (SLAS)	**CLARITY & ACCOUNTABILITY**	**AGREEMENT DRAFTING**	**MEASUREMENT**
Establishing expectations for operational efficiency.	Defines clear performance expectations and commitments.	Develop formal SLAs for internal processes and external vendor services.	Difficulty in precisely measuring and monitoring certain service levels.
KEY PERFORMANCE INDICATORS (KPIS)	**PERFORMANCE VISIBILITY**	**KPI DEFINITION**	**METRIC OVERLOAD**
Tracking productivity, cost management, and service quality.	Provides objective metrics for operational health and efficiency.	Define measurable KPIs (e.g., OEE, lead time, fill rate, cost per unit).	Too many KPIs leading to complexity and difficulty focusing.
BENCHMARKING AGAINST INDUSTRY STANDARDS	**COMPETITIVE IMPROVEMENT**	**DATA GATHERING**	**DATA ACCESS**
Comparing operational KPIs with industry leaders.	Identifies best practices and areas for competitive advantage.	Research industry reports, competitor data, and best practice studies.	Difficulty obtaining reliable and comparable industry benchmark data.

For a business leader, meticulously structuring a strategic roadmap for operations is a critical imperative for ensuring long-term sustainable success and guiding continuous improvement. This roadmap provides a clear vision, outlining the evolution of your operational capabilities and investments over time, aligning them with overall business strategy. Central to this is thoughtfully balancing short-term vs. long-term planning. This involves identifying opportunities for immediate efficiency gains and quick wins while simultaneously laying the groundwork for future scalability, technological advancements, and shifts in operational models.

Furthermore, strategic investment and resource allocation are crucial. Prioritizing funding and distributing resources (human, technological, financial) to meet evolving operational needs and strategic initiatives ensures that critical projects are adequately supported and deliver maximum ROI. Crucially, proactively integrating innovation and transformation strategies is vital. Incorporating new technologies (e.g., AI, automation), methodologies (e.g., Lean, Industry 4.0), and process improvements into your long-term plan ensures continuous operational advancements, maintaining a competitive edge and fostering a culture of continuous evolution. Neglecting a strategic operations roadmap can lead to reactive decision-making, fragmented initiatives, inefficient resource deployment, and an inability to scale or adapt to future market demands. By championing a comprehensive, forward-looking roadmap, you empower your organization to systematically enhance its operational capabilities, drive efficiency, and achieve sustainable growth, ultimately securing your business's long-term competitive strength and market leadership through operational excellence.

WHAT?	WHY?	HOW?	BEWARE!
SHORT-TERM VS. LONG-TERM PLANNING	**BALANCED GROWTH**	**PHASED APPROACH**	**PRIORITIZATION**
Balancing immediate efficiency with future scalability.	Ensures both immediate gains and sustainable future development.	Develop short-term sprints for quick wins and long-term milestones.	Balancing immediate operational pressures with long-term strategic investments.
INVESTMENT AND RESOURCE ALLOCATION	**OPTIMAL UTILIZATION**	**ROI ANALYSIS**	**COMPETITION**
Prioritizing funding and resource distribution.	Ensures critical projects are adequately funded and resourced.	Conduct cost-benefit analysis for operational projects and initiatives.	Internal competition for limited resources across departments.
INNOVATION AND TRANSFORMATION STRATEGIES	**COMPETITIVE EDGE**	**TECHNOLOGY ADOPTION**	**RESISTANCE TO CHANGE**
Incorporating technology and new methodologies.	Drives continuous operational advancements and market leadership.	Research and integrate new technologies (AI, IoT, automation).	Employee and cultural resistance to new technologies and processes.

Ai6x6™ Transformation Blueprint

INITIATION | Operations Planning

"Planning is bringing the future into the present so that you can do something about it now." - Alan Lakein.

An American author and influential time management expert. Best known for "How to Get Control of Your Time and Your Life," he introduced practical strategies for personal productivity and goal setting, emphasizing prioritizing tasks and effective planning.

The Initiation level is your Operational Planning, beginning with the first concrete allocation of resources to meet the mission. Establishing capacity and contingency ensures the operation is functional and prepared for the first stage of action.

Detailing the specific resources, precise schedules, and clear procedures required to execute operational activities effectively and efficiently, translating strategic objectives into actionable plans. This phase ensures smooth workflow, optimal resource allocation, and preparedness for execution, minimizing bottlenecks and maximizing output.

It involves forecasting demand, managing inventory, scheduling production, and allocating personnel to ensure that all necessary components are in place for successful and timely operational execution, supporting overall business objectives.

The "**Operations Planning**" chapter is crucial for business strategy maturity as it focuses on structuring workflows, establishing processes, and defining key execution frameworks. A mature business strategy translates high-level operational goals into actionable plans, ensuring that resources are effectively allocated and processes are optimized for efficiency. This activation phase is vital for mobilizing the operational machinery, ensuring that every step is clearly defined and aligned with strategic objectives. It's about setting up the operational engine to run smoothly and predictably, enabling consistent and reliable strategic execution.

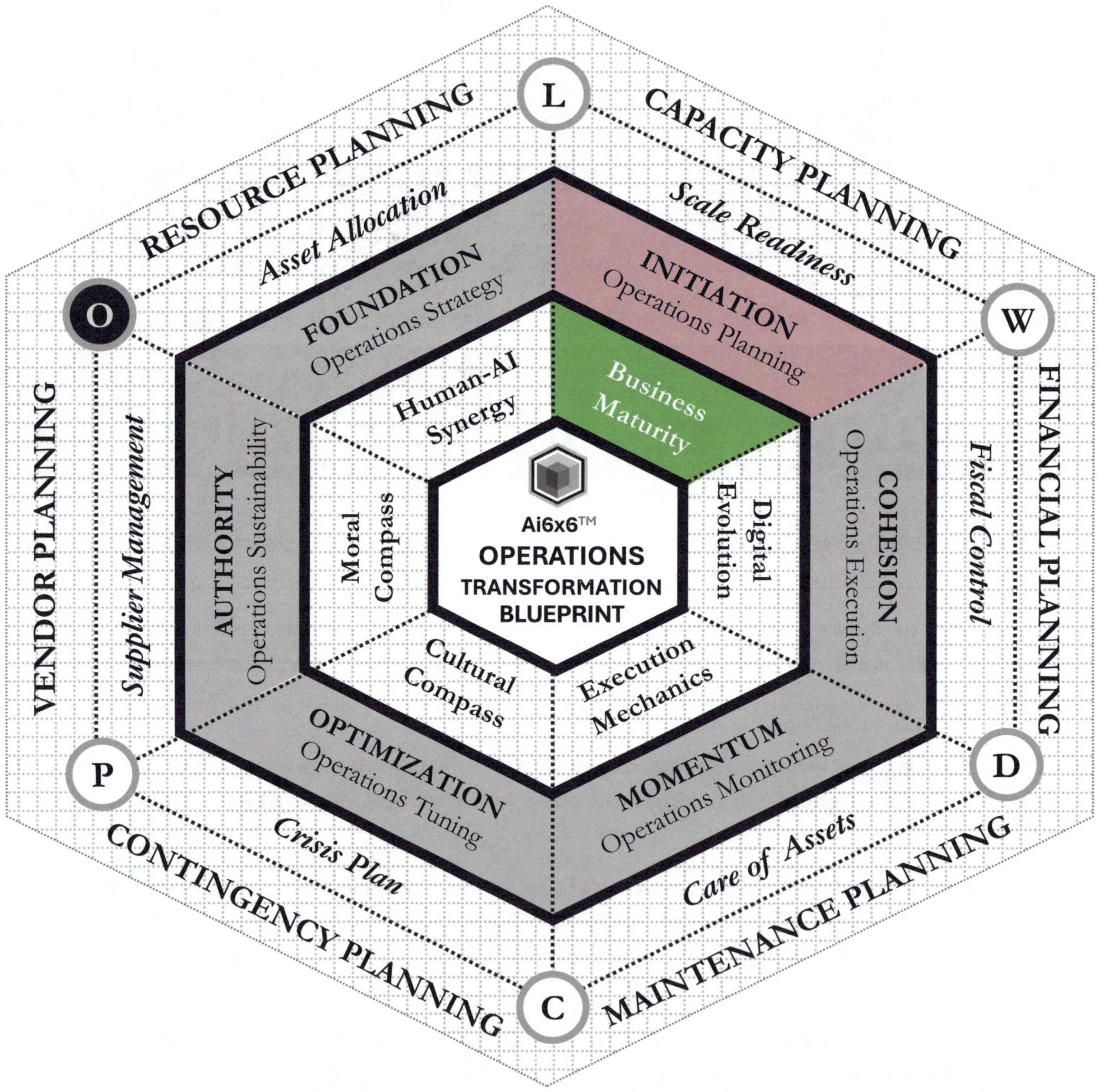

Ai6x6 BUSINESS MATURITY FRAMEWORK
OPERATIONS TRANSFORMATION | OPERATIONS PLANNING

1. Resource planning

For a business leader, effective resource planning is a critical strategic imperative for optimizing the allocation of workforce, materials, and technology to maximize operational efficiency and achieve business objectives. In an environment of finite resources, intelligent planning is key to minimizing waste, reducing costs, and ensuring smooth operations. Central to this is precise workforce scheduling. This involves aligning staffing levels with demand fluctuations and workload priorities, ensuring you have the right number of people with the right skills available at the right time, avoiding both overstaffing and understaffing.

Furthermore, rigorous inventory and material management are crucial. Ensuring the availability of necessary components and raw materials while minimizing excess stock, carrying costs, and waste is a delicate balancing act that directly impacts cash flow and production efficiency. Crucially, strategic technology resource optimization is vital. This means not just acquiring cutting-edge tools but strategically deploying and managing your technology infrastructure to enhance productivity, automate tasks, and support operational workflows, ensuring maximum ROI from your tech investments. Neglecting comprehensive resource planning can lead to operational bottlenecks, increased labor or material costs, inefficient use of technology, and a struggle to meet production or service demands. By championing a disciplined and integrated approach to planning your workforce, materials, and technology, you empower your organization to operate with greater agility, cost-efficiency, and responsiveness, ultimately contributing significantly to your business's sustained success and profitability.

WHAT?	WHY?	HOW?	BEWARE!
WORKFORCE SCHEDULING	**LABOR OPTIMIZATION**	**DEMAND-BASED SCHEDULING**	**COMPLEXITY**
Aligning staffing levels with demand and priorities.	Ensures optimal staffing to meet demand and minimize labor costs.	Use demand forecasts to create dynamic employee schedules.	Managing schedules for large, diverse workforces with varying skills.
INVENTORY AND MATERIAL MANAGEMENT	**COST REDUCTION**	**INVENTORY SYSTEMS**	**SUPPLY CHAIN DISRUPTIONS**
Ensuring availability while minimizing excess costs.	Prevents stockouts and reduces carrying costs of excess inventory.	Implement inventory management systems and just-in-time (JIT) principles.	Unforeseen disruptions impacting material availability.
TECHNOLOGY RESOURCE OPTIMIZATION	**ENHANCED PRODUCTIVITY**	**RESOURCE MAPPING**	**TECHNOLOGY ADOPTION**
Strategically deploying tools and infrastructure.	Maximizes ROI from technology investments.	Assess technology usage and allocate tools based on operational needs.	Ensuring employees effectively adopt and utilize new technologies.

For a business leader, rigorous capacity planning is a critical strategic imperative for ensuring your operational readiness to meet fluctuating demand and enabling seamless scalability for future growth. It's about proactively ensuring you have the necessary resources—production capacity, service capability, workforce—to fulfill customer needs without over-investing or facing bottlenecks. Central to this is robust demand forecasting integration. Aligning your resource availability directly with expected market trends, sales forecasts, and seasonal variations allows you to proactively adjust capacity levels, ensuring you are neither under-resourced nor carrying excessive idle capacity.

Furthermore, meticulously engaging in scalability modeling is crucial. Structuring expansion frameworks that can accommodate growth opportunities—whether through adding new production lines, expanding service teams, or leveraging cloud infrastructure—ensures your operations can scale efficiently and cost-effectively without disrupting current performance. Crucially, implementing load balancing strategies is vital for distributing workloads efficiently across available resources. This ensures optimal utilization of your existing capacity, preventing bottlenecks, maximizing throughput, and maintaining service quality even during peak demand periods. Neglecting comprehensive capacity planning can lead to missed sales opportunities, customer dissatisfaction due to unfulfilled orders, increased operational costs from inefficient resource utilization, and a hindrance to scaling your business. By championing a proactive and adaptive approach to capacity management, you empower your organization to meet demand effectively, achieve sustained growth, and maintain operational efficiency, ultimately contributing significantly to your business's profitability and market responsiveness.

WHAT?	WHY?	HOW?	BEWARE!
DEMAND FORECASTING INTEGRATION	**PROACTIVE READINESS**	**DATA INTEGRATION**	**ACCURACY**
Aligning resource availability with market trends.	Ensures resources are available to meet anticipated demand.	Integrate sales forecasts and market trends directly into capacity models.	Ensuring demand forecasts are sufficiently accurate for capacity decisions.
SCALABILITY MODELING	**EFFICIENT GROWTH**	**SCENARIO ANALYSIS**	**PREDICTION**
Structuring expansion frameworks.	Accommodates growth without compromising operational efficiency.	Model various growth scenarios and plan resource increments.	Predicting future growth rates and resource needs accurately.
LOAD BALANCING STRATEGIES	**OPTIMAL UTILIZATION**	**DISTRIBUTION ALGORITHMS**	**REAL-TIME ADAPTATION**
Distributing workloads efficiently.	Maximizes resource usage and prevents bottlenecks.	Implement algorithms to distribute tasks evenly across resources.	Dynamically adjusting load balancing in real-time for fluctuating demand.

For a business leader, meticulous financial planning is a critical strategic imperative for structuring robust budgets, optimizing cost management, and ensuring the long-term operational sustainability of your business. Operations often represent a significant portion of a company's expenses, and intelligent financial oversight is key to profitability and efficient resource utilization. Central to this is accurate expense forecasting. Analyzing historical spending patterns, leveraging predictive data models, and anticipating future operational needs allows for precise budget accuracy, enabling proactive financial management and resource allocation.

Furthermore, implementing comprehensive cost optimization strategies is crucial. This involves systematically identifying areas for efficiency improvements, streamlining processes, negotiating better vendor terms, and reducing unnecessary expenditures across all operational functions without compromising quality or service levels. Crucially, rigorously tracking profitability and ROI (Return on Investment) is vital. Monitoring the financial performance of operational initiatives, investments, and processes ensures alignment with broader organizational goals, providing clear data on which activities generate the most value and where adjustments are needed. Neglecting robust financial planning in operations can lead to budget overruns, inefficient resource allocation, decreased profitability, and a reactive approach to cost management. By championing a data-driven and proactive approach to budgeting and cost control, you empower your organization to operate with greater financial discipline, maximize value from every dollar spent, and ensure the long-term sustainability and profitability of your business. This financial rigor is fundamental to operational excellence.

WHAT?	WHY?	HOW?	BEWARE!
EXPENSE FORECASTING	**BUDGET ACCURACY**	**DATA ANALYSIS**	**VOLATILITY**
Analyzing historical and predictive data for budget accuracy.	Ensures realistic and precise budget allocation.	Use historical financial data and predictive analytics to project future expenses.	Unforeseen market changes or operational disruptions impacting expense accuracy.
COST OPTIMIZATION STRATEGIES	**PROFITABILITY**	**LEAN PRINCIPLES**	**SHORT-TERM FOCUS**
Identifying efficiency improvements.	Reduces operational expenditures without compromising quality.	Implement lean methodologies and process re-engineering to identify savings.	Prioritizing immediate cuts over long-term strategic investments.
PROFITABILITY AND ROI TRACKING	**STRATEGIC ALIGNMENT**	**FINANCIAL REPORTING**	**ATTRIBUTION**
Monitoring financial performance.	Ensures operational investments contribute to overall business goals.	Track profitability metrics and calculate ROI for operational projects.	Accurately attributing operational costs/savings to specific initiatives.

For a business leader, robust maintenance planning is a critical strategic imperative for preserving operational integrity, minimizing downtime, and ensuring the long-term efficiency and reliability of your assets and infrastructure. In any asset-intensive business, reactive maintenance is costly and inefficient; proactive upkeep is key to sustained performance and profitability. Central to this is implementing comprehensive preventive maintenance schedules. Structuring regular servicing, inspections, and component replacements for equipment and systems before failures occur significantly reduces breakdowns, extends lifespan, and minimizes costly emergency repairs.

Furthermore, rigorous asset lifecycle management is crucial. This involves systematically evaluating repair vs. replacement decisions for long-term efficiency, taking into account factors like age, performance degradation, and technological obsolescence. This strategic approach ensures optimal asset utilization and capital expenditure. Crucially, establishing clear emergency maintenance protocols is vital. Developing rapid response mechanisms, defining escalation procedures, and ensuring immediate availability of resources for unexpected disruptions minimizes downtime, contains damage, and ensures swift recovery from unforeseen failures. Neglecting effective maintenance planning can lead to frequent equipment breakdowns, costly emergency repairs, lost productivity due to downtime, and a shorter lifespan for critical assets. By championing a proactive and systematic approach to maintenance, you empower your organization to operate with greater reliability, reduce operational costs, and maximize asset utilization, ultimately contributing significantly to your business's sustained efficiency, profitability, and operational resilience.

WHAT?	WHY?	HOW?	BEWARE!
PREVENTIVE MAINTENANCE SCHEDULES	**DOWNTIME REDUCTION**	**SCHEDULED MAINTENANCE**	**OVER-MAINTENANCE**
Structuring regular servicing to prevent failures.	Minimizes unexpected breakdowns and extends asset lifespan.	Implement routine inspections and servicing based on manufacturer recommendations.	Performing maintenance too frequently, leading to unnecessary costs.
ASSET LIFECYCLE MANAGEMENT	**OPTIMIZED INVESTMENT**	**COST-BENEFIT ANALYSIS**	**DATA GAPS**
Evaluating repair vs. replacement decisions.	Ensures efficient capital expenditure and asset utilization.	Conduct regular assessments of asset performance and repair costs vs. replacement.	Lack of comprehensive data on asset performance and repair history.
EMERGENCY MAINTENANCE PROTOCOLS	**RAPID RECOVERY**	**RESPONSE PLANS**	**RESOURCE AVAILABILITY**
Establishing rapid response mechanisms.	Minimizes impact and ensures swift recovery from disruptions.	Develop clear protocols for identifying, escalating, and resolving critical failures.	Ensuring immediate availability of skilled personnel and spare parts.

For a business leader, rigorous contingency planning is a non-negotiable strategic imperative for preparing for operational risks and ensuring continuous business continuity in the face of unforeseen disruptions. In today's volatile and interconnected world, a reactive approach to crises is simply insufficient; proactive readiness is key to protecting your assets, reputation, and profitability. Central to this is implementing robust risk assessment frameworks. This involves systematically identifying potential vulnerabilities and analyzing the impact of various disruptions—from natural disasters and cyberattacks to supply chain failures and geopolitical events. A comprehensive risk profile is the foundation for effective planning.

Furthermore, developing clear crisis response strategies is crucial. Establishing detailed action plans for how your organization will react to specific incidents—including communication protocols, roles and responsibilities, and decision-making authority—minimizes chaos and accelerates effective response during a crisis. Crucially, investing in comprehensive business continuity solutions is vital. This includes ensuring redundancies in critical systems, implementing robust data recovery processes, and establishing alternative operational sites, guaranteeing that your business can quickly resume critical functions and maintain resilience even after a significant disruption. Neglecting robust contingency planning can lead to catastrophic operational failures, prolonged downtime, severe financial losses, and irreparable damage to your brand's reputation. By championing a proactive, systematic, and well-rehearsed approach to risk and crisis management, you empower your organization to withstand shocks, recover swiftly, and ultimately secure your business's long-term viability and competitive advantage.

WHAT?	WHY?	HOW?	BEWARE!
RISK ASSESSMENT FRAMEWORKS	**PROACTIVE MITIGATION**	**RISK REGISTER**	**UNFORESEEN RISKS**
Identifying vulnerabilities and potential disruptions.	Uncovers weaknesses to prevent or minimize impact of risks.	Develop a comprehensive risk register, categorizing and prioritizing threats.	Difficulty anticipating every possible type of disruption.
CRISIS RESPONSE STRATEGIES	**CONTROLLED REACTION**	**SIMULATION & TRAINING**	**PSYCHOLOGICAL IMPACT**
Establishing action plans for disruptions.	Ensures swift, coordinated, and effective response during crises.	Conduct regular drills and training for crisis response teams.	Managing stress and emotions of personnel during actual crises.
BUSINESS CONTINUITY SOLUTIONS	**RESILIENT OPERATIONS**	**BACKUP SYSTEMS**	**COST & COMPLEXITY**
Ensuring redundancies and recovery processes.	Guarantees core business functions can resume quickly after disruption.	Implement redundant systems, data backups, and off-site recovery plans.	High costs and complexity of maintaining comprehensive BC/DR solutions.

For a business leader, meticulous vendor planning is a critical strategic imperative for effectively managing supplier relationships, optimizing procurement processes, and ensuring a stable and cost-effective supply chain. In today's global economy, your vendors are extensions of your own operations, and their performance directly impacts your product quality, delivery times, and profitability. Central to this is rigorously developing RFI/RFP (Request for Information/Proposal) processes. Structuring clear vendor selection criteria based on quality, reliability, innovation capabilities, ethical standards, and cost-effectiveness ensures that you partner with suppliers who are truly aligned with your operational goals and values.

Furthermore, mastering negotiation techniques is crucial. Securing favorable contract terms, competitive pricing, and advantageous service level agreements with suppliers directly impacts your cost structure and operational efficiency, contributing to your bottom line. Crucially, establishing robust contract management strategies is vital for ongoing performance. Monitoring vendor compliance with agreed-upon terms, continuously assessing their performance against KPIs, and proactively planning for contract renewals or renegotiations ensures sustained value and mitigates risks over the long term. Neglecting effective vendor planning can lead to unreliable supply chains, inflated procurement costs, quality issues, and a vulnerability to supplier failures. By championing a proactive, strategic, and disciplined approach to vendor relationships, you empower your organization to build a resilient, efficient, and high-performing supply chain, ultimately contributing significantly to your business's operational excellence, profitability, and competitive advantage.

WHAT?	WHY?	HOW?	BEWARE!
RFI/RFP DEVELOPMENT	**OPTIMAL PARTNERSHIPS**	**REQUIREMENTS DEFINITION**	**SCOPE CREEP**
Structuring vendor selection criteria.	Ensures selection of the best-fit suppliers.	Clearly define needs, scope, and evaluation criteria for RFIs/RFPs.	Requirements expanding during the RFI/RFP process.
NEGOTIATION TECHNIQUES	**COST SAVINGS**	**NEGOTIATION TRAINING**	**POWER IMBALANCE**
Securing favorable contract terms and cost efficiencies.	Maximizes financial benefits and favorable contract conditions.	Equip procurement teams with advanced negotiation skills.	Negotiating with dominant or sole-source suppliers.
CONTRACT MANAGEMENT STRATEGIES	**VALUE & COMPLIANCE**	**CONTRACT LIFECYCLE**	**LACK OF MONITORING**
Monitoring compliance, performance, and renewal.	Ensures ongoing vendor performance and contractual adherence.	Implement systems for tracking contracts, milestones, and renewals.	Failing to continuously monitor vendor performance post-contract signing.

Ai6x6™ Transformation Blueprint

COHESION | Operations Execution

"Execution is the great unaddressed issue in the business world." - Larry Bossidy.

Former CEO of AlliedSignal and a highly respected business executive. He is renowned for his unwavering focus on operational excellence and disciplined execution, driving performance across large organizations. His book "Execution: The Discipline of Getting Things Done" is a practical guide for leaders.

The Cohesion level is your Operational Execution, occurring when supply chains and internal systems work in perfect harmony. ERP and supply chain integration turn individual tasks into a single, cohesive operational engine.

The actual, hands-on implementation of all planned activities, directly leading to the production of goods or the delivery of services according to established standards and schedules. This is the critical juncture where raw materials and processes are transformed into tangible value; efficient and flawless execution is paramount for consistently meeting customer demand, maintaining high quality standards, and achieving operational targets.

It requires effective coordination, skilled labor, and robust systems to ensure smooth, continuous, and high-quality output.

"Operations Execution" is vital for business strategy maturity because it represents the core active work of operations—carrying out tasks, managing workflows, and ensuring smooth day-to-day operations. A mature business strategy relies on efficient and effective operational execution to translate plans into tangible results. This acceleration phase is critical for achieving productivity targets, meeting customer demands, and delivering on strategic promises. It's about maintaining momentum and ensuring that the operational engine runs at peak performance, directly contributing to the organization's ability to achieve its strategic objectives and competitive advantage. Here is how we approach our learning:

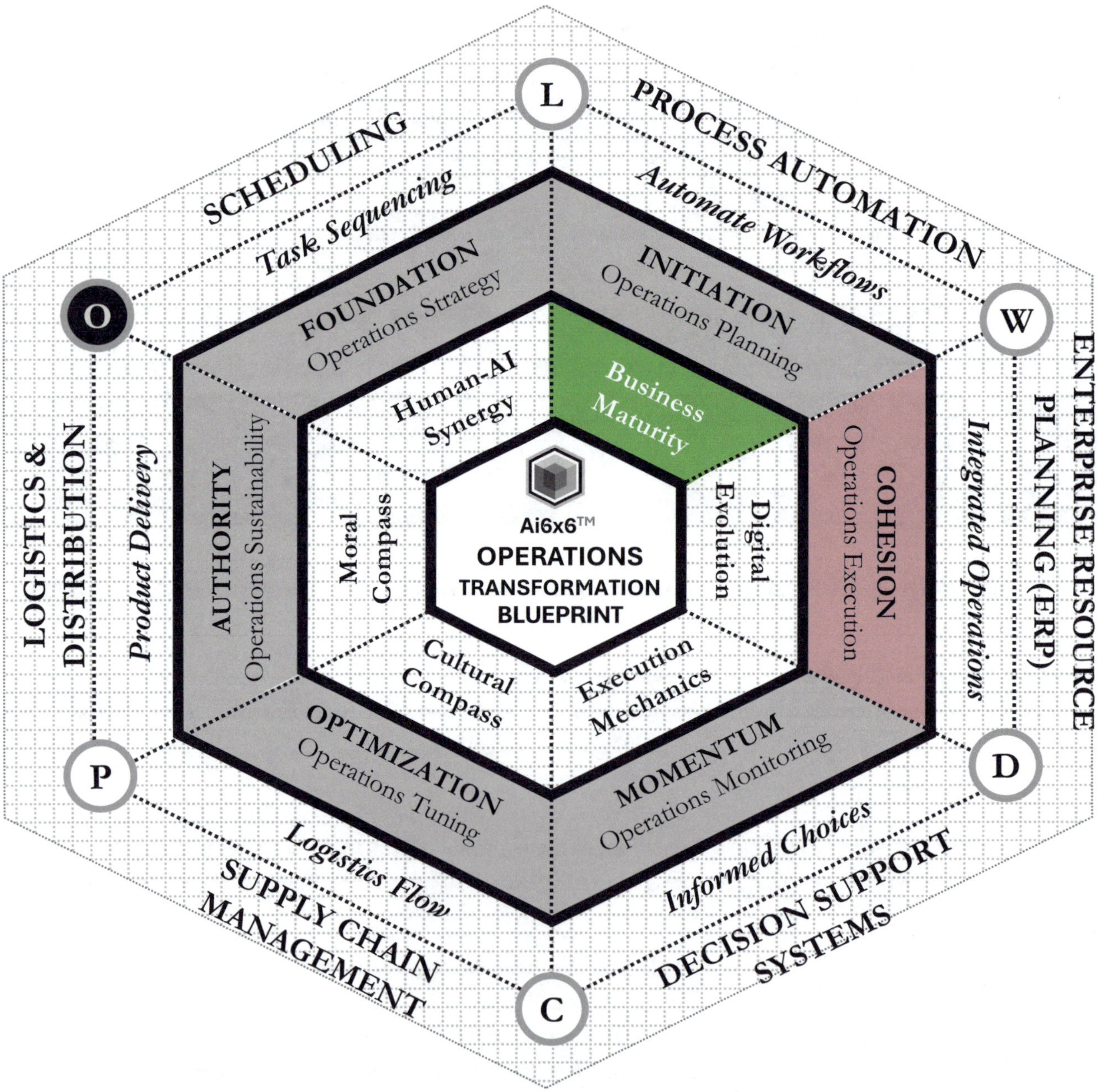

Ai6x6 BUSINESS MATURITY FRAMEWORK
OPERATIONS TRANSFORMATION | OPERATIONS EXECUTION

1. Scheduling

For a business leader, optimizing scheduling is a critical strategic imperative for enhancing task management, streamlining workforce coordination, and ensuring the smooth, efficient execution of day-to-day operations. In any dynamic operational environment, precise scheduling is key to maximizing productivity, minimizing idle time, and meeting deadlines effectively. Central to this is implementing dynamic scheduling systems. By utilizing AI and predictive analytics, these intelligent systems can adjust schedules in real time, responding to unexpected demand fluctuations, resource availability, or urgent priorities, ensuring optimal resource utilization and responsiveness.

Furthermore, developing robust workforce allocation models is crucial. Distributing personnel based on demand patterns, individual skill sets, and priority tasks ensures that your workforce is deployed efficiently, maximizing productivity and minimizing labor costs while maintaining service levels. Crucially, achieving meticulous project timeline synchronization is vital. Ensuring alignment and coordination across different departments and project phases allows for seamless execution, preventing bottlenecks, accelerating project completion, and improving overall operational flow. Neglecting effective scheduling can lead to inefficient resource utilization, missed deadlines, employee burnout, and a struggle to meet operational targets. By championing a data-driven, dynamic, and integrated approach to scheduling, you empower your organization to operate with greater precision, agility, and efficiency, ultimately contributing significantly to your business's sustained success and profitability through optimized operational execution.

WHAT?	WHY?	HOW?	BEWARE!
DYNAMIC SCHEDULING SYSTEMS	**OPTIMIZED ALLOCATION**	**AI-POWERED PLATFORM**	**DATA ACCURACY**
Using AI and predictive analytics.	Adjusts schedules in real time for optimal resource utilization.	Deploy AI/ML-driven scheduling software that responds to changing conditions.	Reliance on accurate real-time data for effective dynamic adjustments.
WORKFORCE ALLOCATION MODELS	**MAXIMIZED PRODUCTIVITY**	**SKILL-BASED MATCHING**	**FLEXIBILITY**
Distributing personnel based on demand and priority.	Ensures the right people are in the right place at the right time.	Develop models that match employee skills to task requirements.	Employee availability and willingness to adapt to dynamic assignments.
PROJECT TIMELINE SYNCHRONIZATION	**SEAMLESS EXECUTION**	**COLLABORATIO N TOOLS**	**COMMUNICATION**
Ensuring alignment across departments.	Prevents bottlenecks and accelerates project completion.	Use project management software to track dependencies and milestones.	Ensuring constant and clear communication across all involved departments.

For a business leader, strategically implementing process automation is a critical imperative for significantly enhancing efficiency, reducing costs, and streamlining workflows throughout your operations. In today's competitive environment, automating repetitive, manual tasks frees up human capital for higher-value activities, driving both productivity and innovation. Central to this is deploying Robotic Process Automation (RPA). Automating repetitive, rule-based tasks—such as data entry, form processing, or report generation—reduces manual workload, minimizes human error, and accelerates execution speed across various operational functions.

Furthermore, comprehensive workflow orchestration is crucial. Integrating automation across different departments and systems streamlines end-to-end operations, ensuring seamless handoffs, reducing bottlenecks, and creating a more cohesive and efficient operational flow across the entire organization. Crucially, leveraging AI-driven optimization enhances process adaptability. By using predictive insights and machine learning, AI can identify opportunities for further automation, optimize process routes, and even adapt workflows in real time based on changing conditions, leading to continuous improvement and greater resilience. Neglecting process automation can lead to inefficient manual processes, high operational costs, increased human error rates, and a struggle to scale operations effectively. By championing a comprehensive and intelligent approach to automation, you empower your organization to operate with unprecedented efficiency, reduce operational overhead, and free up valuable human resources for strategic initiatives, ultimately contributing significantly to your business's sustained success and competitive advantage.

WHAT?	WHY?	HOW?	BEWARE!
ROBOTIC PROCESS AUTOMATION (RPA)	**MANUAL WORKLOAD REDUCTION**	**BOT DEPLOYMENT**	**PROCESS FRAGMENTATION**
Automating repetitive tasks.	Frees up human resources for higher-value activities.	Identify repetitive, rule-based tasks suitable for RPA bots.	RPA bots needing frequent updates if underlying processes change.
WORKFLOW ORCHESTRATION	**STREAMLINED OPERATIONS**	**INTEGRATION PLATFORM**	**COMPLEXITY**
Integrating automation across departments.	Ensures seamless handoffs and reduces bottlenecks.	Use an orchestration platform to connect automated processes across systems.	Orchestrating complex workflows across diverse systems and departments.
AI-DRIVEN OPTIMIZATION	**CONTINUOUS IMPROVEMENT**	**ML INTEGRATION**	**TRUST IN AI**
Using predictive insights to enhance process adaptability.	Enables dynamic adaptation and optimization of automated processes.	Apply AI/ML to analyze process data and suggest optimization opportunities.	Building trust in AI recommendations for critical process changes.

For a business leader, strategically implementing and optimizing Enterprise Resource Planning (ERP) systems is a fundamental imperative for centralizing data, integrating workflows, and achieving seamless operational efficiency across all business units. ERP acts as the nervous system of your organization, unifying disparate functions and providing a single source of truth for critical business data. Central to this is leveraging cloud-based ERP systems. This enhances accessibility for remote and distributed teams, provides inherent scalability to accommodate growth, and reduces the need for expensive on-premise infrastructure, ensuring flexibility and cost-efficiency.

Furthermore, ensuring customizable modules are available is crucial. Adapting ERP functionalities to align with your unique business needs, specific workflows, and industry requirements maximizes its relevance and effectiveness, ensuring the system supports your distinct operational processes rather than forcing them into a rigid template. Crucially, establishing real-time data synchronization is vital. This ensures seamless communication and immediate information flow between departments—from sales and finance to supply chain and HR—eliminating data silos and enabling rapid, informed decision-making across the entire organization. Neglecting a modern, well-integrated ERP system can lead to fragmented data, inefficient manual processes, poor cross-functional visibility, and a struggle to make timely, data-driven decisions. By championing a robust ERP strategy, you empower your organization with unparalleled operational control, enhanced visibility, and streamlined workflows, ultimately contributing significantly to your business's sustained efficiency, scalability, and profitability through integrated enterprise management.

WHAT?	WHY?	HOW?	BEWARE!
CLOUD-BASED ERP SYSTEMS	**FLEXIBILITY & EFFICIENCY**	**CLOUD MIGRATION**	**DATA SECURITY**
Enhancing accessibility and scalability.	Provides scalable and accessible enterprise data management.	Migrate or implement ERP systems on cloud platforms (e.g., SAP S/4HANA Cloud, Oracle Cloud ERP).	Ensuring data security and compliance in a cloud environment.
CUSTOMIZABLE MODULES	**RELEVANCE & ADAPTABILITY**	**CONFIGURATION**	**OVER-CUSTOMIZATION**
Adapting ERP functionalities to unique business needs.	Ensures the ERP system aligns with specific operational workflows.	Configure modules to match unique business processes and industry requirements.	Excessive customization leading to complex upgrades and maintenance.
REAL-TIME DATA SYNCHRONIZATION	**INFORMED DECISIONS**	**INTEGRATION INTERFACES**	**DATA CONSISTENCY**
Ensuring seamless communication between departments.	Provides up-to-date information for accurate, timely decisions.	Implement APIs and connectors for real-time data flow between modules.	Maintaining data consistency and integrity across all integrated modules.

For a business leader, implementing robust decision support systems is a critical strategic imperative for empowering more informed and accurate strategic choices through readily available, data-driven insights. In today's complex business environment, raw data alone is insufficient; contextualized, actionable intelligence is key to effective leadership. Central to this is leveraging predictive analytics tools. By utilizing machine learning models for forecasting and trend analysis, you can anticipate future market shifts, identify emerging opportunities, and proactively prepare for challenges, moving beyond reactive problem-solving to strategic foresight.

Furthermore, providing intuitive business intelligence dashboards is crucial. These real-time reporting tools transform complex data into easily digestible visual formats, allowing leaders and managers to monitor key performance indicators, identify trends, and make rapid, informed decisions without needing deep technical analytical skills. Crucially, integrating comprehensive risk assessment frameworks helps identify operational vulnerabilities and informs mitigation strategies. By providing data on potential disruptions and their likely impact, these systems enable leaders to make risk-aware decisions that protect the business from unforeseen challenges. Neglecting robust decision support systems can lead to reactive decision-making, missed opportunities, inefficient resource allocation, and a struggle to adapt swiftly to market changes. By championing data-driven insights and analytical tools, you empower your organization to make more precise, impactful, and proactive strategic choices, ultimately contributing significantly to your business's sustained success, profitability, and competitive advantage in a data-centric world.

WHAT?	WHY?	HOW?	BEWARE!
PREDICTIVE ANALYTICS TOOLS	**STRATEGIC FORESIGHT**	**ML MODEL DEVELOPMENT**	**ACCURACY**
Leveraging machine learning models for forecasting.	Anticipates future trends and opportunities for proactive decision-making.	Build and deploy machine learning models for sales, demand, or risk forecasting.	Ensuring predictive models provide reliable and actionable forecasts.
BUSINESS INTELLIGENCE DASHBOARDS	**PERFORMANCE VISIBILITY**	**BI PLATFORM DEPLOYMENT**	**DASHBOARD FATIGUE**
Providing real-time reports for informed decision-making.	Offers quick, visual insights into key operational performance metrics.	Implement BI tools (e.g., Tableau, Power BI) to create interactive dashboards.	Overwhelm from too many dashboards or irrelevant metrics.
RISK ASSESSMENT FRAMEWORKS	**RISK-AWARE DECISIONS**	**RISK MAPPING**	**COMPLEXITY**
Identifying operational vulnerabilities and mitigation strategies.	Informs decisions by highlighting potential threats and their impact.	Develop tools or processes to map and quantify operational risks.	Identifying and quantifying all potential operational vulnerabilities.

For a business leader, mastering supply chain management is a critical strategic imperative for ensuring smooth coordination across all stages, from procurement to distribution. In today's globalized and complex economy, an efficient and resilient supply chain is fundamental to operational excellence, customer satisfaction, and sustained profitability. Central to this is rigorous supplier performance tracking. Continuously monitoring vendors for efficiency, reliability, quality, and adherence to sustainability standards ensures that your procurement processes are supported by high-performing partners, minimizing risks and maximizing value.

Furthermore, optimizing inventory is crucial for efficiency. By reducing waste, preventing stockouts, and improving stock availability based on precise demand forecasting, you can balance the need for product availability with minimizing carrying costs, ensuring lean and responsive operations. Crucially, adopting demand-driven supply chain models is vital. This involves dynamically adjusting procurement and distribution strategies in real-time based on actual customer demand, rather than fixed forecasts. This agility allows for greater responsiveness to market changes, reduced overproduction, and improved customer satisfaction. Neglecting effective supply chain management can lead to costly delays, stockouts, inefficient inventory, inflated operational costs, and customer dissatisfaction. By championing an integrated, data-driven, and agile approach to supply chain coordination, you empower your organization to operate with seamless efficiency, build inherent resilience against disruptions, and ultimately contribute significantly to your business's sustained profitability and competitive advantage.

WHAT?	WHY?	HOW?	BEWARE!
SUPPLIER PERFORMANCE TRACKING	**OPTIMAL VENDORS**	**KPI MONITORING**	**DATA INTEGRATION**
Monitoring vendors for efficiency and reliability.	Ensures consistent quality and timely delivery from suppliers.	Track supplier KPIs (e.g., on-time delivery, defect rate, responsiveness).	Integrating performance data from diverse supplier systems.
INVENTORY OPTIMIZATION	**COST EFFICIENCY**	**INVENTORY SOFTWARE**	**DEMAND VOLATILITY**
Reducing waste and improving stock availability.	Minimizes carrying costs and prevents stockouts.	Implement advanced inventory management systems and forecasting tools.	Managing inventory levels for highly unpredictable demand.
DEMAND-DRIVEN SUPPLY CHAIN MODELS	**AGILE RESPONSIVENESS**	**REAL-TIME DATA**	**INFRASTRUCTURE**
Adjusting procurement and distribution dynamically.	Ensures the supply chain adapts swiftly to market changes.	Use real-time sales data to adjust production and logistics schedules.	Requires highly flexible and interconnected supply chain infrastructure.

For a business leader, strategically structuring efficient logistics and distribution networks is a critical imperative for meeting market demands, optimizing delivery speeds, and ensuring customer satisfaction. In today's e-commerce driven world, a robust and agile distribution system is a direct competitive advantage, impacting everything from delivery times to shipping costs and customer experience. Central to this is implementing advanced route optimization. By utilizing AI-based logistics planning, you can identify the most cost-effective transportation paths, minimize fuel consumption, and reduce delivery times, enhancing efficiency and environmental sustainability.

Furthermore, optimizing warehouse and fulfillment strategies is crucial. This involves enhancing inventory flow, improving order accuracy, and streamlining picking and packing processes within your warehouses to accelerate order fulfillment and ensure precise delivery, directly impacting customer satisfaction. Crucially, adopting an omnichannel distribution approach coordinates both physical and digital sales channels for seamless delivery. Ensuring that customers can receive products efficiently whether they order online for home delivery, click-and-collect, or purchase in-store, creates a unified and friction-less buying experience that builds loyalty. Neglecting efficient logistics and distribution can lead to high shipping costs, delayed deliveries, customer dissatisfaction, and a struggle to scale with market demand. By championing a data-driven, automated, and integrated approach to your delivery networks, you empower your organization to fulfill orders faster, more accurately, and at a lower cost, ultimately contributing significantly to your business's profitability, customer loyalty, and competitive edge in the marketplace.

WHAT?	WHY?	HOW?	BEWARE!
ROUTE OPTIMIZATION	**COST-EFFECTIVE DELIVERY**	**AI-POWERED SOFTWARE**	**REAL-TIME DATA**
Implementing AI-based logistics planning.	Reduces transportation costs and improves delivery times.	Deploy AI-driven route planning and optimization software.	Reliance on accurate real-time traffic and delivery data.
WAREHOUSE AND FULFILLMENT STRATEGIES	**ORDER EFFICIENCY**	**AUTOMATED SYSTEMS**	**INFRASTRUCTURE**
Enhancing inventory flow and order accuracy.	Accelerates order fulfillment and minimizes errors.	Implement warehouse management systems (WMS) and automation.	High initial investment in warehouse automation and technology.
OMNI-CHANNEL DISTRIBUTION	**SEAMLESS CUSTOMER EXPERIENCE**	**INTEGRATED SYSTEMS**	**COMPLEXITY**
Coordinating physical and digital sales channels.	Provides flexible and convenient delivery options.	Ensure inventory and order management are unified across all sales channels.	Managing inventory and fulfillment across diverse physical and digital channels.

Ai6x6™ Transformation Blueprint

MOMENTUM | Operations Monitoring

"Measurement is the first step that leads to control and eventually to improvement." - H. James Harrington.

A globally recognized expert in business process improvement and quality management. He has authored numerous influential books on performance improvement and business process re-engineering, significantly shaping how organizations approach operational efficiency and quality control worldwide.

Systematically tracking and meticulously measuring operational performance against established key metrics, performance indicators (KPIs), and predefined goals. This provides real-time insights into process health, enabling timely adjustments, proactive problem-solving, and ensuring that all operational processes remain precisely on track and aligned with objectives.

Effective monitoring identifies inefficiencies, quality issues, and deviations, allowing for immediate corrective actions and continuous improvement in operational effectiveness and efficiency.

The "**Operations Monitoring**" chapter is paramount for business strategy maturity as it focuses on continuously optimizing current operations and adapting to issues through data-driven insights. A mature business strategy employs robust monitoring systems to track performance, identify inefficiencies, and make data-driven improvements. This proactive approach ensures that operational processes are continuously refined, risks are mitigated, and resources are utilized optimally. It's about creating a feedback loop that allows the organization to learn from its operations and adjust its strategic execution in real-time, ensuring sustained efficiency and effectiveness. Here is how we approach our learning:

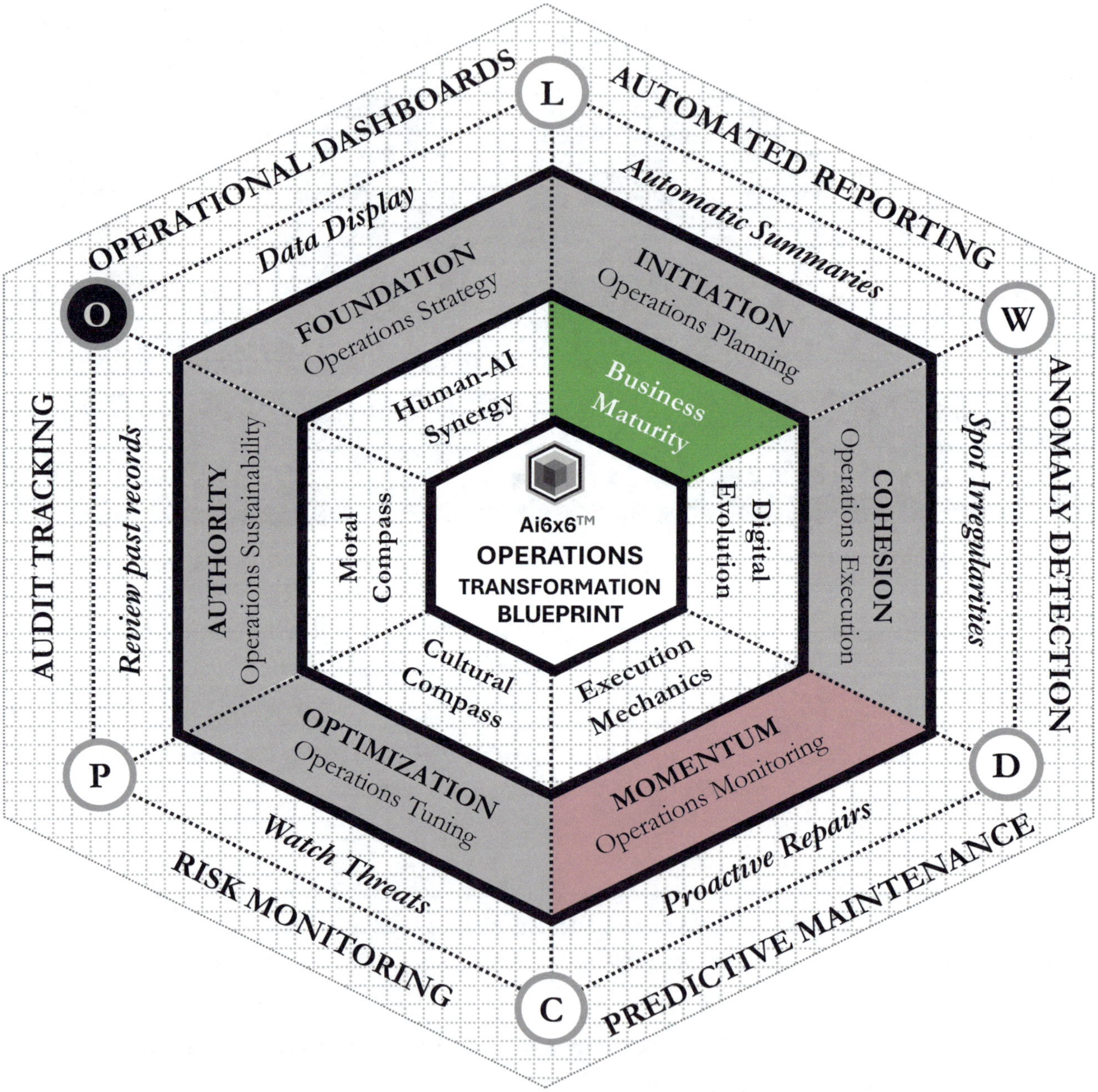

Ai6x6 BUSINESS MATURITY FRAMEOWORK
OPERATIONS TRANSFORMATION | OPERATIONS MONITORING

1. Operational dashboards

For a business leader, establishing robust operational dashboards is a critical strategic imperative for providing real-time visibility into key performance metrics, enabling immediate insights, and fostering agile decision-making across all operational functions. In today's fast-paced environment, static reports are insufficient; dynamic, actionable dashboards are key to maintaining control and optimizing performance. Central to this is enabling customizable data visualization. Tailoring dashboards to display only the most relevant operational insights for specific roles or departments ensures that users are not overwhelmed by irrelevant data, allowing for focused analysis and quicker decision-making.

Furthermore, rigorous KPI tracking and analytics are crucial. Continuously monitoring productivity, efficiency, and performance trends against predefined KPIs provides an objective and quantitative view of operational health, allowing you to identify strengths, weaknesses, and areas needing immediate attention. Crucially, implementing multi-source data integration consolidates information from various disparate systems (e.g., ERP, CRM, IoT sensors) into a single, comprehensive monitoring view. This eliminates data silos and provides a holistic understanding of end-to-end operational performance. Neglecting real-time operational dashboards can lead to delayed problem-solving, missed opportunities for optimization, and a reactive approach to performance management. By championing intuitive, data-integrated dashboards, you empower your organization with unparalleled visibility, accelerating insights, and enabling proactive adjustments that ultimately drive the sustained efficiency, profitability, and competitive advantage of your business. This visual intelligence is a cornerstone of operational leadership.

WHAT?	WHY?	HOW?	BEWARE!
CUSTOMIZABLE DATA VISUALIZATION	**FOCUSED INSIGHTS**	**BI PLATFORM**	**DESIGN COMPLEXITY**
Tailoring dashboards to display relevant insights.	Ensures users get pertinent information without data overload.	Utilize BI tools (e.g., Power BI, Tableau) to design role-specific dashboards.	Designing intuitive and effective visualizations for complex data.
KPI TRACKING AND ANALYTICS	**PERFORMANCE OPTIMIZATION**	**KPI DEFINITION**	**METRIC OVERLOAD**
Monitoring productivity, efficiency, and performance trends.	Provides objective measures for identifying areas of improvement.	Define and continuously monitor key performance indicators relevant to operations.	Too many KPIs making it difficult to discern critical information.
MULTI-SOURCE DATA INTEGRATION	**COMPREHENSIVE VIEW**	**DATA WAREHOUSING**	**DATA GOVERNANCE**
Consolidating information from various systems.	Eliminates data silos, providing a holistic operational overview.	Integrate data from ERP, MES, CRM, and IoT into a central data warehouse.	Ensuring data consistency and quality across disparate sources.

For a business leader, strategically implementing automated reporting is a critical imperative for significantly enhancing efficiency, ensuring real-time data collection and analysis, and empowering stakeholders with consistent, timely performance insights. Manual report generation is slow, prone to error, and cannot keep pace with the demands of modern operations. Central to this is leveraging AI-powered report generation. By utilizing artificial intelligence to analyze data, identify key trends, and summarize findings, you can reduce manual workload, accelerate report delivery, and ensure greater accuracy, freeing up valuable human resources for more strategic analysis.

Furthermore, establishing scheduled reporting workflows is crucial. This ensures that consistent, up-to-date performance updates are automatically delivered to relevant stakeholders—from executives to frontline managers—on a predefined cadence, fostering a culture of informed decision-making without manual intervention. Crucially, incorporating interactive reporting tools empowers users to conduct drill-down analysis for deeper operational insights. Beyond static reports, these tools allow users to explore data dynamically, investigate anomalies, and uncover root causes, facilitating more proactive problem-solving and continuous improvement. Neglecting automated reporting can lead to outdated information, delayed insights, inefficient manual processes, and a reactive approach to operational challenges. By championing comprehensive automated reporting, you empower your organization with timely, accurate, and actionable data, accelerating decision-making, improving operational visibility, and ultimately contributing significantly to your business's sustained efficiency, profitability, and competitive advantage.

WHAT?	WHY?	HOW?	BEWARE!
REPORT GENERATION	**EFFICIENCY & ACCURACY**	**ML INTEGRATION**	**CONTEXT & NUANCE**
Reducing manual workload with AI powered automated insights.	Accelerates report delivery and minimizes human error.	Deploy AI/ML models to analyze data and auto-generate summaries and insights.	AI struggling to provide nuanced qualitative insights or context.
SCHEDULED REPORTING WORKFLOWS	**TIMELY INSIGHTS**	**AUTOMATION PLATFORM**	**RECIPIENT FATIGUE**
Delivering consistent performance updates.	Ensures stakeholders receive consistent, up-to-date information.	Implement a reporting platform that supports scheduled report distribution.	Over-saturating stakeholders with too many automated reports.
INTERACTIVE REPORTING	**DEEPER UNDERSTANDING**	**BI PLATFORM**	**DATA LITERACY**
Enabling drill-down analysis for deeper insights.	Empowers users to explore data and uncover root causes.	Utilize BI tools (e.g., Power BI, Looker) with drill-down and filter capabilities.	Users lacking the skills to effectively use interactive features for insights.

For a business leader, proactively implementing anomaly detection systems is a critical strategic imperative for identifying irregularities and operational risks before they escalate into major disruptions. In complex operational environments, manually sifting through vast amounts of data to spot unusual patterns is impossible; automated, intelligent detection is key to preventing costly failures and maintaining continuity. Central to this is leveraging AI-driven pattern recognition. By deploying machine learning algorithms that continuously monitor data streams, these systems can detect subtle deviations from expected performance metrics in real time, signaling potential issues like equipment malfunction, fraudulent activity, or supply chain disruptions.

Furthermore, establishing robust real-time alerts and notifications is crucial. Flagging unusual trends for immediate investigation ensures that operational teams are notified instantly when an anomaly is detected, enabling rapid response and preventing minor issues from escalating into major crises. Crucially, integrating systematic root cause analysis is vital. Pinpointing the underlying factors behind anomalies ensures that you don't just react to symptoms but develop sustainable resolutions to prevent future occurrences, driving continuous improvement and long-term resilience. Neglecting proactive anomaly detection can lead to unexpected operational failures, significant financial losses, compromised data, and a reactive approach to problem-solving. By championing AI-powered anomaly detection, you empower your organization to anticipate and mitigate risks, enhance operational stability, and respond swiftly to threats, ultimately contributing significantly to your business's sustained efficiency, profitability, and competitive advantage.

WHAT?	WHY?	HOW?	BEWARE!
AI-DRIVEN PATTERN RECOGNITION	**EARLY WARNING**	**ML ALGORITHMS**	**FALSE POSITIVES**
Detecting deviations from expected performance.	Identifies potential issues before they escalate into major problems.	Deploy AI/ML models to monitor real-time operational data for anomalies.	Distinguishing genuine anomalies from normal variations or data noise.
REAL-TIME ALERTS AND NOTIFICATIONS	**RAPID RESPONSE**	**ALERTING SYSTEMS**	**ALERT FATIGUE**
Flagging unusual trends for immediate investigation.	Ensures operational teams are notified instantly for swift action.	Configure automated alerts via email, SMS, or dashboard notifications.	Overwhelming users with too many non-critical alerts.
ROOT CAUSE ANALYSIS	**PREVENTION**	**ANALYTICAL FRAMEWORKS**	**COMPLEXITY**
Pinpointing underlying factors behind anomalies.	Develop sustainable resolutions to prevent future occurrences.	Use methodologies like Five Whys or Fishbone diagrams to investigate incidents.	Identifying root causes in highly complex and interconnected systems.

For a business leader, strategically implementing predictive maintenance is a critical imperative for preventing operational disruptions, minimizing downtime, and optimizing asset utilization through proactive upkeep strategies. Moving beyond reactive repairs or time-based maintenance, predictive approaches leverage data to anticipate failures, ensuring more efficient and cost-effective operations. Central to this is utilizing sensor-based equipment monitoring. Leveraging IoT data from machinery and infrastructure allows for real-time collection of performance metrics (e.g., temperature, vibration, pressure), enabling early detection of potential issues before critical failures.

Furthermore, applying AI-powered failure predictions is crucial. By using analytics and machine learning to forecast potential breakdowns based on historical data and real-time sensor readings, you can shift from scheduled maintenance to condition-based maintenance, performing repairs precisely when needed, optimizing asset lifespan and reducing unnecessary costs. Crucially, automating maintenance scheduling is vital. Implementing repair and servicing plans based on predictive insights ensures that maintenance activities are triggered proactively and efficiently, minimizing human error and optimizing resource allocation for maintenance teams. Neglecting predictive maintenance can lead to costly unexpected equipment failures, prolonged downtime, reduced asset lifespan, and inefficient maintenance schedules. By championing a data-driven and proactive approach to asset upkeep, you empower your organization to enhance operational reliability, reduce maintenance costs, and maximize productivity, ultimately contributing significantly to your business's sustained efficiency, profitability, and competitive advantage through smarter asset management.

WHAT?	WHY?	HOW?	BEWARE!
SENSOR-BASED EQUIPMENT MONITORING	**EARLY WARNING**	**IOT DEPLOYMENT**	**DATA VOLUME**
Leveraging IoT data for early issue detection.	Detects potential equipment issues before they cause failures.	Install sensors on critical equipment to collect real-time performance data.	Managing and processing large volumes of sensor data.
AI-POWERED FAILURE PREDICTIONS	**OPTIMIZED MAINTENANCE**	**ML MODEL DEVELOPMENT**	**ACCURACY**
Using analytics to forecast potential breakdowns.	Schedules maintenance precisely when needed, reducing costs.	Build AI/ML models to analyze sensor data and predict failures.	Ensuring predictive models accurately identify failure patterns.
AUTOMATED MAINTENANCE SCHEDULING	**EFFICIENCY & TIMELINESS**	**INTEGRATION**	**COMPLEXITY**
Implementing repair and servicing plans.	Streamlines maintenance workflows and optimizes resource allocation.	Integrate predictive insights into automated maintenance scheduling systems.	Integrating predictive insights with existing maintenance management systems.

For a business leader, continuously engaging in robust risk monitoring is a critical strategic imperative for identifying and mitigating operational vulnerabilities, safeguarding your business against potential disruptions and crises. In an unpredictable global environment, proactive vigilance is key to maintaining stability, protecting assets, and ensuring business continuity. Central to this is conducting continuous risk assessment. This involves regularly evaluating both external factors (e.g., market volatility, geopolitical events, natural disasters) and internal factors (e.g., system vulnerabilities, human error, supply chain dependencies) that could affect operational stability. A dynamic risk profile is essential.

Furthermore, meticulous compliance and regulatory tracking are crucial. Ensuring strict adherence to industry standards, legal requirements, and internal policies minimizes legal liabilities, avoids penalties, and protects your brand's reputation by demonstrating responsible operations. Crucially, developing comprehensive crisis response modeling prepares your organization for high-impact risk scenarios. Designing and rehearsing contingency plans for various potential disruptions—including communication strategies, resource allocation, and recovery protocols—ensures your teams can react effectively and minimize damage when a crisis strikes. Neglecting continuous risk monitoring can lead to being blindsided by unexpected events, suffering prolonged operational downtime, significant financial losses, and irreparable damage to your brand's credibility. By championing a proactive, systematic approach to identifying and mitigating operational risks, you empower your organization to build inherent resilience, respond swiftly to threats, and ultimately secure your business's long-term viability.

WHAT?	WHY?	HOW?	BEWARE!
CONTINUOUS RISK ASSESSMENT	**PROACTIVE MITIGATION**	**RISK REGISTER**	**UNFORESEEN RISKS**
Evaluating external and internal factors.	Uncovers weaknesses to prevent or minimize impact of risks.	Develop a comprehensive risk register, categorizing and prioritizing threats.	Difficulty anticipating every possible type of disruption.
COMPLIANCE AND REGULATORY TRACKING	**LEGAL PROTECTION**	**MONITORING TOOLS**	**EVOLVING REGULATIONS**
Ensuring adherence to industry standards.	Minimizes legal liabilities and enhances brand reputation.	Use tools to track changes in regulations and assess internal adherence.	Keeping pace with frequent changes in legal and industry compliance.
CRISIS RESPONSE MODELING	**RAPID RECOVERY**	**SIMULATION & TRAINING**	**REALISM**
Developing contingency plans for high-impact scenarios.	Ensures swift and effective response to unexpected disruptions.	Conduct regular crisis drills and develop clear communication plans.	Creating realistic crisis simulations that challenge existing protocols.

For a business leader, establishing robust audit tracking is a critical strategic imperative for strengthening transparency, fostering accountability, and ensuring the integrity of your operations. In an environment of increasing scrutiny and regulatory demands, having clear, traceable records of all operational activities is fundamental for compliance, performance evaluation, and continuous improvement. Central to this is maintaining meticulous digital audit logs. Securely capturing detailed and traceable records of all operational activities—including user actions, system changes, data access, and process completions—provides an immutable historical record essential for forensic analysis, troubleshooting, and demonstrating adherence to policies.

Furthermore, implementing automated compliance checks is crucial. Conducting systematic evaluations to meet internal policies and external regulations ensures continuous adherence, reducing the risk of human error and minimizing legal or reputational exposure. This proactive approach ensures operational integrity. Crucially, conducting regular performance assessment reviews allows for deep learning from operational history. Analyzing audit trails and performance data over time helps identify areas for improvement, uncover inefficiencies, and validate the effectiveness of changes, fostering a culture of continuous learning and optimization. Neglecting robust audit tracking can lead to a lack of accountability, difficulties in tracing errors or breaches, non-compliance with regulations, and an inability to pinpoint areas for operational improvement. By championing comprehensive audit tracking, you empower your organization with unparalleled transparency, strengthens accountability, and drives continuous operational refinement, ultimately contributing significantly to your business's sustained efficiency.

WHAT?	WHY?	HOW?	BEWARE!
DIGITAL AUDIT LOGS	**TRANSPARENCY & INVESTIGATION**	**LOGGING SYSTEMS**	**STORAGE & ANALYSIS**
Maintaining secure and traceable records of activities.	Provides clear, immutable records for accountability and forensics.	Implement centralized logging systems (SIEM) for all operational activities.	Managing and analyzing vast volumes of audit log data.
AUTOMATED COMPLIANCE CHECKS	**CONTINUOUS ADHERENCE**	**RULE-BASED MONITORS**	**FALSE POSITIVES**
Conducting systematic evaluations.	Ensures ongoing compliance with internal and external regulations.	Deploy automated tools to check processes against compliance rules.	Automated checks generating false positives, requiring manual review.
PERFORMANCE ASSESSMENT REVIEWS	**LEARNING & OPTIMIZATION**	**RETROSPECTIVES**	**CORRELATION**
Analyzing operational history to identify improvements.	Translates historical data into actionable insights for improvement.	Conduct regular reviews of audit logs and performance data to identify trends.	Difficulty in correlating specific log entries with broader performance outcomes.

Ai6x6™ Transformation Blueprint

OPTIMIZATION | Operations Tuning

"The most dangerous phrase in the language is, 'We've always done it this way.'" - Grace Hopper.

An American computer scientist and U.S. Navy rear admiral. A pioneering figure in computer programming, she developed the first compiler and was instrumental in COBOL's development. Her innovative contributions significantly advanced computer science and software development.

The Optimization level is your Operational Nirvana, where process tuning identifies and removes every millisecond of waste. This root cause refinement achieves the ultimate operational hum by eliminating the final frictions within the machine.

The continuous and systematic process of improving operational processes, workflows, and resource utilization to enhance overall efficiency, significantly reduce costs, eliminate waste, and consistently improve quality and speed. This relentless pursuit of improvement drives substantial gains in productivity, profitability, and overall operational excellence.

Optimization leverages data from monitoring, implements best practices, and embraces technological advancements to refine every aspect of operations, ensuring the organization remains competitive and agile in a dynamic market.

"Operations Tuning" is critical for business strategy maturity because it focuses on transitioning operations to a higher level of efficiency or adapting to new demands. A mature business strategy continuously seeks opportunities for optimization, whether through expanding operations, integrating new technologies, or adjusting to market demands. This adaptability ensures that operational processes remain agile, cost-effective, and capable of supporting evolving strategic objectives. It's about proactively refining operational models to enhance competitiveness and long-term viability, ensuring the business can pivot and grow in dynamic environments. Here is how we approach our learning:

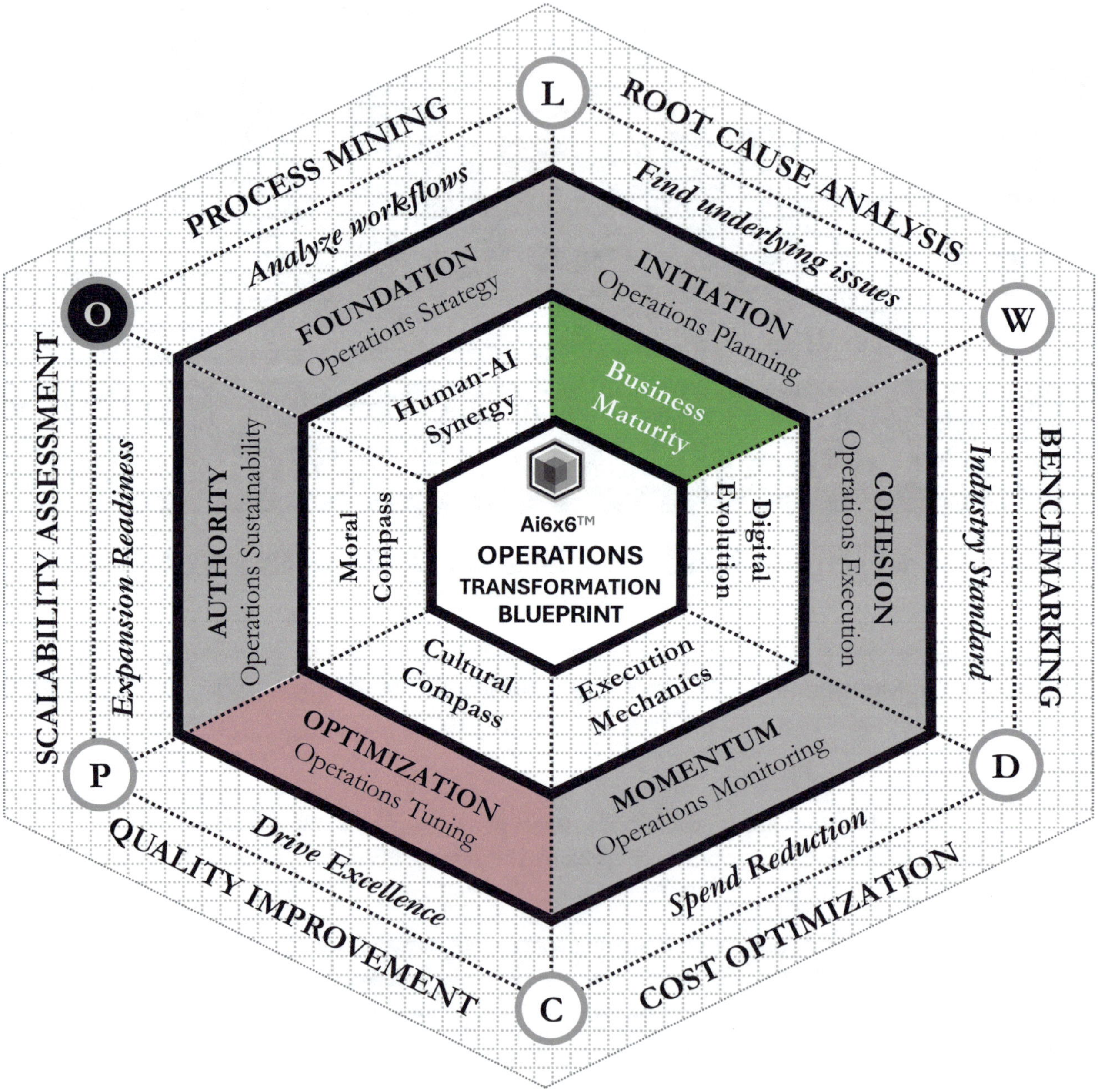

Ai6x6 BUSINESS MATURITY FRAMEWORK
OPERATIONS TRANSFORMATION | OPERATIONS OPTIMIZATION

For a business leader, strategically leveraging process mining is a critical imperative for using data-driven analysis to identify inefficiencies, streamline workflows, and unlock significant operational improvements. In complex organizational processes, hidden bottlenecks and suboptimal paths often go unnoticed; process mining provides objective, empirical insights into how work truly flows. Central to this is implementing automated process mapping. By tracking event logs from IT systems, you can automatically uncover and visualize real workflows, identifying bottlenecks, deviations from ideal paths, and optimization opportunities that might be invisible through traditional manual process mapping. This provides a clear, data-backed blueprint for improvement.

Furthermore, deriving real-time operational insights is crucial. By leveraging continuous analytics on process data, you gain immediate visibility into performance, allowing for prompt identification of issues and agile adjustments to ongoing operations, ensuring continuous optimization rather than reactive fixes. Crucially, integrating workflow automation is a natural next step. Implementing AI-driven solutions directly into identified opportunities for automation enhances efficiency, reduces manual effort, and ensures that streamlined processes are consistently executed, maximizing the impact of your process mining efforts. Neglecting process mining can lead to persistent inefficiencies, wasted resources, slow operational cycles, and a struggle to identify fundamental areas for improvement. By championing this data-driven approach, you empower your organization to gain unprecedented visibility into its operations, systematically eliminate waste, and continuously optimize its workflows, ultimately contributing significantly to your business's sustained efficiency, profitability, and competitive advantage.

WHAT?	WHY?	HOW?	BEWARE!
AUTOMATED PROCESS MAPPING	**VISIBILITY & IDENTIFICATION**	**EVENT LOG ANALYSIS**	**DATA QUALITY**
Tracking workflows to uncover bottlenecks.	Uncovers hidden inefficiencies and optimization opportunities.	Use process mining software to analyze system logs and reconstruct process flows.	Relying on accurate and complete event log data for mapping.
REAL-TIME OPERATIONAL INSIGHTS	**AGILE OPTIMIZATION**	**DASHBOARD INTEGRATION**	**ACTIONABLE INSIGHTS**
Leveraging analytics for continuous performance assessment.	Provides immediate visibility for prompt adjustments.	Display process mining insights on real-time operational dashboards.	Translating complex process insights into clear, actionable steps.
WORKFLOW AUTOMATION INTEGRATION	**STREAMLINED OPERATIONS**	**RPA/BPM INTEGRATION**	**COMPLEXITY**
Implementing AI-driven solutions to enhance efficiency.	Automates identified inefficiencies for improved flow.	Connect process mining tools with RPA or BPM suites for automation.	Integrating automation into highly variable or human-centric processes.

For a business leader, mastering root cause analysis is a critical strategic imperative for identifying the underlying factors that impact operational performance, ensuring that you solve problems fundamentally rather than just treating symptoms. In complex operational environments, surface-level fixes are often temporary; addressing root causes is key to sustainable improvement and preventing recurrence. Central to this is meticulously conducting failure mode analysis. Systematically assessing recurring issues allows you to develop targeted solutions that address the core reasons for operational problems, rather than repeatedly applying quick fixes. This preventive approach minimizes future disruptions.

Furthermore, applying systematic troubleshooting methodologies is crucial. Utilizing structured frameworks like the Five Whys (asking "why" repeatedly to dig deeper) or Fishbone diagrams (categorizing potential causes) ensures a comprehensive and logical approach to problem-solving, leading to more accurate root cause identification. Crucially, developing robust corrective action planning is vital. Once a root cause is identified, creating sustainable resolutions ensures that issues do not recur, embedding process improvements and preventing future operational inefficiencies or failures. This closes the loop on problem-solving. Neglecting thorough root cause analysis can lead to persistent operational issues, wasted resources on ineffective solutions, and a reactive approach to problem-solving that hinders continuous improvement. By championing a disciplined, analytical approach to identifying underlying factors, you empower your organization to solve problems effectively, drive lasting improvements, and ultimately enhance your business's long-term operational stability, efficiency, and profitability.

WHAT?	WHY?	HOW?	BEWARE!
FAILURE MODE ANALYSIS	**PROBLEM RESOLUTION**	**FMEA SESSIONS**	**COMPLEXITY**
Assessing recurring issues to develop targeted solutions.	Addresses core reasons for operational problems, preventing recurrence.	Conduct Failure Mode and Effects Analysis to identify potential failure points.	Analyzing intricate systems with multiple interdependent failure points.
SYSTEMATIC TROUBLESHOOTING	**ACCURATE IDENTIFICATION**	**DIAGRAMS & QUESTIONS**	**BIAS**
Applying structured methodologies.	Ensures a comprehensive and logical approach to problem-solving.	Use Five Whys, Fishbone diagrams, and fault tree analysis.	Jumping to conclusions without thorough investigation.
CORRECTIVE ACTION PLANNING	**PREVENTION**	**CAPA FRAMEWORK**	**IMPLEMENTATION**
Developing sustainable resolutions.	Ensures issues do not recur and drives continuous improvement.	Implement Corrective and Preventive Action (CAPA) plans with clear owners.	Ensuring consistent implementation and follow-through of corrective actions.

For a business leader, rigorously engaging in benchmarking is a critical strategic imperative for continuously improving your operations and enhancing your competitive positioning. It's about systematically comparing your operational performance against industry leaders and best-in-class organizations to identify strengths, weaknesses, and opportunities for significant improvement. Central to this is meticulous performance metric comparison. Analyzing your key performance indicators (KPIs) (e.g., cost per unit, cycle time, quality defect rate, customer service response time) relative to industry standards provides an objective measure of your operational efficiency and effectiveness, highlighting where you excel or fall short.

Furthermore, actively adopting best practices is crucial. Identifying and implementing proven strategies, processes, and technologies from high-performing organizations allows you to learn from their successes, accelerate your own improvements, and avoid reinventing the wheel. This external learning is a powerful catalyst for change. Crucially, engaging in internal benchmarking can drive optimization across different departments or business units within your own organization. Evaluating performance across various internal teams helps identify internal best practices that can be replicated, fostering a culture of continuous learning and improvement from within. Neglecting benchmarking can lead to operational stagnation, a failure to recognize competitive disadvantages, and a reactive approach to market demands. By championing a systematic approach to comparing and learning from the best, you empower your organization to continuously elevate its operational standards, enhance efficiency, and ultimately secure a stronger competitive position and sustained profitability for your business.

WHAT?	WHY?	HOW?	BEWARE!
PERFORMANCE METRIC COMPARE	**GAP IDENTIFICATION**	**KPI ANALYSIS**	**DATA ACCESS**
Analyzing KPIs relative to industry standards.	Pinpoints areas where operational performance lags or excels.	Collect and compare your operational KPIs with published industry benchmarks.	Difficulty in obtaining accurate and comparable industry data.
BEST PRACTICE ADOPTION	**ACCELERATED IMPROVEMENT**	**RESEARCH & PILOTS**	**CULTURAL RESISTANCE**
Implementing proven strategies from high-performing organizations.	Drives efficiency by leveraging external successes.	Research best practices, conduct pilot programs, and adapt for your context.	Employees or teams resisting adoption of new, externally-sourced practices.
INTERNAL BENCHMARKING	**INTERNAL OPTIMIZATION**	**CROSS-DEPARTMENTAL ANALYSIS**	**RESISTANCE**
Evaluating performance across departments.	Identifies internal best practices for replication and overall improvement.	Compare operational KPIs and processes across different internal units.	Departments being reluctant to share data or reveal inefficiencies.

For a business leader, relentless cost optimization is a critical strategic imperative for reducing operational expenses while simultaneously maintaining or even enhancing efficiency. In a competitive economic climate, managing costs effectively is directly linked to profitability and the ability to invest in growth. Your ability to extract maximum value from every dollar spent is paramount. Central to this is meticulous resource allocation refinement. This involves strategically distributing assets—workforce, materials, technology, capital—for maximum productivity and minimum waste. Identifying and reallocating underutilized resources can lead to significant savings without compromising output.

Furthermore, implementing robust expense reduction strategies is crucial. This entails systematically identifying areas where costs can be cut without compromising quality, safety, or core business functions. This might involve renegotiating contracts, optimizing energy consumption, or streamlining administrative overhead. Crucially, adopting Lean process implementation methodologies is vital. Streamlining operations to eliminate unnecessary overhead, reduce waste (e.g., excess inventory, waiting times, overproduction), and improve workflow efficiency directly impacts your cost structure, leading to sustained savings and improved profitability. Neglecting cost optimization can lead to bloated operations, decreased profitability, a reduced capacity for innovation, and a struggle to compete on price. By championing a continuous, data-driven approach to cost management, you empower your organization to operate with greater financial discipline, achieve higher levels of efficiency, and ultimately secure your business's sustained profitability and competitive advantage through smarter resource utilization.

WHAT?	WHY?	HOW?	BEWARE!
RESOURCE ALLOCATION REFINEMENT	**EFFICIENCY GAINS**	**RESOURCE AUDITS**	**RESISTANCE**
Distributing assets for maximum productivity and minimum waste.	Ensures optimal use of all operational resources.	Conduct regular audits of labor, materials, and technology usage.	Departmental resistance to resource reallocation.
EXPENSE REDUCTION STRATEGIES	**PROFITABILITY**	**COST ANALYSIS**	**SHORT-TERMISM**
Identifying areas to cut costs.	Directly improves the bottom line without compromising quality.	Conduct detailed spend analysis and renegotiate vendor contracts.	Focusing on short-term cuts that harm long-term operational health.
LEAN PROCESS IMPLEMENTATION	**WASTE REDUCTION**	**LEAN PRINCIPLES**	**CULTURAL SHIFT**
Streamlining operations to eliminate unnecessary overhead.	Improves efficiency by eliminating non-value-added activities.	Apply Lean methodologies (e.g., value stream mapping, 5S) to workflows.	Requires significant cultural change and continuous training.

5. Quality improvement

For a business leader, relentless quality improvement is a critical strategic imperative for enhancing standards, significantly increasing customer satisfaction, and bolstering operational reliability. In today's competitive landscape, product or service quality is a direct differentiator and a key driver of loyalty and brand reputation. Your commitment to excellence is paramount. Central to this is incorporating continuous improvement methodologies. Embracing principles from frameworks like Six Sigma (reducing defects), Kaizen (continuous small improvements), and Total Quality Management (TQM) fosters a culture where every employee is committed to ongoing enhancement, leading to systemic, sustained improvements.

Furthermore, ensuring rigorous compliance and regulatory enhancements is crucial. Adhering to industry quality standards, certifications (e.g., ISO), and legal requirements not only prevents penalties but also builds trust with customers and stakeholders, demonstrating your commitment to high-quality outputs. Crucially, developing robust defect prevention frameworks leverages predictive analytics to minimize errors and inefficiencies before they occur. By analyzing data to anticipate potential quality issues, you can implement proactive measures that save time, reduce waste, and ensure a higher standard of output. Neglecting quality improvement can lead to customer dissatisfaction, increased returns or complaints, wasted resources on rework, and a damaged brand reputation. By championing a systematic, data-driven, and culture-wide approach to quality, you empower your organization to deliver superior products and services, enhance operational reliability, and ultimately secure your business's long-term success, profitability, and competitive advantage through a reputation for excellence.

WHAT?	WHY?	HOW?	BEWARE!
CONTINUOUS IMPROVEMENT METHODOLOGIES	**SYSTEMIC ENHANCEMENT**	**METHODOLOGY ADOPTION**	**CULTURAL RESISTANCE**
Incorporating Six Sigma, Kaizen, TQM.	Fosters a culture of ongoing quality improvement.	Train teams in Lean, Six Sigma, Kaizen principles and tools.	Employees resisting change or seeing quality as a function.
COMPLIANCE AND REGULATORY ENHANCEMENTS	**TRUST & LEGITIMACY**	**CERTIFICATION & AUDITS**	**EVOLVING STANDARDS**
Ensuring adherence to industry quality standards.	Builds customer trust and ensures market acceptance.	Pursue industry certifications (e.g., ISO) and conduct regular compliance audits.	Keeping up with constantly changing quality standards and regulations.
DEFECT PREVENTION FRAMEWORKS	**PROACTIVE QUALITY**	**PREDICTIVE MODELS**	**DATA ACCURACY**
Using predictive analytics to minimize errors.	Reduces errors and inefficiencies before they impact output.	Implement AI/ML models to identify potential defects in processes or products.	Reliance on high-quality data for accurate defect prediction.

For a business leader, rigorously conducting scalability assessments is a critical strategic imperative for structuring operations to support growth effectively while simultaneously maintaining efficiency. As your business expands, an unprepared operational infrastructure can quickly become a bottleneck, hindering growth and eroding profitability. Your ability to grow gracefully is paramount. Central to this is meticulous capacity planning for expansion. Ensuring that your resources—production capabilities, workforce, IT infrastructure—can scale effectively with anticipated demand growth prevents bottlenecks and ensures you can meet increased customer needs without compromising service quality.

Furthermore, developing adaptive infrastructure models is crucial. Implementing flexible systems, leveraging cloud-native architectures, and adopting modular components allows your operations to accommodate future needs and technological advancements without major overhauls, ensuring agility and cost-efficiency during growth. Crucially, comprehensive performance stress testing evaluates operational readiness under high-load conditions. Simulating peak demand scenarios and extreme usage allows you to identify system weaknesses, bottlenecks, and breaking points before real growth impacts performance, ensuring stability and reliability as your business scales. Neglecting scalability assessment can lead to operational chaos, system failures during growth spurts, spiraling costs, and customer dissatisfaction due to inability to deliver. By championing a proactive and systematic approach to ensuring operational scalability, you empower your organization to grow confidently, maintain efficiency at scale, and ultimately secure your business's long-term market leadership and profitability by handling demand seamlessly.

WHAT?	WHY?	HOW?	BEWARE!
CAPACITY PLANNING FOR EXPANSION	**GROWTH SUPPORT**	**FORECASTING & MODELING**	**PREDICTION ACCURACY**
Ensuring resources scale effectively with demand.	Prevents bottlenecks and ensures demand can be met efficiently.	Use demand forecasts to project future resource needs and plan increments.	Difficulty in accurately forecasting long-term growth rates.
ADAPTIVE INFRASTRUCTURE MODELS	**AGILITY & RESILIENCE**	**CLOUD-NATIVE DESIGN**	**COMPLEXITY**
Implementing flexible systems to accommodate future needs.	Allows operations to evolve without major overhauls.	Build or migrate to cloud-native, microservices-based architectures.	High initial investment and technical complexity of adaptive systems.
PERFORMANCE STRESS TESTING	**SYSTEM STABILITY**	**LOAD SIMULATION**	**REALISM**
Evaluating operational readiness under high-load conditions.	Identifies weaknesses before real growth impacts performance.	Use tools to simulate high user traffic and data processing loads.	Creating test environments that accurately mimic real-world peak conditions.

Ai6x6™ Transformation Blueprint

AUTHORITY | Operations Sustainability

"Sustainability is no longer a niche issue; it's a business imperative." - Paul *Polman.*

Former CEO of Unilever. He is recognized for pioneering the integration of sustainability into Unilever's core business strategy, demonstrating that responsible practices drive both profit and positive impact. He is a prominent global advocate for sustainable development and corporate responsibility.

The Authority level is your Operational Sustainability, reached when systems are so efficient they become self-sustaining and waste-neutral. Efficiency and waste reduction ensure the organization is a model of long-term endurance and market leadership.

Integrating comprehensive environmental, social, and economic considerations into all operational practices and decision-making, aiming for long-term viability and positive impact. This ensures the long-term viability of operations, significantly reduces risks (e.g., regulatory, reputational), enhances brand reputation, and aligns operational activities with broader societal values and ethical responsibilities, fostering responsible growth.

It involves minimizing environmental footprint, ensuring fair labor practices, and contributing positively to communities, building a resilient and ethically sound operational framework for the future.

The "**Operations Sustainability**" chapter is fundamental for business strategy maturity as it focuses on implementing responsible business practices for long-term success. A mature business strategy embeds sustainability into its operational framework, ensuring ethical, environmental, and economic viability. This commitment amplifies operational resilience, enhances brand reputation, and attracts socially conscious stakeholders. It's about ensuring that operations not only meet current strategic goals but also contribute to the organization's enduring value and societal impact, securing its position for future growth and competitive advantage. Here is how we approach our learning:

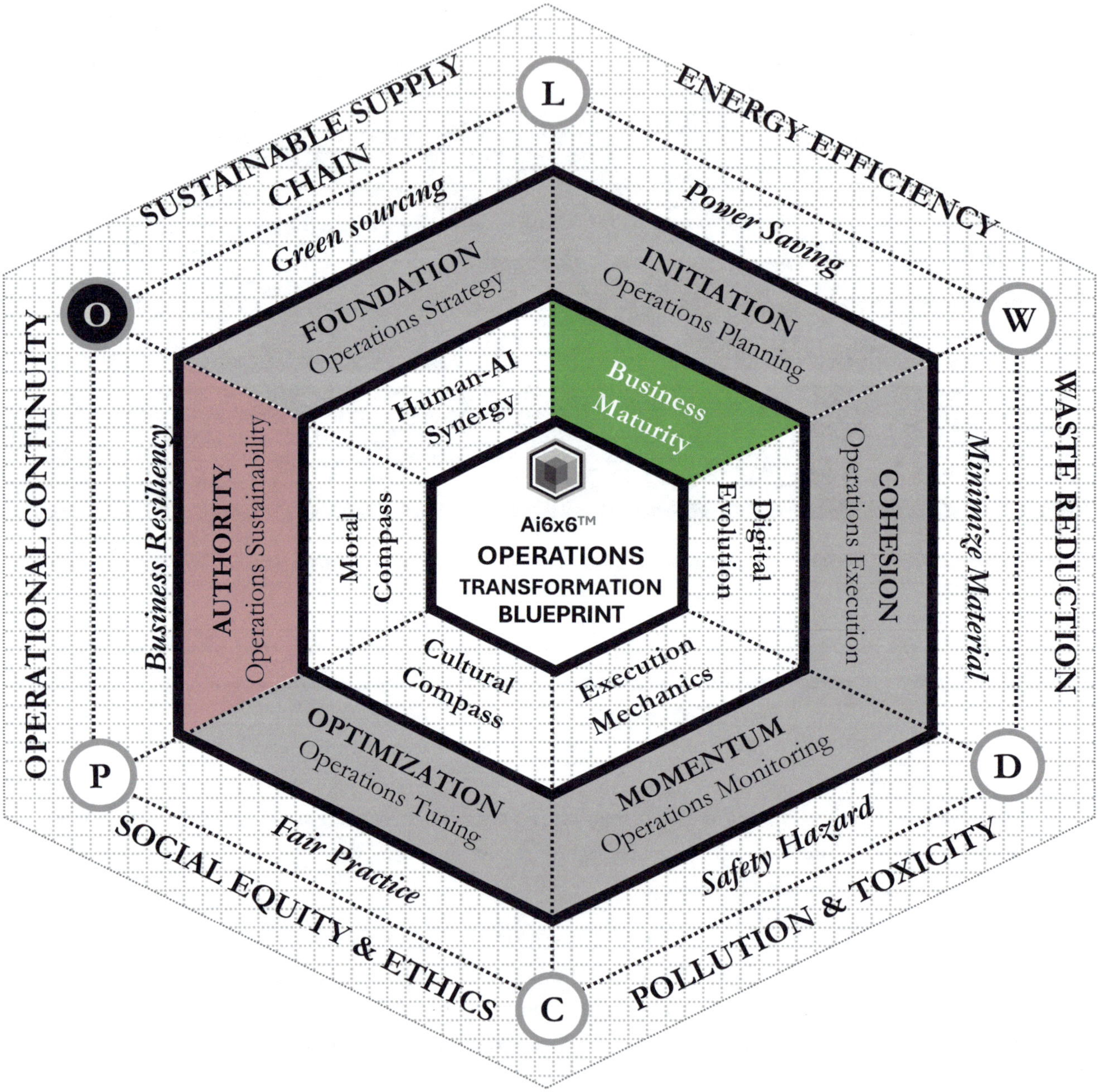

BUSINESS MATURITY FOR SUSTAINABILITY | OPERATIONS TRANSFORMATION

1. Sustainable supply chain

For a business leader, establishing a sustainable supply chain is a critical strategic imperative for implementing environmentally and socially responsible sourcing practices, enhancing brand reputation, and ensuring long-term operational resilience. In an era of increasing consumer and regulatory scrutiny, a sustainable supply chain is no longer a niche concern but a core competitive advantage. Central to this is adopting ethical procurement strategies. Partnering with suppliers who adhere to stringent sustainability and fair labor standards ensures that your entire value chain operates responsibly, mitigating risks of human rights violations or environmental degradation and enhancing brand integrity.

Furthermore, rigorously pursuing carbon footprint reduction is crucial. Optimizing logistics, streamlining transportation routes, and promoting green transportation methods minimize emissions across your supply chain, contributing to climate goals and reducing operational costs. Crucially, integrating circular economy principles encourages product lifecycle extensions through refurbishment, reuse, and recycling. Designing products and processes with their end-of-life in mind ensures materials stay in use longer, reducing waste and reliance on virgin resources. Neglecting a sustainable supply chain can lead to reputational damage from unethical practices, supply chain disruptions due to environmental factors, and a failure to meet evolving consumer and regulatory expectations. By championing a comprehensive, ethical, and environmentally conscious approach to your supply chain, you empower your organization to operate responsibly, build stronger supplier relationships, and ultimately secure your business's long-term viability, brand leadership, and competitive advantage in a world.

WHAT?	WHY?	HOW?	BEWARE!
ETHICAL PROCUREMENT	**REPUTATION & ESG**	**SUPPLIER AUDITS**	**TRANSPARENCY**
Partnering with suppliers who adhere to stringent sustainability and fair labor standards.	Enhances brand reputation and meets Environmental, Social, Governance goals.	Conduct regular ethical and social compliance audits of suppliers.	Ensuring full transparency across complex, multi-tiered supply chains.
CARBON FOOTPRINT REDUCTION	**ENVIRONMENTAL IMPACT**	**GREEN LOGISTICS**	**INFRASTRUCTURE**
Optimizing logistics to minimize emissions.	Reduces greenhouse gas emissions and operational costs.	Optimize transportation routes, use eco-friendly fleets, & reduce air freight.	Limited availability of green transportation infrastructure.
CIRCULAR ECONOMY INTEGRATION	**RESOURCE EFFICIENCY**	**DESIGN FOR CIRCULARITY**	**CONSUMER BEHAVIOR**
Encouraging product lifecycle extensions through refurbishment, reuse, and recycling.	Minimizes waste and reduces reliance on virgin materials.	Design products for easy repair, reuse, and recycling.	Shifting consumer habits towards circular consumption.

For a business leader, prioritizing energy efficiency is a critical strategic imperative for reducing operational energy consumption, enhancing environmental sustainability, and driving significant cost-effectiveness. In an era of rising energy costs and climate concerns, optimizing energy use is not just a green initiative but a direct contributor to your bottom line and brand reputation. Central to this is accelerating renewable energy adoption. Utilizing solar, wind, and other alternative energy sources in your operations (e.g., on-site generation, purchasing renewable energy credits) reduces reliance on fossil fuels, lowers long-term energy costs, and aligns your business with global sustainability goals.

Furthermore, implementing smart energy management systems is crucial. By deploying AI-driven tools and IoT sensors to monitor and optimize power usage across your facilities and equipment, you can identify inefficiencies, automate energy savings, and gain real-time insights into consumption patterns. Crucially, continuously improving facility and equipment efficiency is vital. Upgrading to energy-saving machinery, optimizing HVAC systems, improving insulation, and investing in sustainable infrastructure reduces overall energy demand, leading to long-term cost reductions and a smaller carbon footprint. Neglecting energy efficiency can lead to escalating operational costs, a larger environmental footprint, and a failure to meet investor and consumer expectations for sustainability. By championing a comprehensive approach to energy optimization, you empower your organization to reduce expenses, enhance environmental stewardship, and ultimately secure your business's long-term profitability and competitive advantage through sustainable operations.

WHAT?	WHY?	HOW?	BEWARE!
RENEWABLE ENERGY ADOPTION	**CARBON REDUCTION**	**INVESTMENT**	**INITIAL COST**
Utilizing solar, wind, and other alternative energy sources.	Lowers operational carbon footprint and long-term energy costs.	Invest in on-site renewables or purchase renewable energy credits (RECs).	High upfront investment for renewable energy infrastructure.
SMART ENERGY MANAGEMENT	**OPTIMIZED CONSUMPTION**	**IOT SENSORS**	**INTEGRATION**
Implementing AI-driven tools to monitor and optimize power usage.	Identifies inefficiencies and automates energy savings.	Deploy smart meters and AI-powered building management systems.	Integrating smart energy systems with existing legacy infrastructure.
FACILITY AND EQUIPMENT EFFICIENCY	**COST REDUCTION**	**TECHNOLOGY UPGRADES**	**CAPITAL EXPENDITURE**
Upgrading to energy-saving machinery.	Reduces overall energy demand and associated operational expenses.	Invest in energy-efficient lighting, HVAC, and machinery.	Significant capital expenditure required for major upgrades.

For a business leader, prioritizing waste reduction is a critical strategic imperative for minimizing environmental impact, optimizing resource utilization, and driving significant cost savings across operations. In an era of finite resources and growing environmental awareness, a commitment to reducing waste is both an ecological necessity and a smart business practice. Central to this is implementing zero-waste initiatives. By developing comprehensive strategies to divert waste from landfills through improved material recovery, composting, and rigorous recycling programs, you reduce your environmental footprint and align with circular economy principles.

Furthermore, transitioning to sustainable packaging solutions is crucial. Replacing traditional packaging materials with biodegradable, recyclable, or reusable alternatives minimizes waste generated by your products throughout their lifecycle, enhancing your brand's eco-friendly credentials and appealing to conscious consumers. Crucially, optimizing production waste through Lean process implementation is vital. Applying Lean methodologies to identify and eliminate waste streams (e.g., overproduction, defects, unnecessary transportation) within your manufacturing and operational processes maximizes resource efficiency, reduces costs, and improves overall productivity. Neglecting waste reduction can lead to increased disposal costs, a larger environmental footprint, negative brand perception, and a failure to meet evolving regulatory requirements. By championing a comprehensive approach to waste management, you empower your organization to operate more efficiently, reduce its environmental impact, and build a reputation for responsible stewardship, ultimately securing your business's long-term viability and competitive advantage in a sustainability-driven market.

WHAT?	WHY?	HOW?	BEWARE!
ZERO-WASTE INITIATIVES	**ENVIRONMENTAL IMPACT**	**RECYCLING & COMPOSTING**	**BEHAVIORAL CHANGE**
Reducing landfill contributions.	Minimizes waste sent to landfills and promotes resource recovery.	Implement comprehensive recycling, composting, and material recovery programs.	Encouraging employee and customer participation in waste sorting.
SUSTAINABLE PACKAGING SOLUTIONS	**ECO-FRIENDLY BRANDING**	**MATERIAL SUBSTITUTION**	**COST & AVAILABILITY**
Transitioning to biodegradable and reusable materials.	Reduces environmental impact of products and enhances brand image.	Research and switch to recyclable, compostable, or reusable packaging.	Higher costs or limited availability of sustainable packaging materials.
PRODUCTION WASTE OPTIMIZATION	**RESOURCE EFFICIENCY**	**LEAN PRINCIPLES**	**CULTURAL RESISTANCE**
Implementing lean practices to maximize efficiency.	Eliminates waste within manufacturing processes and reduces costs.	Apply Lean tools (e.g., 5S, value stream map) to production workflows.	Employees resisting changes to established production methods.

For a business leader, rigorously managing pollution and toxicity is a critical strategic imperative for ensuring environmental protection, mitigating hazards, and safeguarding your organization's reputation and compliance. In an era of heightened environmental awareness and stricter regulations, minimizing harmful emissions and hazardous substances is paramount to responsible operations. Central to this is meticulous air and water pollution management. Establishing regulatory-compliant controls for emissions, discharges, and waste disposal ensures that your operations do not contaminate natural resources, preventing legal penalties and protecting public health.

Furthermore, prioritizing non-toxic material selection is crucial. Systematically substituting harmful substances with safer, more sustainable alternatives in manufacturing processes and final products reduces environmental risk, protects employee health, and aligns your brand with eco-conscious values. Crucially, unwavering compliance with environmental regulations is vital. Aligning all operational processes with national and international sustainability laws and industry standards ensures legal adherence, avoids costly fines, and builds trust with regulators and the public. Neglecting pollution and toxicity management can lead to severe legal penalties, significant reputational damage from environmental incidents, health risks for employees and communities, and increased operational costs from remediation efforts. By championing a proactive and comprehensive approach to environmental protection, you empower your organization to operate responsibly, reduce its ecological footprint, and ultimately secure your business's long-term viability, brand integrity, and competitive advantage through genuine commitment to environmental stewardship.

WHAT?	WHY?	HOW?	BEWARE!
AIR AND WATER POLLUTION MGMT.	**ENVIRONMENTAL PROTECTION**	**EMISSION CONTROLS**	**INVESTMENT**
Establishing regulatory-compliant controls.	Minimizes harm to ecosystems and public health.	Implement air scrubbers, water treatment systems, and strict waste disposal protocols.	High capital investment for pollution control technologies.
NON-TOXIC MATERIAL SELECTION	**RISK REDUCTION**	**SUSTAINABLE SOURCING**	**AVAILABILITY**
Substituting harmful substances with safer alternatives.	Reduces environmental and health risks from hazardous materials.	Research and source certified non-toxic raw materials.	Limited availability or higher cost of non-toxic alternatives.
COMPLIANCE WITH ENVIRONMENTAL REGULATIONS	**LEGAL ADHERENCE**	**REGULATORY MONITORING**	**COMPLEXITY**
Aligning operations with local, state and federal aws.	Avoids fines and legal actions, enhances corporate reputation.	Regularly monitor and adapt to evolving environmental laws and standards.	Navigating complex and often conflicting international environmental regulations.

5. Social equity & ethics

For a business leader, rigorously upholding social equity and ethics is a critical strategic imperative for ensuring responsible operations, fostering a positive workplace culture, and building a trusted brand. Beyond legal compliance, a genuine commitment to fair labor practices and equitable business standards attracts talent, enhances reputation, and resonates with socially conscious consumers. Central to this is establishing fair wage and workplace policies. Ensuring employee well-being, promoting safe working conditions, and guaranteeing ethical treatment throughout your supply chains (both direct and extended) is fundamental to responsible business conduct and human rights.

Furthermore, actively engaging in community engagement programs is crucial. Supporting local initiatives that contribute to sustainability, social progress, and economic development in the communities where you operate enhances your social license to operate and builds goodwill. Crucially, implementing comprehensive Diversity and Inclusion (DEI) strategies fosters equitable opportunities within operations and leadership. Creating a workplace that values and celebrates differences in background, perspective, and identity not only enhances innovation but also ensures fair treatment and a sense of belonging for all employees. Neglecting social equity and ethics can lead to severe reputational damage, consumer boycotts, talent flight, legal challenges, and a fundamental erosion of trust. By championing these ethical and equitable business standards, you empower your organization to operate with integrity, build a strong and diverse workforce, and contribute positively to society, ultimately securing your business's long-term viability, brand strength, and a sustainable competitive advantage.

WHAT?	WHY?	HOW?	BEWARE!
FAIR WAGE AND WORKPLACE POLICIES	**ETHICAL TREATMENT**	**LIVING WAGE**	**GLOBAL STANDARDS**
Ensuring employee well-being.	Fosters employee satisfaction, loyalty, and brand reputation.	Implement fair compensation, safe working conditions, and non-discriminatory policies.	Applying consistent fair labor standards across diverse international supply chains.
COMMUNITY ENGAGEMENT PROGRAMS	**SOCIAL LICENCE**	**CSR INITIATIVES**	**IMPACT MEASUREMENT**
Supporting local initiatives.	Builds goodwill and enhances company's social license to operate.	Partner with local non-profits, support community development, and volunteer programs.	Quantifying the real social impact of community programs.
DIVERSITY AND INCLUSION STRATEGIES	**INNOVATION & BELONGING**	**DEI TRAINING**	**CULTURAL RESISTANCE**
Fostering equitable opportunities.	Drives innovation and creates an inclusive workplace.	Implement DEI training, diverse hiring practices, and equitable promotion.	Overcoming deep-seated biases or resistance to DEI.

For a business leader, prioritizing operational continuity is a non-negotiable strategic imperative for safeguarding business stability against a wide range of disruptions and crises. In an increasingly unpredictable global environment, the ability to maintain essential functions and recover swiftly from unforeseen events is paramount to protecting assets, reputation, and profitability. Central to this is meticulous disaster recovery planning. Establishing clear protocols and rehearsing responses for climate-related emergencies (e.g., floods, extreme weather) and other critical events (e.g., fires, power outages) ensures that your organization can protect its people, data, and infrastructure, and resume operations rapidly.

Furthermore, cultivating robust cybersecurity resilience is crucial. Implementing breach prevention measures, sophisticated ransomware protection, and comprehensive data recovery systems safeguards your digital assets from cyberattacks, minimizing downtime and data loss. This proactive defense is vital for modern operations. Crucially, developing comprehensive force majeure response strategies prepares contingency measures for unpredictable global events beyond your control (e.g., pandemics, geopolitical conflicts, major economic downturns). Having flexible plans allows your business to adapt and minimize disruption when external forces hit. Neglecting operational continuity can lead to catastrophic downtime, severe financial losses, loss of customer trust, and a fundamental threat to your business's survival. By championing a proactive, multi-faceted approach to operational continuity, you empower your organization to withstand shocks, recover swiftly, and ultimately secure your business's long-term viability, resilience, and competitive advantage in a volatile world.

WHAT?	WHY?	HOW?	BEWARE!
DISASTER RECOVERY	**RAPID RESUMPTION**	**ACTIVE DR SITE**	**COST & COMPLEXITY**
Establishing protocols for climate-related and emergency risks.	Ensures quick recovery of critical operations after disasters.	Test disaster recovery plans, including off-site backups and recovery sites.	High costs and complexity of maintaining robust DR infrastructure.
CYBERSECURITY RESILIENCE	**DATA PROTECTION**	**LAYERED SECURITY**	**EVOLVING THREATS**
Implementing breach prevention and data recovery systems.	Safeguards digital assets from cyberattacks and data loss.	Implement firewalls, anti-malware, data encryption, and regular security audits.	Keeping pace with rapidly evolving and sophisticated cyber threats.
FORCE MAJEURE RESPONSE	**ADAPTABILITY**	**SCENARIO PLANNING**	**UN-PREDICTABILITY**
Preparing contingency measures for unpredictable global events.	Enables the business to adapt and minimize disruption from unforeseen external events.	Develop contingency plans for major global disruptions (pandemics, geopolitical crises).	The inherent unpredictability of true force majeure events.

BLUEPRINT SERIES

SECTION D

Digital Evolution

Ai6x6™ Transformation Blueprint

Story 1 — The Birth of COBOL and Business Computing (1959–1965): Early enterprises struggled to translate business logic into machine-readable form, limiting the usefulness of emerging computers. COBOL introduced a standardized language that mirrored human reasoning, enabling organizations to encode processes consistently. This shift allowed systems to scale beyond isolated experiments. Human analysts defined rules while machines executed them with precision. The partnership created the first bridge between business intent and digital execution.

Story 2 — ARPANET and Packet Switching (1969–1972): Communication networks were fragile, centralized, and prone to failure. Packet switching introduced a decentralized architecture that allowed information to flow even when parts of the network failed. Humans designed protocols while machines routed data autonomously. This resilience transformed digital communication from a technical novelty into a strategic infrastructure. ARPANET demonstrated how distributed systems could scale without collapsing. It laid the foundation for the modern internet.

Story 3 — Xerox PARC and the Graphical User Interface (1973–1983): Computers were powerful but inaccessible to most users, requiring specialized knowledge to operate. Xerox PARC introduced the graphical user interface, enabling humans to interact with machines visually rather than through code. This shift democratized computing and redefined usability. Machines handled complexity behind the scenes while humans engaged through intuitive metaphors. The interface became a catalyst for mass digital adoption. It showed that digital evolution accelerates when technology adapts to human cognition.

Story 4 — The Rise of Enterprise Resource Planning (1980s–1990s): Organizations operated in silos, with fragmented systems and inconsistent data. ERP platforms unified finance, operations, and supply chains into integrated digital environments. Machines enforced standardization while humans redesigned processes around shared information. This integration improved visibility, coordination, and control. ERP became the backbone of global enterprises. It demonstrated how digital evolution transforms not just tools but entire operating models.

Story 5 — The Commercial Internet and E-Commerce (1994–2005): The internet shifted from academic infrastructure to commercial ecosystem, enabling new forms of interaction, distribution, and value creation. Websites became digital storefronts, and payment systems matured into secure transactional platforms. Machines automated logistics while humans reimagined customer experience. The digital marketplace expanded global reach and lowered barriers to entry. It marked a turning point where digital capability became essential for competitiveness.

Story 6 — Cloud Computing and Elastic Infrastructure (2006–2015): Organizations struggled with rigid, capital-intensive infrastructure that limited experimentation and scale. Cloud computing introduced elastic resources that could expand or contract instantly. Machines automated provisioning while humans focused on architecture and innovation. This shift accelerated deployment cycles, reduced cost, and enabled global reach. Cloud became the foundation for modern digital ecosystems. It demonstrated how digital evolution thrives when infrastructure becomes fluid, on-demand, and programmable.

LESSONS FROM AN OPERATIONS TRANSFORMATION PERSPECTIVE

Operations transform as digital evolution replaces manual coordination with integrated, automated, and elastic systems. Early digital tools standardized processes, while later platforms unified data, synchronized workflows, and enabled real-time decision-making. Cloud infrastructure introduced flexibility that allowed operations to scale without friction. Across decades, the pattern is consistent: machines handle volume, speed, and repeatability; humans manage exceptions, interpretation, and strategic alignment. Operational maturity emerges when digital systems enhance resilience, reduce variability, and support continuous improvement. Digital evolution ultimately creates operations that are adaptive, transparent, and capable of responding to complexity with precision and stability.

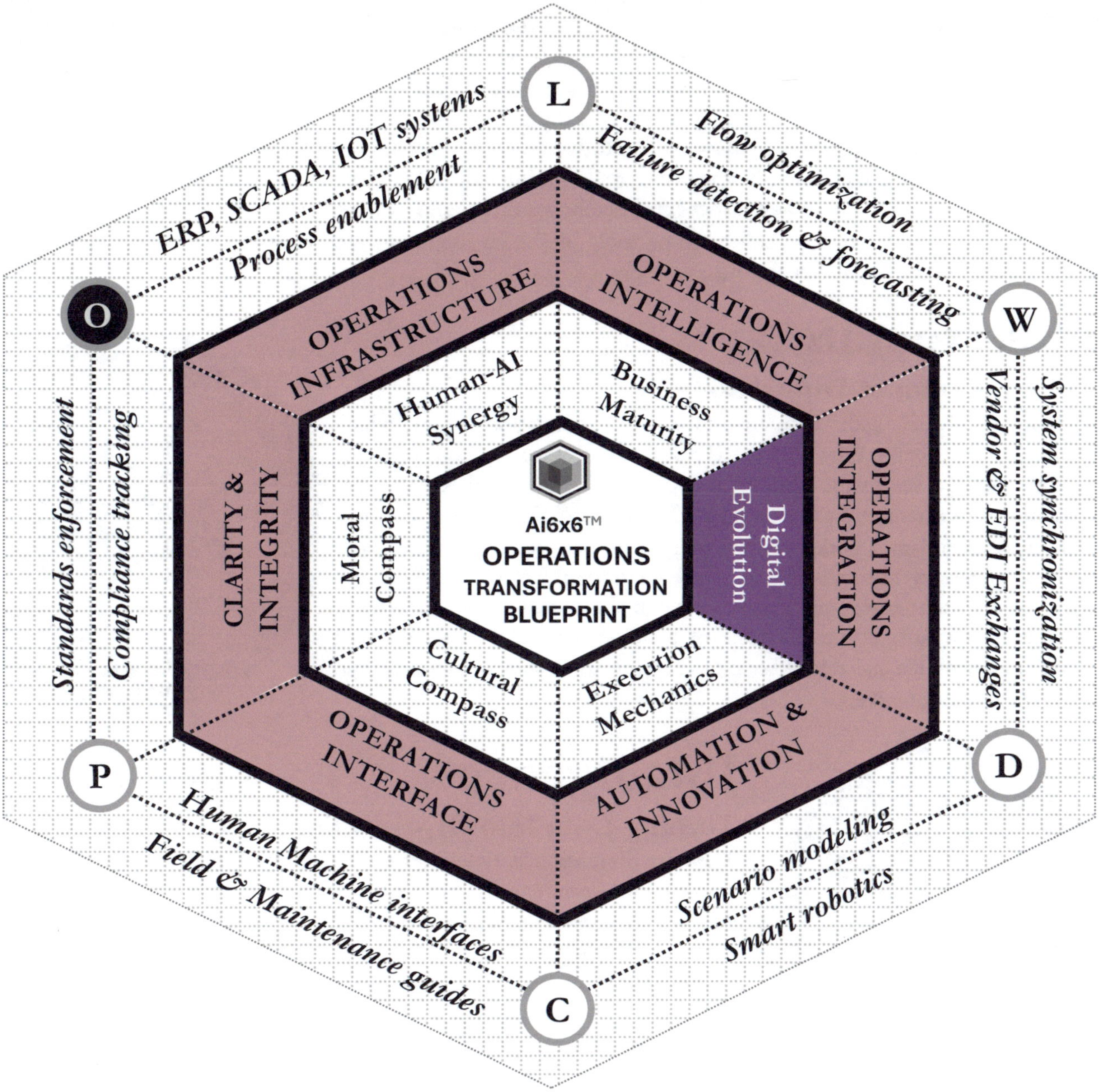

The **Ai6x6 Research Lab** chose this lens to solve the "visibility gap" that often leaves leaders blind to the "Machines Behind the Mission." We believe an operation must possess **Digital Gravitas**—the quiet, unshakeable confidence that comes from "Thinking Through Movement" and being **Connected at Every Step**. This isn't about the number of sensors in a warehouse; it is about the mental shift toward creating a "UI for Everything" that makes complex data actionable. It is vital because a leader cannot "Automate What Matters" if they are constantly overwhelmed by the noise of disconnected systems.

By mastering this evolution, the leader moves from a state of operational anxiety to a state of absolute strategic command over the physical world. This lens ensures that the **Infrastructure** is built on a foundation of **Integrity**, where every process is "Built to Comply" by design. When a leader understands the "Interface" of their movement, they gain the freedom to lead with an undistracted focus on **Innovation**. This lens provides the psychological peace of mind that comes from knowing the backbone of the mission is secure, reliable, and designed for human-AI synergy.

Digital Evolution is no longer a technical initiative. It is the structural rewiring of how an organization thinks, operates, and competes. For years, leaders treated digital transformation as a project—something with a start date, a roadmap, and a finish line. But the world has moved beyond that framing. Digital Evolution is not a destination; it is a permanent state of adaptation. It is the recognition that technology is no longer an enabler of strategy but a **shaper** of it. The organizations that thrive are those that treat digital capability as a living system—continuously upgraded, continuously integrated, and continuously aligned with the realities of the market. This lens reframes Digital Evolution as a leadership discipline rather than a technology program. It challenges leaders to see digital capability not as a collection of tools but as an interconnected ecosystem that influences every decision, every workflow, and every customer interaction.

Infrastructure

The Foundation of Digital Stability

The journey begins with **Infrastructure**, the foundation upon which everything else depends. Infrastructure is not just hardware or cloud capacity; it is the stability, scalability, and reliability of the systems that carry the organization's intelligence. Leaders who underestimate infrastructure discover that innovation collapses under its own weight. Without a strong foundation, every new capability becomes fragile, every integration becomes brittle, and every attempt at modernization becomes a patch rather than a transformation. Infrastructure maturity is the discipline of building for endurance—ensuring that the organization's digital backbone can support the speed, complexity, and volatility of modern markets.

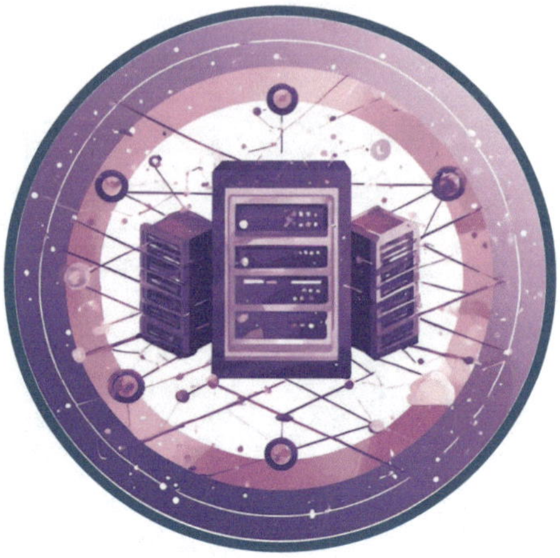

Intelligence

Turning Data Into Organizational Foresight

The next dimension is **Intelligence**, representing the organization's ability to convert data into insight and insight into action. Intelligence is not about dashboards or reports; it is about building a system that learns. It requires data that flows freely, models that are trained responsibly, and decisions that are informed by evidence rather than assumption. Leaders who embrace intelligence understand that it is not a technical asset but a strategic one. Intelligence shapes priorities, exposes risks, and reveals opportunities that would otherwise remain invisible. It is the cognitive engine of Digital Evolution—the mechanism through which the organization perceives, interprets, and responds to its environment.

Integration

Making the Enterprise Whole

Integration is where Digital Evolution becomes real. Organizations often accumulate technologies faster than they can connect them. Systems become islands, data becomes siloed, and teams become fragmented. Integration is the discipline that ensures the enterprise operates as a coherent whole. It is the work of connecting platforms, harmonizing data, and designing workflows that move across boundaries. Without integration, digital investments become isolated experiments rather than enterprise capabilities. Integration maturity is the art of coherence—ensuring that the organization's digital landscape functions as a single, unified system rather than a patchwork of disconnected tools.

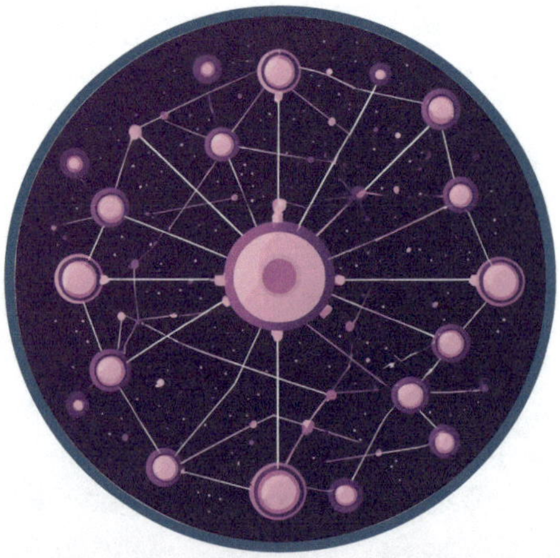

Innovation

Expanding the Field of Possibility

The fourth dimension, **Innovation**, expands the field of possibility. Innovation is not about chasing trends or adopting technology for its own sake. It is about using digital capability to reimagine how value is created, delivered, and experienced. Innovation demands that leaders challenge assumptions, explore alternatives, and test ideas quickly. It requires a culture where experimentation is encouraged, failure is reframed as learning, and progress is measured by insight rather than perfection. Innovation maturity is the discipline of curiosity—using digital tools not just to optimize the present but to invent the future.

Interface

Where Technology Meets Human Behavior

Interface represents the human experience of Digital Evolution. It is the point where technology meets behavior, where systems meet expectations, and where design meets trust. Leaders who ignore interface discover that even the most powerful systems fail if people cannot or will not use them. Interface is not cosmetic; it is strategic. It shapes adoption, accelerates learning, and determines whether digital capability becomes embedded or abandoned. Interface maturity is the discipline of empathy—designing digital experiences that are intuitive, accessible, and aligned with human needs.

Integrity

The Guardrail of Responsible Evolution

The final dimension, **Integrity**, grounds Digital Evolution in responsibility. Integrity ensures that systems are secure, data is protected, and decisions are ethical. It prevents speed from becoming recklessness and innovation from becoming harm. Leaders who prioritize integrity build organizations that are trusted, resilient, and prepared for the scrutiny that accompanies digital capability. Integrity maturity is the discipline of stewardship—ensuring that digital evolution strengthens the organization's moral and operational foundations rather than compromising them.

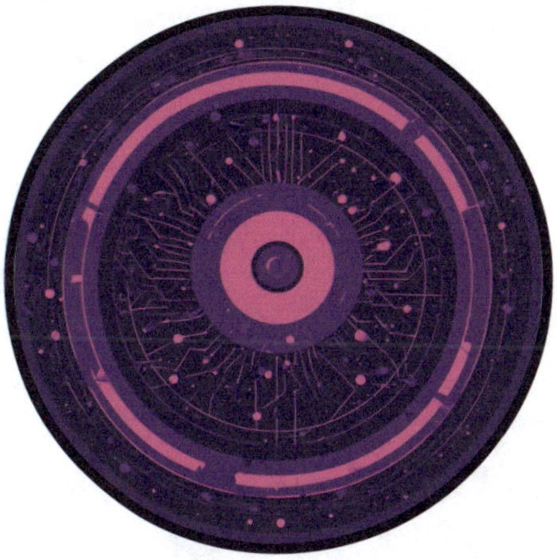

Digital Evolution is not a technology story. It is a leadership story. It is the recognition that the future belongs to organizations that treat digital capability as a core competency, not a side project. It is the shift from building tools to building systems, from adopting technology to evolving with it, and from reacting to change to shaping it. In this future, Digital Evolution is not a department, a project, or a budget line. It is the operating philosophy of the enterprise. It shapes how decisions are made, how systems are built, how teams collaborate, and how value is created. It turns technology from a source of disruption into a source of stability. It transforms the organization from a follower of change into a shaper of it.

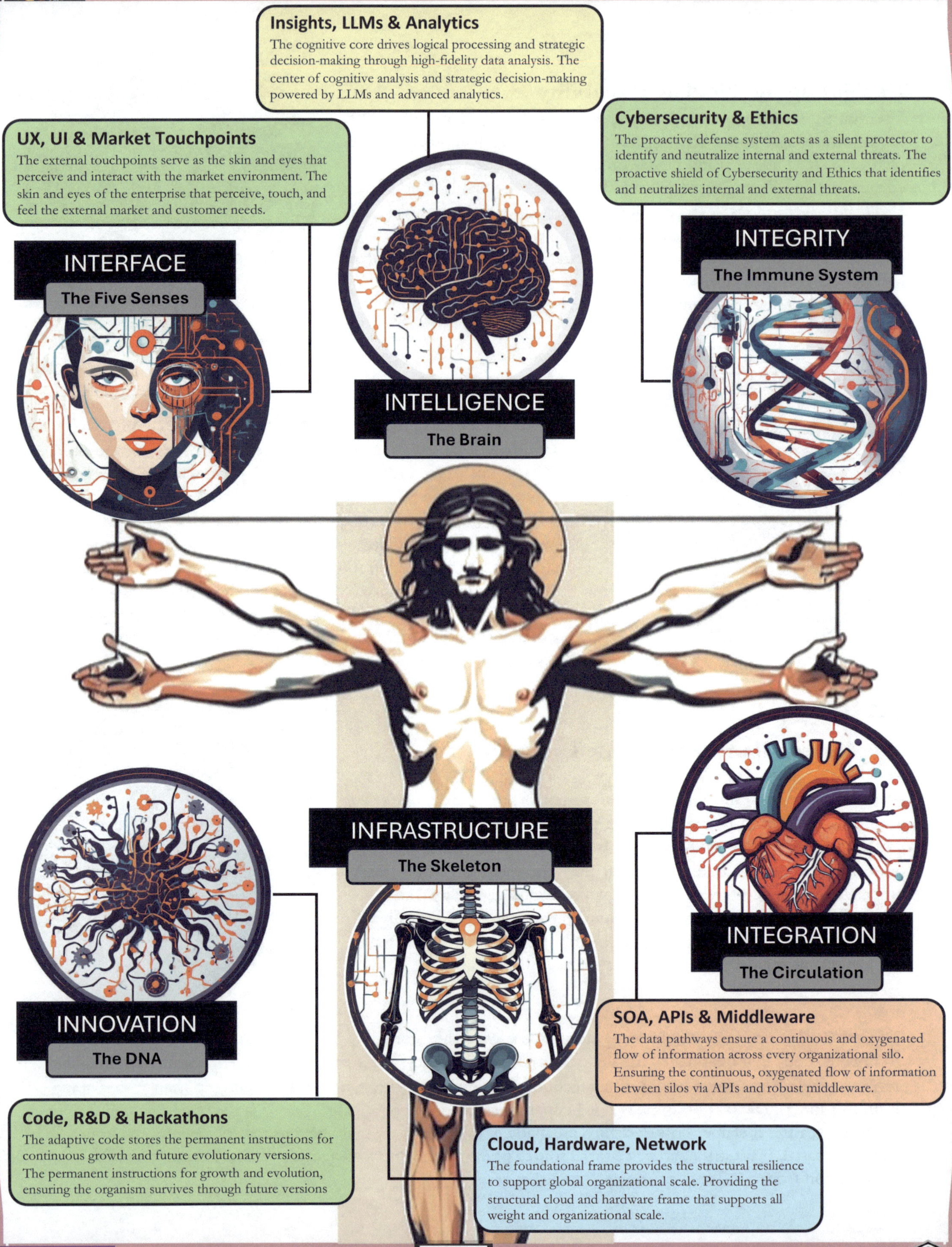
Insights, LLMs & Analytics
The cognitive core drives logical processing and strategic decision-making through high-fidelity data analysis. The center of cognitive analysis and strategic decision-making powered by LLMs and advanced analytics.
UX, UI & Market Touchpoints
The external touchpoints serve as the skin and eyes that perceive and interact with the market environment. The skin and eyes of the enterprise that perceive, touch, and feel the external market and customer needs.
Cybersecurity & Ethics
The proactive defense system acts as a silent protector to identify and neutralize internal and external threats. The proactive shield of Cybersecurity and Ethics that identifies and neutralizes internal and external threats.
INTERFACE
The Five Senses
INTELLIGENCE
The Brain
INTEGRITY
The Immune System
INFRASTRUCTURE
The Skeleton
INTEGRATION
The Circulation
INNOVATION
The DNA
SOA, APIs & Middleware
The data pathways ensure a continuous and oxygenated flow of information across every organizational silo. Ensuring the continuous, oxygenated flow of information between silos via APIs and robust middleware.
Code, R&D & Hackathons
The adaptive code stores the permanent instructions for continuous growth and future evolutionary versions. The permanent instructions for growth and evolution, ensuring the organism survives through future versions
Cloud, Hardware, Network
The foundational frame provides the structural resilience to support global organizational scale. Providing the structural cloud and hardware frame that supports all weight and organizational scale.

Ai6x6™ Transformation Blueprint

Operations 'Infrastructure'

"Digital operations are the nervous system of modern business." — Thomas Siebel , Founder of Siebel Systems and C3.ai; influential in building enterprise digital infrastructure.

Operations don't run on machines—they run on orchestration. ERP, SCADA, and control towers form the digital spine that turns ambition into execution. "Operations Infrastructure" recognizes that infrastructure must be expressive, not just efficient. Mature transformation moves fast, but without friction. Visibility into asset flows, edge gateways, and MES dashboards ensures alignment without micromanagement. Infrastructure isn't technical debt—it's strategic credit. When systems respond, predict, and adapt in real time, operations stop being reactive—they start rehearsing precision.

For operations transformation, the "Operations Infrastructure" chapter is foundational for digital maturity because it establishes the technological backbone that automates, connects, and optimizes core operational processes. A digitally mature organization relies on advanced systems like ERP, Manufacturing Execution Systems (MES), and IoT edge gateways to gain real-time control and visibility over its physical assets and workflows. This robust infrastructure ensures operational efficiency, reduces manual intervention, and provides the essential data foundation for intelligent decision-making, driving productivity, resilience, and scalability across the entire operational landscape.

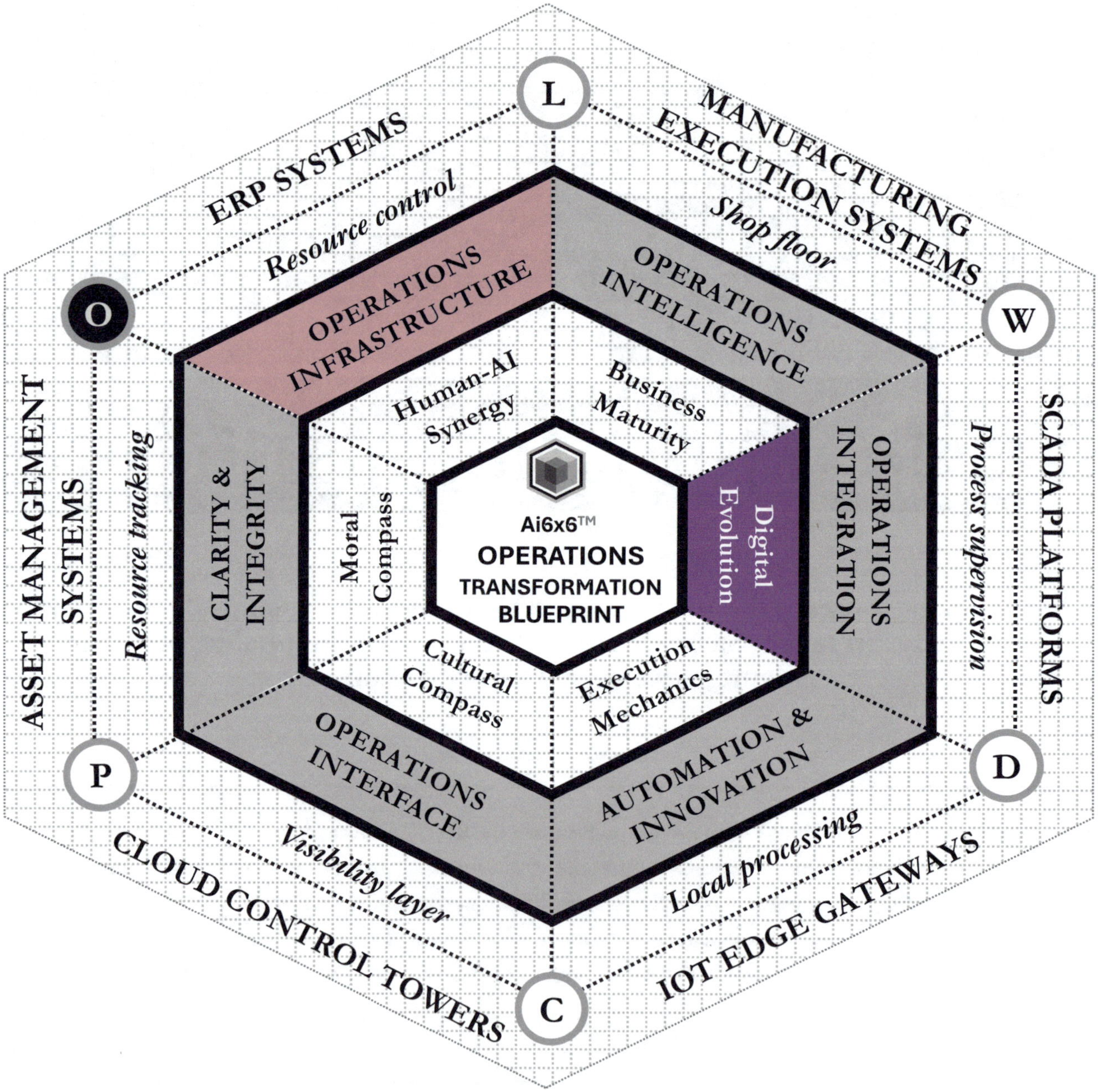

Ai6x6 DIGITAL EVOLUTION FRAMEWORK FOR **OPERATIONS LIFECYCLE**

Enterprise Resource Planning (ERP) Systems are integrated software suites that manage an organization's core business processes across various departments, including finance, human resources, manufacturing, supply chain, services, and procurement. By centralizing data and standardizing workflows, ERP systems provide a single, comprehensive view of the entire operation, enabling real-time insights and improved decision-making. This holistic approach streamlines business processes, enhances operational efficiency, reduces costs, and improves cross-functional collaboration. Implementing an ERP system is fundamental for organizations seeking to optimize resource utilization, improve data accuracy, and achieve greater agility in response to market demands, laying the groundwork for digital transformation across the enterprise.

Without a well-implemented ERP system, organizations typically suffer from fragmented data, siloed departments, and inefficient manual processes across their core operations. This leads to inconsistencies in information, delayed decision-making, redundant data entry, and a lack of real-time visibility into critical business functions like inventory, production, or finance. The absence of an integrated system results in higher operational costs, reduced productivity, poor resource utilization, and an inability to adapt quickly to changing market conditions. This ultimately hinders growth, limits scalability, and makes it challenging to achieve overall operational excellence and strategic alignment across the organization.

Implementation	Challenges
* Define business processes & requirements * Select appropriate ERP software * Migrate existing data & cleanse * Configure modules & customize workflows * Integrate with legacy systems & third-parties * Train employees & manage change	* High implementation costs & complexity * Significant change management & user adoption * Data migration & quality issues * Customization vs. standard process trade-offs * Integration with diverse existing systems * Measuring clear ROI on efficiency gains

Products	Platform Cost	Project Cost	Duration
SAP S/4HANA	$50-300K	$500K-5M+	12-36 m
Oracle ERP Cloud	$30-200K	$300K-3M+	10-30 m
Microsoft Dynamics 365	$20-150K	$200K-2M+	8-24 m
NetSuite (Oracle)	$10-80K	$100K-800K	6-18 m
Infor CloudSuite	$20-100K	$150K-1M	8-20 m
Acumatica	$5-50K	$50K-500K	4-12 m

Manufacturing Execution Systems (MES) are comprehensive software solutions that monitor, control, and manage work-in-progress on the factory floor, from raw materials to finished goods. They bridge the gap between enterprise-level planning (ERP) and shop-floor control systems (SCADA/PLC), providing real-time data on production activities, equipment status, quality control, and labor management. By enabling real-time visibility and control over manufacturing operations, MES improves production efficiency, reduces scrap and rework, enhances product quality, and ensures regulatory compliance. This is critical for optimizing resource utilization, accelerating throughput, and achieving operational excellence in complex manufacturing environments by providing the granular control needed to execute production schedules effectively.

Without a dedicated Manufacturing Execution System (MES), factory operations often suffer from a lack of real-time visibility and control over the production floor. This leads to manual data collection, delayed decision-making, inconsistent processes, and difficulties in tracking work-in-progress, quality, and material flow. The absence of an MES results in inefficiencies, increased waste, higher operational costs, and an inability to respond quickly to production issues or changing demands. This ultimately impacts product quality, delivery schedules, and overall competitiveness by hindering operational efficiency and responsiveness on the factory floor.

Implementation	Challenges
* Define shop floor processes & data points * Select an MES platform * Integrate with ERP, SCADA/PLC systems * Configure production workflows & rules * Train operators & production supervisors * Plan for continuous data collection & analysis	* Complex integration with diverse machinery * Managing real-time data acquisition from devices * User adoption on the factory floor * Ensuring data accuracy & reliability * Customization for unique production processes * Measuring clear ROI on operational improvements

Products	Platform Cost	Project Cost	Duration
Siemens Opcenter MES	$20-100K	$150K-800K	8-24 m
Rockwell Automation FactoryTalk ProductionCentre	$15-80K	$120K-700K	7-20 m
AVEVA MES	$10-70K	$100K-600K	6-18 m
SAP Digital Manufacturing	$10-60K	$80K-500K	6-18 m
GE Digital Plant Applications	$10-50K	$80K-450K	6-16 m
Plex Systems (Cloud MES)	$5-40K	$50K-300K	4-12 m

SCADA (Supervisory Control and Data Acquisition) Platforms are industrial control systems used to monitor and control large-scale, geographically dispersed facilities and processes, such as power generation, oil and gas pipelines, water treatment plants, and smart factories. They collect real-time data from remote devices (RTUs, PLCs, sensors), transmit it to a central control room, and allow operators to remotely control equipment. By providing real-time operational visibility, remote control capabilities, and alarm management, SCADA Platforms enhance operational efficiency, improve safety, and reduce downtime in critical infrastructure and industrial environments. This is essential for maintaining continuous operations, optimizing resource allocation, and ensuring the reliability and security of complex industrial processes.

Without SCADA Platforms, industrial and critical infrastructure operations rely on manual monitoring and control of dispersed equipment, leading to significant delays in detecting and responding to issues, inefficient resource allocation, and increased risk of operational failures. The absence of centralized real-time data acquisition and remote control capabilities makes it difficult to maintain continuous operations, diagnose problems quickly, or optimize performance across a wide geographical area. This ultimately impacts safety, operational efficiency, and system reliability, increasing costs and exposing critical infrastructure to greater risks of downtime and environmental incidents due to uncoordinated and reactive management.

Implementation	Challenges
* Define control points & data tags * Select a SCADA platform * Integrate with PLCs, RTUs, sensors * Configure HMIs & alarm systems * Develop control logic & data historians * Plan for cybersecurity & redundancy	* Complex integration with diverse industrial hardware * Ensuring real-time data accuracy & latency * Cybersecurity risks of connected systems * High cost of implementation & maintenance * Training operators on complex systems * Managing system scalability & upgrades

Products	Platform Cost	Project Cost	Duration
AVEVA PI System (OSIsoft PI)	$20-100K	$150K-800K	8-24 m
Siemens SIMATIC WinCC	$10-60K	$80K-500K	6-18 m
Rockwell Automation FactoryTalk View SE	$10-50K	$70K-400K	6-16 m
GE Digital iFIX	$5-40K	$50K-300K	4-12 m
Ignition (Inductive Automation)	$5-30K	$40K-250K	3-10 m
Wonderware InTouch (AVEVA)	$5-30K	$40K-250K	3-10 m

IoT Edge Gateways are physical devices or software platforms that serve as intermediaries between IoT devices (sensors, machines, actuators) and the cloud or enterprise systems. They perform functions such as data aggregation, protocol translation, data filtering, and local processing at the network's edge, close to the data source. By processing data locally before sending it to the cloud, IoT Edge Gateways reduce network latency, minimize bandwidth consumption, enhance security, and enable real-time decision-making for critical operational processes. This is essential for deploying scalable, reliable, and secure IoT solutions in industrial, smart city, and remote monitoring applications, optimizing data flow and enabling intelligent operations where connectivity may be limited or real-time response is crucial.

Without IoT Edge Gateways, organizations face significant challenges in managing large volumes of raw data from distributed IoT devices, leading to high bandwidth costs, increased network latency, and security vulnerabilities due to direct cloud connections. The absence of local data processing and protocol translation capabilities means that all data must be sent to the cloud, hindering real-time decision-making for critical operational tasks and making IoT deployments less resilient to connectivity interruptions. This ultimately impacts the scalability, cost-effectiveness, and responsiveness of IoT solutions, limiting their ability to drive real-time insights and automation in operational environments.

Implementation	Challenges
* Identify edge data processing needs * Select appropriate hardware/software gateways * Develop/deploy edge applications * Configure device connectivity & protocols * Establish security & data encryption * Plan for remote management & updates	* Managing diverse IoT device protocols * Ensuring robust edge computing capabilities * Securing edge devices & data * Remote management & updates of gateways * Integrating with cloud platforms * Scalability for large-scale deployments

Products	Platform Cost	Project Cost	Duration
AWS IoT Greengrass	$0.1-5K	$2-20K	1-3 m
Azure IoT Edge	$0.1-5K	$2-20K	1-3 m
Google Cloud IoT Edge	$0.1-5K	$2-20K	1-3 m
Cisco IoT Gateway Solutions	$1-10K	$5-30K	1-4 m
Dell Edge Gateway 5000 Series	$1-5K	$2-15K	1-3 m
Eclipse Kura (Open Source)	N/A	$5-20K	1-3 m

Cloud Control Towers are centralized, cloud-based platforms that provide end-to-end visibility, real-time monitoring, and proactive management across complex supply chains and logistics operations. They integrate data from various sources (ERP, WMS, TMS, IoT devices, external partners) into a single, unified dashboard, leveraging advanced analytics and AI to identify anomalies, predict disruptions, and recommend corrective actions. By offering a comprehensive and real-time view of the entire operational landscape, Cloud Control Towers enable organizations to make faster, more informed decisions, optimize inventory, improve delivery performance, and enhance supply chain resilience. This is crucial for navigating dynamic global markets, reducing operational risks, and ensuring efficient and agile execution of complex logistical processes.

Without Cloud Control Towers, organizations struggle with fragmented visibility across their complex supply chains and logistics networks. Data is siloed in disparate systems, leading to delayed information, reactive decision-making, and an inability to proactively identify or mitigate disruptions. The absence of real-time, end-to-end operational transparency hinders the ability to optimize inventory levels, manage transportation efficiently, or respond quickly to unexpected events like supplier delays or weather disruptions. This ultimately results in higher operational costs, increased risk of stockouts, slower delivery times, and reduced customer satisfaction due to a lack of coordinated and intelligent oversight of the entire supply chain.

Implementation	Challenges
* Define supply chain scope & data points	* Integrating diverse & siloed data sources
* Select a cloud control tower platform	* Ensuring real-time data freshness & accuracy
* Integrate with ERP, TMS, WMS, IoT	* Overcoming organizational silos & data ownership
* Configure real-time dashboards & alerts	* Developing accurate predictive models
* Develop predictive analytics & AI models	* User adoption & change management
* Plan for data governance & security	* Measuring clear ROI on improved visibility

Products	Platform Cost	Project Cost	Duration
Kinaxis RapidResponse	$20-100K	$150K-800K	8-24 m
One Network Enterprises	$15-80K	$120K-700K	7-20 m
E2open	$10-70K	$100K-600K	6-18 m
Coupa Supply Chain (LLamasoft)	$10-60K	$80K-500K	6-18 m
FourKites (Real-time Visibility)	$5-40K	$50K-300K	4-12 m
Blue Yonder Luminate Platform	$10-50K	$80K-450K	6-16 m

Asset Management Systems (AMS), often integrated within Enterprise Asset Management (EAM) platforms, are software solutions used to manage and track the lifecycle of physical assets within an organization, from acquisition and operation to maintenance and disposal. This includes machinery, vehicles, infrastructure, and IT equipment. These systems help organizations maintain detailed records of asset location, condition, maintenance history, and depreciation. By providing centralized control and insights into asset performance, AMS optimizes asset utilization, extends asset lifespan, reduces maintenance costs, and improves operational reliability. This is critical for industries with high capital investments like manufacturing, utilities, and transportation, ensuring maximum value extraction from assets and minimizing unexpected downtime.

Without dedicated Asset Management Systems, organizations face significant challenges in tracking, maintaining, and optimizing their physical assets. This leads to inefficient maintenance schedules, frequent unexpected breakdowns, inaccurate asset valuations, and difficulty in planning capital expenditures. The absence of a centralized system for asset data results in increased operational costs due to reactive repairs, reduced asset lifespan, poor resource utilization, and a higher risk of non-compliance with regulatory requirements. This ultimately impacts productivity, profitability, and the overall stability of operations by hindering the effective management of critical physical infrastructure and equipment.

Implementation	Challenges
* Define asset types & hierarchies * Select an Asset Management System * Collect & input asset data * Configure maintenance schedules & workflows * Integrate with ERP, IoT, and GIS * Train maintenance & operations teams	* Accurate data collection for asset inventory * Integrating with diverse legacy systems * User adoption in maintenance departments * Managing complex maintenance schedules * Ensuring data integrity & real-time updates * Measuring clear ROI on asset optimization

Products	Platform Cost	Project Cost	Duration
IBM Maximo	$15-80K	$100K-800K	8-24 m
SAP Enterprise Asset Management (EAM)	$10-70K	$80K-600K	6-18 m
Infor EAM	$10-60K	$70K-500K	6-16 m
Oracle EAM	$10-50K	$60K-450K	5-16 m
FMX (Facilities/Maintenance)	$2-15K	$10K-70K	1-4 m
Fiix (CMMS/EAM)	$1-10K	$5K-50K	1-3 m

Ai6x6™ Transformation Blueprint

Operational 'Intelligence'

"The factory of the future is driven by data, not walls." — Henrik von Scheel , Originator of Industry 4.0; shaped how physical operations merge with intelligent systems.

Movement requires memory. Predictive maintenance, process mining, and throughput analytics represent a shift from managing tasks to choreographing flow. "Operational Intelligence" captures operational maturity as continuous insight—where systems suggest, not report. Transformation replaces observation with orchestration. Every machine, workflow, and energy node feeds strategy. Intelligence moves beyond visibility—it becomes directional. Operations no longer rely on schedules—they respond to situational relevance, making performance dynamically intelligent.

The "Operational Intelligence" chapter is crucial for operations transformation's digital maturity as it focuses on leveraging data to gain deep insights into operational performance and predict future states. A digitally mature organization utilizes predictive maintenance platforms, process mining tools, and demand forecasting systems to move beyond reactive problem-solving to proactive optimization. This intelligence, enhanced by quality control and throughput analytics, enables real-time adjustments, reduces downtime, and enhances overall operational efficiency, ensuring that resources are utilized optimally and the organization can respond swiftly and intelligently to dynamic conditions.

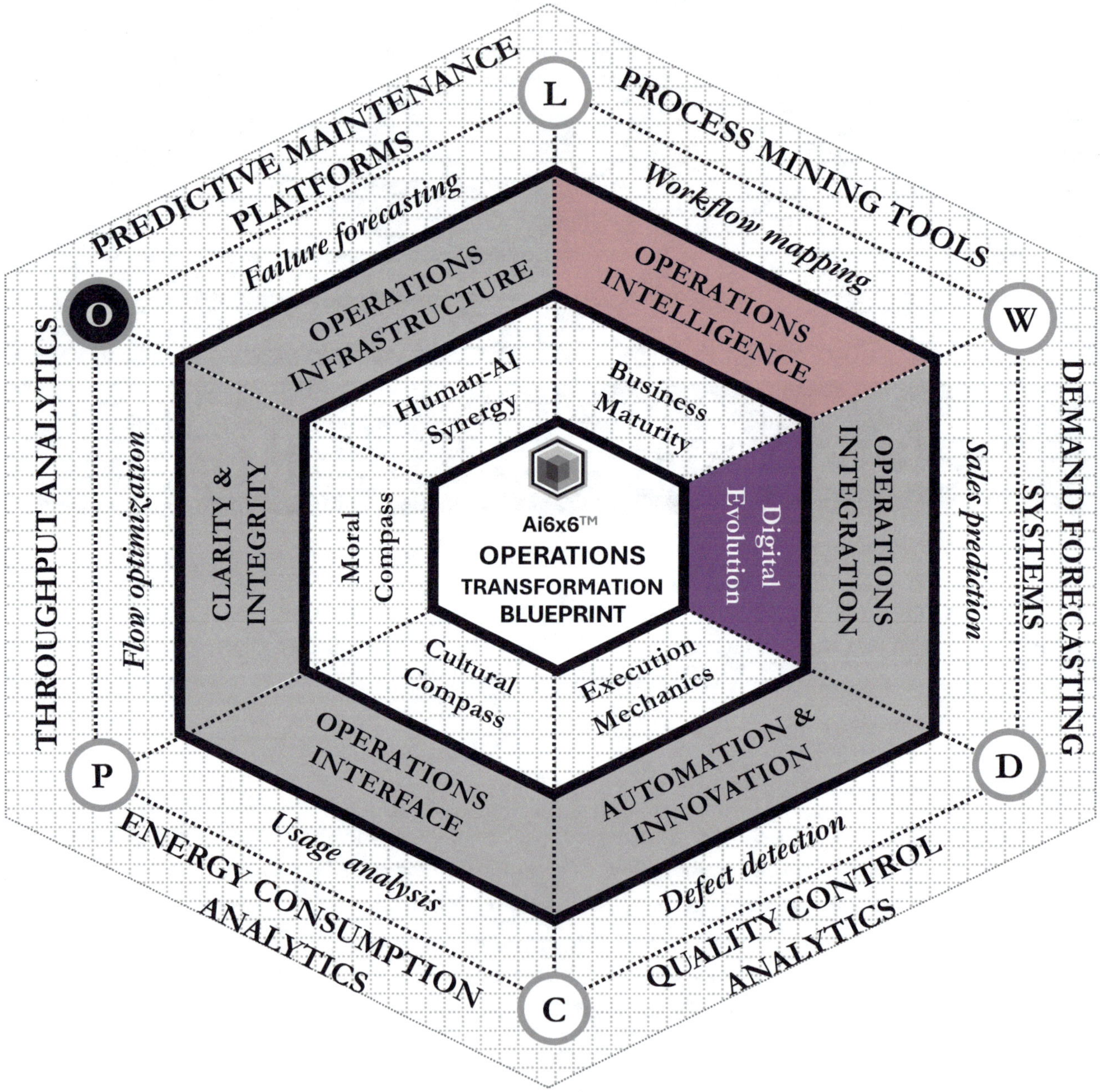

Ai6x6 DIGITAL EVOLUTION FOR **OPERATIONS LIFECYCLE**

Predictive Maintenance Platforms leverage data analytics, machine learning, and IoT sensor data to forecast equipment failures before they occur. By continuously monitoring asset health parameters (e.g., vibration, temperature, pressure, fluid levels), these platforms identify anomalies and predict the optimal time for maintenance interventions, moving beyond reactive or time-based maintenance schedules. This approach significantly reduces unplanned downtime, extends asset lifespan, optimizes maintenance scheduling, and lowers operational costs by preventing catastrophic failures and allowing for just-in-time repairs. Predictive Maintenance Platforms are crucial for industries relying on heavy machinery and continuous operations, ensuring maximum asset uptime and improved operational efficiency.

Without Predictive Maintenance Platforms, organizations primarily rely on reactive maintenance (fixing equipment after it breaks) or time-based preventative maintenance (servicing equipment on a fixed schedule regardless of actual need). This leads to frequent unexpected breakdowns, high repair costs, prolonged downtime, and inefficient resource allocation. The absence of data-driven insights into asset health means that maintenance is performed too early (wasting resources) or too late (leading to costly failures). This ultimately impacts production schedules, operational efficiency, and overall profitability, as critical assets are not utilized optimally and are prone to unpredictable disruptions.

Implementation	Challenges
* Identify critical assets & failure modes * Install IoT sensors for data collection * Select a predictive maintenance platform * Develop/train predictive models * Integrate with CMMS/EAM systems * Plan for continuous data analysis & alerts	* High initial investment in sensors & platforms * Data quality & volume from diverse assets * Developing accurate predictive models (ML expertise) * Integrating with legacy operational systems * Change management for maintenance teams * Proving clear ROI on avoided downtime

Products	Platform Cost	Project Cost	Duration
IBM Maximo Predictive Maintenance	$10-50K	$80K-400K	6-16 m
Siemens MindSphere (Predictive Maint. apps)	$10-60K	$70K-350K	5-14 m
GE Digital APM (Asset Performance Mgmt)	$10-50K	$60K-300K	5-14 m
AVEVA Predictive Analytics	$5-40K	$50K-250K	4-12 m
PTC ThingWorx (with Analytics)	$5-30K	$40K-200K	3-10 m
Augury (Machine Health Monitoring)	$10-60K	$70K-350K	5-14 m

Process Mining Tools are analytical software solutions that use event logs generated by information systems (e.g., ERP, CRM, workflow systems) to discover, monitor, and improve actual business processes. By visualizing process flows, identifying bottlenecks, deviations, and inefficiencies, these tools provide objective, data-driven insights into how processes are truly executed, rather than how they are assumed to be executed. This enables organizations to pinpoint areas for automation, optimize resource allocation, enhance compliance, and ultimately streamline operational workflows. Process Mining is crucial for achieving continuous process improvement, reducing operational costs, and increasing efficiency across various business functions by revealing the hidden realities of process execution.

Without Process Mining Tools, organizations rely on manual process mapping, interviews, or anecdotal evidence to understand and improve their business operations. This often results in incomplete, inaccurate, or outdated process documentation, leading to suboptimal improvements, missed bottlenecks, and continued inefficiencies. The absence of data-driven insights into actual process execution makes it difficult to identify root causes of delays, deviations, or non-compliance. This ultimately hinders effective process optimization, slows down digital transformation initiatives, and prevents organizations from achieving maximum operational efficiency and cost savings through targeted process improvements.

Implementation	Challenges
* Identify target processes & data sources * Extract & prepare event logs * Select a process mining platform * Analyze process maps & identify deviations * Validate findings with process owners * Plan for continuous monitoring & improvement	* Extracting clean & consistent event data * Interpreting complex process maps * Gaining organizational buy-in for changes * Integrating with diverse source systems * Managing data volume for large processes * Measuring clear ROI on process optimizations

Products	Platform Cost	Project Cost	Duration
Celonis EMS (Execution Management System)	$20-100K	$150K-800K	8-24 m
UiPath Process Mining (ProcessGold)	$15-70K	$100K-600K	6-18 m
ABBYY Timeline	$10-60K	$80K-500K	6-16 m
IBM Process Mining	$10-50K	$70K-400K	5-14 m
Disco (Fluxicon)	$2-15K	$10K-70K	1-4 m
QPR ProcessMining	$5-30K	$40K-200K	3-10 m

Demand Forecasting Systems are analytical software solutions that use historical data, statistical models, and increasingly, machine learning algorithms to predict future customer demand for products or services. These systems consider various factors such as seasonality, trends, promotions, economic indicators, and external events. By providing accurate demand predictions, these systems enable organizations to optimize inventory levels, plan production schedules efficiently, improve supply chain responsiveness, and reduce stockouts or excess inventory. This is crucial for minimizing operational costs, maximizing sales opportunities, and ensuring that products are available when and where customers need them, leading to improved customer satisfaction and financial performance.

Without Demand Forecasting Systems, organizations rely on manual methods, intuition, or simplistic historical averages to predict future demand. This inevitably leads to significant inaccuracies, resulting in either costly overstocking (excess inventory, waste) or revenue-losing stockouts (lost sales, customer dissatisfaction). The absence of data-driven, systematic demand prediction hinders effective planning across the supply chain, from procurement and manufacturing to logistics and sales. This ultimately impacts operational efficiency, increases carrying costs, reduces customer service levels, and limits an organization's ability to respond strategically to market fluctuations and capitalize on sales opportunities.

Implementation	Challenges
* Define forecasting scope & data sources * Collect & prepare historical sales data * Select a forecasting system/tool * Develop/train forecasting models * Integrate with ERP, SCM, CRM systems * Plan for continuous model monitoring & recalibration	* Data quality & availability for historical trends * Handling external factors & market volatility * Selecting appropriate forecasting models * Integrating with diverse internal systems * Gaining trust in automated forecasts * Measuring clear ROI on inventory optimization

Products	Platform Cost	Project Cost	Duration
SAP IBP (Integrated Business Planning)	$20-100K	$150K-800K	8-24 m
Oracle SCM Cloud (Planning Central)	$15-70K	$100K-600K	6-18 m
Blue Yonder Demand Planning	$10-60K	$80K-500K	6-16 m
Kinaxis RapidResponse (Demand Planning)	$10-50K	$70K-400K	5-14 m
SAS Demand Planning	$10-50K	$60K-350K	5-14 m
Anaplan (with Demand Planning apps)	$5-30K	$40K-200K	3-10 m

Quality Control Analytics involves the use of data analysis, statistical process control (SPC), and machine learning to monitor, measure, and improve product quality throughout the manufacturing and operational processes. These systems collect data from production lines, inspection points, and sensory inputs to identify defects, predict quality deviations, and pinpoint root causes of quality issues. By providing real-time insights into quality performance, Quality Control Analytics enables proactive adjustments, reduces scrap and rework, minimizes warranty claims, and ensures consistent product quality. This is vital for maintaining brand reputation, reducing operational waste, and achieving regulatory compliance by making data-driven decisions that elevate the overall standard of products delivered to market.

Without Quality Control Analytics, organizations rely on manual inspections, limited sampling, or reactive measures to address product quality issues. This leads to delayed detection of defects, higher rates of scrap and rework, increased warranty costs, and customer dissatisfaction due to inconsistent product quality. The absence of data-driven insights into process variations and defect root causes makes it difficult to implement effective preventative measures or achieve continuous quality improvement. This ultimately impacts production efficiency, profitability, and brand reputation, as quality issues become a reactive problem rather than a proactively managed aspect of operations.

Implementation	Challenges
* Define critical quality parameters & KPIs * Install sensors/data collection points * Select a quality analytics platform * Develop/configure SPC charts & dashboards * Train machine learning models for defect prediction * Integrate with MES/ERP & inspection systems	* Data quality & consistency from production lines * Integrating with diverse manufacturing equipment * Developing accurate defect detection models * Gaining user adoption on the factory floor * Real-time processing for immediate feedback * Measuring clear ROI on defect reduction

Products	Platform Cost	Project Cost	Duration
InfinityQS (SPC Software)	$5-30K	$40K-200K	3-10 m
Minitab (Statistical Software)	$1-10K	$5K-50K	1-3 m
SAS Quality Management	$10-50K	$60K-350K	5-14 m
AVEVA Manufacturing Operations Management (Quality)	$10-60K	$70K-350K	5-14 m
Plex Systems (Quality Management)	$5-40K	$50K-250K	4-12 m
Custom BI/Analytics for Quality (e.g., Power BI, Tableau)	$2-15K	$10K-70K	1-4 m

5. ENERGY CONSUMPTION ANALYTICS

Energy Consumption Analytics involves collecting, analyzing, and visualizing data related to energy usage across an organization's operations, including facilities, machinery, and production lines. These systems typically integrate with smart meters, sensors, and existing building management systems to provide real-time insights into consumption patterns, peak usage, and potential areas of waste. By identifying energy inefficiencies, optimizing operational schedules, and enabling proactive energy management, Energy Consumption Analytics helps organizations reduce utility costs, improve sustainability, and comply with environmental regulations. This is crucial for minimizing operational expenses, achieving corporate social responsibility goals, and enhancing overall resource efficiency in energy-intensive industries.

Without Energy Consumption Analytics, organizations have limited visibility into their energy usage patterns, leading to undetected waste, suboptimal operational schedules, and higher utility costs. Relying on monthly bills or sporadic meter readings prevents real-time identification of energy inefficiencies, equipment malfunctions that cause spikes, or opportunities for optimization. The absence of granular data and analytical insights makes it difficult to pinpoint energy drains, implement effective conservation strategies, or accurately measure the impact of efficiency initiatives. This ultimately impacts profitability, environmental footprint, and the ability to meet sustainability targets, as energy consumption remains an unmanaged operational expense.

Implementation	Challenges
* Install smart meters/sensors * Select an energy analytics platform * Integrate with BMS, SCADA, IoT * Configure real-time dashboards & alerts * Develop anomaly detection models * Plan for energy audits & optimization strategies	* Integrating with diverse legacy energy systems * Accurate data collection from various sources * Interpreting complex energy consumption patterns * Ensuring data security & privacy * Gaining buy-in for energy-saving initiatives * Measuring clear ROI on energy reduction

Products	Platform Cost	Project Cost	Duration
Siemens Energy Management	$5-30K	$40K-200K	3-10 m
Schneider Electric EcoStruxure Power	$5-25K	$30K-180K	3-9 m
IBM Maximo APM for Energy Optimization	$10-40K	$50K-250K	4-12 m
AVEVA Insight (Energy apps)	$2-15K	$10K-70K	1-4 m
Accruent EMS (Energy Management Software)	$5-20K	$20K-100K	2-6 m
Enel X (Demand Response, Energy Mgmt)	$2-15K	$10K-70K	1-4 m

Throughput Analytics involves the systematic measurement, analysis, and optimization of the rate at which a system, process, or production line processes work or produces output. These analytical tools collect data from various operational touchpoints (e.g., manufacturing stations, logistics hubs, service queues) to identify bottlenecks, idle times, and constraints that limit overall efficiency. By providing real-time insights into flow rates, cycle times, and capacity utilization, Throughput Analytics enables organizations to optimize resource allocation, balance workloads, and streamline operational processes. This is crucial for maximizing productivity, reducing lead times, and increasing overall operational capacity, leading to significant improvements in efficiency and profitability across manufacturing, logistics, and service operations.

Without Throughput Analytics, organizations often operate with limited visibility into the actual flow and efficiency of their core processes, relying on anecdotal evidence or aggregated metrics that don't reveal bottlenecks. This leads to suboptimal resource utilization, unexplained delays, and an inability to accurately predict output or identify limiting factors in production or service delivery. The absence of granular data and real-time insights into process flow makes it difficult to pinpoint constraints, optimize workflows, or make informed decisions about capacity planning. This ultimately impacts productivity, increases operational costs, and hinders an organization's ability to meet delivery commitments and scale operations efficiently.

Implementation	Challenges
* Define key process stages & metrics * Install sensors/data collection points * Select a throughput analytics platform * Configure real-time dashboards & alerts * Develop bottleneck identification algorithms * Plan for continuous process optimization	* Data quality & consistency from diverse sources * Integrating with heterogeneous operational systems * Identifying true bottlenecks in complex processes * Gaining buy-in for process changes from teams * Real-time processing & visualization needs * Measuring clear ROI on efficiency gains

Products	Platform Cost	Project Cost	Duration
AVEVA System Platform (Operations Analytics)	$10-60K	$70K-350K	5-14 m
Siemens Opcenter Intelligence	$10-50K	$60K-300K	5-14 m
IBM Maximo Application Suite (Operations)	$10-50K	$60K-300K	5-14 m
TrakSYS (MES & OEE)	$5-40K	$50K-250K	4-12 m
Lighthouse Systems (MES & Performance)	$5-30K	$40K-200K	3-10 m
Custom BI/Analytics for Ops (e.g., Tableau, Power BI)	$2-15K	$10K-70K	1-4 m

Ai6x6™ Transformation Blueprint

Operations 'Integration'

"The future is integrated—not just connected." — Shantanu Narayen, CEO of Adobe; responsible for seamless experiences across vast product ecosystems.

Integration in operations isn't convenience—it's consequence. EDI layers, IoT syncs, and digital twin connectors craft a choreography where every step informs the next. "Operations Integration" means decisions carry downstream context and upstream accountability. Maturity lies in the disappearance of delays and the emergence of shared rhythm. Systems should not just share data—they should validate direction. Integration is no longer a technical project—it's the operational language of awareness.

"Operations Integration" is vital for operations transformation's digital maturity because it ensures seamless data flow and interoperability across the entire operational value chain, from suppliers to customers. A digitally mature organization breaks down silos between systems like EDI, vendor portals, and production-inventory sync tools, creating a unified and transparent operational ecosystem. This deep integration, supported by IoT device connectors and digital twin integration, fosters real-time visibility, improves coordination, and enables end-to-end process optimization, which is essential for agile and responsive operations, minimizing friction and maximizing efficiency in a complex digital environment.

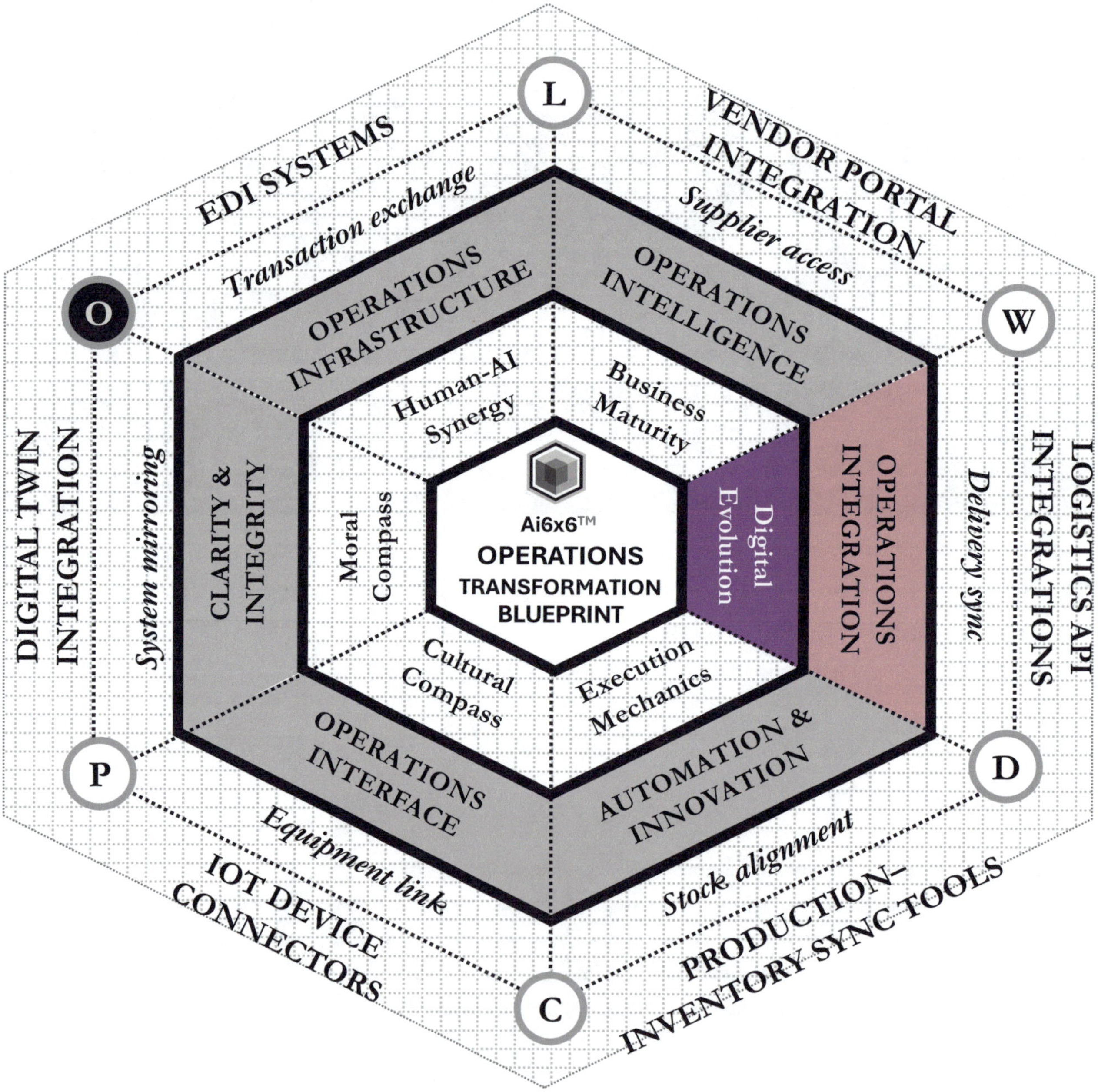

Ai6x6 DIGITAL EVOLUTION FRAMEOWORK FOR **OPERATIONS LIFECYCLE**

Electronic Data Interchange (EDI) Systems are standardized electronic communication platforms that enable the rapid and secure exchange of business documents (e.g., purchase orders, invoices, shipping notices, inventory reports) between different companies' computer systems. By replacing paper-based processes and manual data entry with automated digital exchange, EDI systems significantly improve efficiency, reduce errors, and accelerate transaction cycles across the supply chain. This enables seamless collaboration with trading partners, streamlines procurement and logistics, and enhances overall operational responsiveness. Implementing EDI is fundamental for organizations to automate B2B interactions, reduce operational costs, and improve data accuracy in high-volume transaction environments, fostering robust and efficient supply chain partnerships.

Without EDI Systems, organizations rely heavily on manual processes, faxes, emails, or proprietary data formats for exchanging business documents with trading partners. This leads to significant delays, high rates of human error, increased administrative costs, and a lack of real-time visibility into transaction statuses. The absence of standardized, automated data exchange hinders efficient communication, complicates order processing and fulfillment, and creates bottlenecks in the supply chain. This ultimately impacts operational speed, increases disputes, reduces data accuracy, and limits an organization's ability to scale efficiently and maintain strong, integrated relationships with its business partners in a fast-paced global economy.

Implementation	Challenges
* Identify key trading partners & document types * Select an EDI solution/provider * Map internal data to EDI standards * Configure communication protocols (AS2, SFTP) * Integrate with ERP/WMS systems * Plan for testing with trading partners	* Complex data mapping & transformation * Onboarding diverse trading partners * Managing various EDI standards & versions * Ensuring data security & non-repudiation * Debugging connectivity & data errors * High cost of initial setup & maintenance

Products	Platform Cost	Project Cost	Duration
OpenText B2B Integration	$10-60K	$80K-500K	6-18 m
TrueCommerce EDI Solutions	$5-30K	$40K-200K	3-10 m
SPS Commerce EDI	$2-15K	$10K-70K	1-4 m
IBM Sterling B2B Integrator	$10-50K	$70K-400K	5-14 m
Dell Boomi (EDI integration)	$5-30K	$30K-180K	3-9 m
Cleo Integration Cloud	$5-40K	$50K-250K	4-12 m

Vendor Portal Integration involves establishing a secure, centralized online platform where an organization and its suppliers can collaboratively manage various aspects of their business relationship. These portals typically allow vendors to view purchase orders, submit invoices, update order status, share product information, and manage their profiles. By digitizing and streamlining communication and data exchange with suppliers, Vendor Portal Integration significantly improves supply chain visibility, reduces administrative overhead, accelerates procurement cycles, and enhances supplier collaboration. This is crucial for optimizing supplier performance, reducing errors, and building stronger, more efficient relationships across the entire supply network, transforming reactive vendor management into a proactive, integrated partnership.

Without Vendor Portal Integration, organizations often struggle with fragmented communication channels, manual data entry, and a lack of real-time visibility into their supplier interactions. This leads to delays in order fulfillment, inaccurate data, increased administrative costs for both parties, and strained vendor relationships due to inefficient communication. The absence of a centralized platform for supplier collaboration hinders proactive issue resolution, complicates data sharing, and limits the ability to effectively manage vendor performance. This ultimately impacts supply chain efficiency, increases operational risks, and prevents the optimization of procurement processes due to a lack of seamless, integrated interaction with key suppliers.

Implementation	Challenges
* Define scope of vendor interactions * Select a vendor portal solution * Integrate with ERP, procurement, finance * Configure user roles & access permissions * Onboard vendors & provide training * Plan for ongoing maintenance & support	* Encouraging vendor adoption & participation * Managing diverse vendor capabilities & tech stacks * Data synchronization & consistency across systems * Ensuring data security & privacy * Customization for specific vendor workflows * Measuring clear ROI on improved collaboration

Products	Platform Cost	Project Cost	Duration
SAP Ariba Supplier Management	$10-60K	$80K-500K	6-18 m
Oracle Supplier Relationship Mgmt (SRM)	$10-50K	$70K-400K	5-14 m
Coupa Supplier Portal	$5-40K	$50K-300K	4-12 m
Jaggaer Supplier Management	$5-30K	$40K-200K	3-10 m
Custom Portal (built on existing tech)	N/A	$30K-150K	3-9 m
Workday (Supplier Accounts)	$5-30K	$40K-200K	3-10 m

3. LOGISTICS API INTEGRATIONS

Logistics API (Application Programming Interface) Integrations involve connecting an organization's internal systems (e.g., ERP, WMS, e-commerce platforms) directly with external logistics service providers (LSPs) and carriers using their APIs. These integrations enable real-time exchange of critical logistics data, such as shipping rates, tracking information, delivery status updates, and order fulfillment requests. By automating data flow and communication with logistics partners, API integrations eliminate manual data entry, reduce errors, provide real-time shipment visibility, and enable dynamic routing and optimization. This is crucial for streamlining transportation, improving delivery efficiency, enhancing customer satisfaction through accurate tracking, and achieving greater agility in managing complex supply chain movements.

Without Logistics API Integrations, organizations rely on manual data entry, batch file transfers, or static communication methods to interact with their logistics service providers and carriers. This leads to delays in shipment processing, inaccurate tracking information, increased administrative overhead, and limited visibility into the real-time status of goods in transit. The absence of automated, real-time data exchange hinders efficient routing, complicates order fulfillment, and prevents proactive issue resolution. This ultimately impacts delivery speed, increases transportation costs, reduces customer satisfaction, and limits an organization's ability to optimize its supply chain and respond quickly to dynamic logistics challenges.

Implementation	Challenges
* Identify key logistics partners & data needs * Select relevant APIs (shipping, tracking, customs) * Develop/configure API connectors * Map data fields & ensure compatibility * Implement error handling & retry mechanisms * Plan for monitoring & performance optimization	* Managing diverse API standards & versions * Ensuring data security & authentication * Handling high volume of real-time data * Debugging complex integration failures * Maintaining connections with external systems * Measuring clear ROI on efficiency gains

Products	Platform Cost	Project Cost	Duration
Celigo (iPaaS for Logistics)	$2-15K	$10K-70K	1-4 m
Boomi (Integration Platform)	$5-30K	$30K-180K	3-9 m
MuleSoft Anypoint Platform	$10-50K	$70K-400K	5-14 m
ShipStation / Shippo (Shipping APIs)	$0.1-1K	$0.5K-5K	0.5-1.5 m
Custom API Development	N/A	$20K-100K	2-6 m
Flexport (Logistics Platform with APIs)	$5-30K	$20K-100K	2-6 m

4. PRODUCTION–INVENTORY SYNC TOOLS

Production–Inventory Sync Tools are software solutions that enable real-time or near real-time synchronization of production schedules with inventory levels of raw materials, work-in-progress, and finished goods. These tools integrate data from MES, ERP, and WMS systems to provide a unified view of production progress against material availability and demand. By automatically adjusting production plans based on inventory fluctuations and material shortages, or conversely, updating inventory based on production output, these tools optimize material flow, reduce excess inventory, prevent stockouts, and improve production scheduling accuracy. This is crucial for achieving just-in-time manufacturing, minimizing carrying costs, and ensuring efficient resource utilization across the entire supply chain and manufacturing operations.

Without Production–Inventory Sync Tools, organizations often operate with disjointed production schedules and inventory levels, leading to material shortages that halt production, or excess inventory that incurs high carrying costs. Relying on manual updates or periodic batch processes for inventory-production data exchange results in outdated information, inefficient resource allocation, and a lack of responsiveness to real-time changes in supply or demand. The absence of seamless synchronization hinders the ability to achieve lean manufacturing, optimize working capital, or respond quickly to supply chain disruptions. This ultimately impacts production efficiency, increases operational costs, and limits an organization's ability to meet customer demand reliably and profitably.

Implementation	Challenges
* Map production & inventory data points * Select a synchronization tool/platform * Integrate with ERP, MES, WMS systems * Define synchronization rules & frequency * Implement alerts for discrepancies * Plan for continuous monitoring & adjustments	* Ensuring data accuracy & consistency across systems * Managing complex data transformations * Real-time data processing for large operations * Overcoming legacy system integration hurdles * Gaining user adoption across production & warehouse * Measuring clear ROI on inventory reduction

Products	Platform Cost	Project Cost	Duration
SAP SCM (Supply Chain Management)	$15-80K	$100K-600K	6-18 m
Oracle SCM Cloud (Inventory/Manufacturing)	$10-70K	$80K-500K	6-16 m
Blue Yonder Dispatcher Paragon	$10-60K	$70K-400K	5-14 m
Infor Nexus (Supply Chain Orchestration)	$5-40K	$50K-300K	4-12 m
Custom Integration Middleware (e.g., Boomi, MuleSoft)	$5-30K	$40K-250K	3-10 m
NetSuite (Inventory & Manufacturing)	$5-30K	$40K-200K	3-10 m

5. IOT DEVICE CONNECTORS

IoT Device Connectors are software components or platforms that facilitate the secure and reliable communication between diverse Internet of Things (IoT) devices (sensors, machines, actuators) and cloud-based IoT platforms or enterprise applications. They handle various communication protocols, data formats, authentication mechanisms, and often provide device management capabilities. By abstracting the complexities of device connectivity and ensuring seamless data ingestion from heterogeneous IoT devices, IoT Device Connectors enable organizations to rapidly deploy and scale their IoT initiatives without extensive custom development. This is crucial for harnessing real-time operational data, driving data-driven insights, and enabling automation across industrial, smart city, and asset monitoring applications.

Without IoT Device Connectors, organizations attempting to deploy IoT solutions face significant challenges in integrating diverse devices that speak different protocols and generate varied data formats. This leads to extensive custom development for each device type, increased integration costs, security vulnerabilities due to insecure connections, and difficulty in scaling IoT deployments. The absence of standardized and robust device connectivity solutions hinders the ability to efficiently collect real-time operational data, effectively manage a growing fleet of IoT devices, or build scalable IoT applications. This ultimately slows down IoT adoption, increases development complexity, and limits an organization's ability to leverage the full potential of connected operations.

Implementation	Challenges
* Identify IoT device types & protocols * Select appropriate device connectors/platforms * Configure device provisioning & registration * Implement data filtering & transformation * Establish secure communication channels * Plan for device lifecycle management	* Supporting diverse communication protocols * Ensuring device security & authentication * Managing device scalability & remote updates * Handling high volume & velocity of data * Integrating with backend IoT platforms * Debugging connectivity & data issues

Products	Platform Cost	Project Cost	Duration
AWS IoT Core	$0.1-5K	$2-20K	1-3 m
Azure IoT Hub	$0.1-5K	$2-20K	1-3 m
Google Cloud IoT Core	$0.1-5K	$2-20K	1-3 m
PTC ThingWorx (Connectivity)	$5-30K	$20K-100K	2-6 m
Bosch IoT Suite (Connectivity)	$2-15K	$10K-70K	1-4 m
Eclipse IoT Projects (Open Source)	N/A	$5K-30K	1-3 m

Digital Twin Integration involves creating and connecting virtual replicas (digital twins) of physical assets, processes, or systems to real-world operational data. These digital twins receive real-time data from sensors, integrate with historical data, and often incorporate simulation models to provide a comprehensive, dynamic representation of their physical counterparts. By synchronizing the virtual and physical worlds, Digital Twin Integration enables real-time monitoring, predictive analytics, what-if scenario planning, and remote control of complex operations. This is crucial for optimizing asset performance, conducting predictive maintenance, simulating process changes without disruption, and improving decision-making across manufacturing, smart cities, and critical infrastructure, offering unparalleled insights and control over physical realities.

Without Digital Twin Integration, organizations lack a comprehensive, real-time virtual representation of their physical assets or processes, relying instead on fragmented data, static models, or manual observations. This hinders the ability to conduct predictive analysis, simulate operational changes, or diagnose issues remotely with full context. The absence of a continuously synchronized digital twin makes it difficult to optimize asset performance proactively, identify potential failures before they occur, or test process improvements without impacting live operations. This ultimately limits operational efficiency, increases risk, and prevents organizations from leveraging the full power of advanced analytics and AI for real-time, data-driven operational decision-making.

Implementation	Challenges
* Identify assets/processes for digital twinning * Select a digital twin platform * Integrate with IoT, SCADA, ERP, PLM * Develop digital twin models & simulations * Configure real-time data synchronization * Plan for ongoing model refinement & validation	* Data quality & integration from diverse sources * Building accurate & complex twin models * Ensuring real-time synchronization & low latency * High computational requirements for simulations * User adoption & trust in twin insights * Measuring clear ROI on complex twin deployments

Products	Platform Cost	Project Cost	Duration
Siemens Teamcenter (with Digital Twin)	$20-100K	$150K-800K	8-24 m
PTC ThingWorx (Digital Twin capabilities)	$15-70K	$100K-600K	6-18 m
Azure Digital Twins (Microsoft)	$5-30K	$20K-100K	2-6 m
AWS IoT TwinMaker	$5-25K	$20K-90K	2-5 m
GE Digital Predix (Digital Twin Focus)	$10-50K	$70K-400K	5-14 m
Dassault Systèmes 3DEXPERIENCE	$20-90K	$120K-700K	6-24 m

Ai6x6™ Transformation Blueprint

Automation & 'Innovation'

"Automation is not about elimination. It's about elevation." — Rodney Brooks , Robotics pioneer and co-founder of iRobot; shaped real-world automation narratives.

Not everything should be automated—only what accelerates purpose. RPA engines, dispatch algorithms, and smart robotics must align with operational intent. "Automation & Innovation" turns innovation from disruption into dignity. Transformation maturity is measured in friction removed and clarity gained. 3D printing and autonomous vehicles don't just replace—they rethink. Laboratories for automation must prototype balance—between intelligence, trust, and throughput. Automation becomes less technical and more philosophical: are we elevating the mission or merely optimizing the machinery?

The "Automation & Innovation" chapter is paramount for operations transformation's digital maturity as it focuses on leveraging cutting-edge technologies to achieve unprecedented levels of automation and efficiency. A digitally mature organization explores Robotic Process Automation (RPA), autonomous vehicle systems, and smart robotics platforms to automate repetitive tasks, optimize logistics, and enhance production capabilities. This commitment to operational innovation, including 3D printing and AI-driven dispatch systems, not only reduces costs and improves throughput but also frees up human capital for higher-value activities, driving significant competitive advantage and transforming the very nature of work.

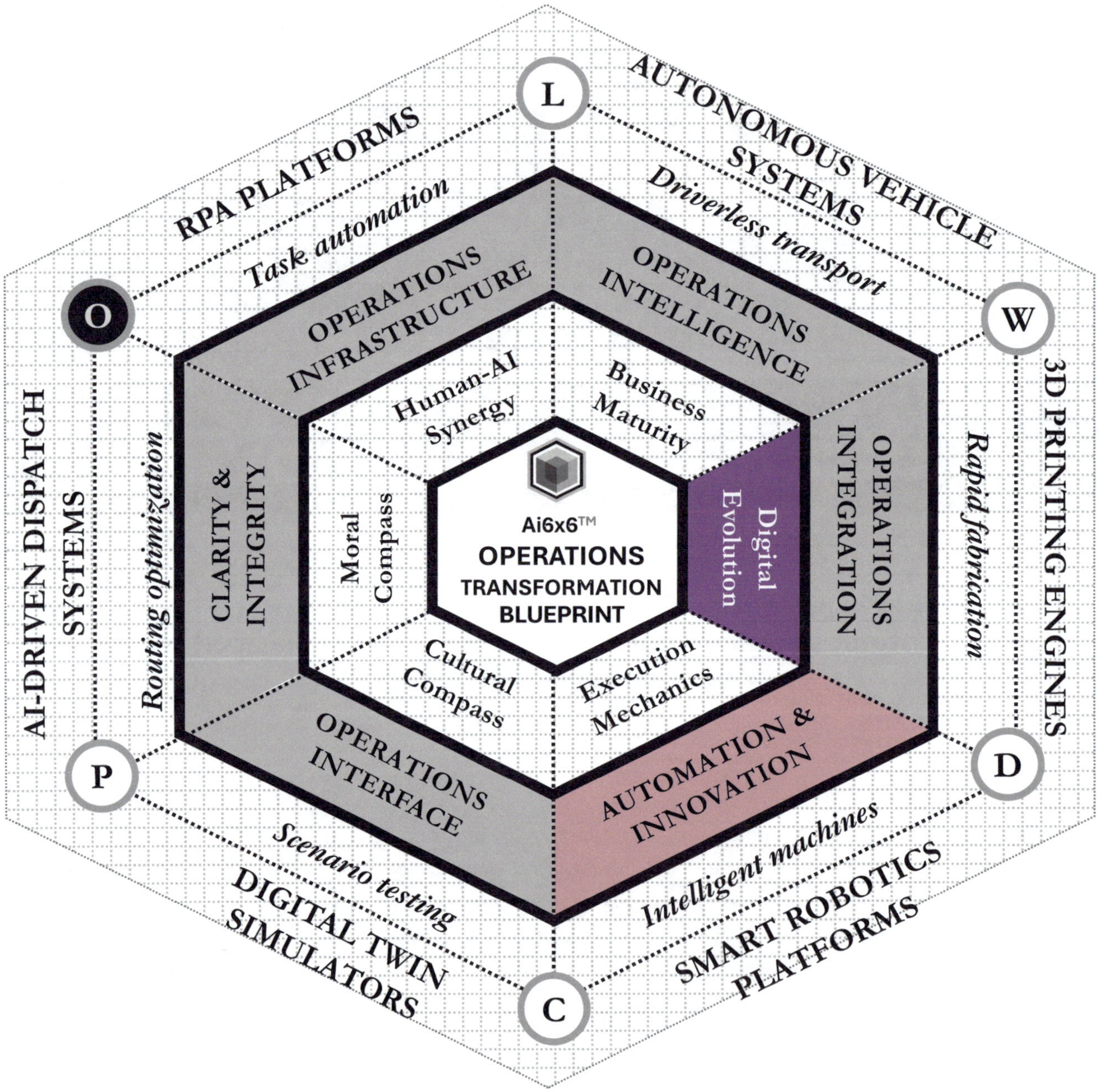

Ai6x6 DIGITAL EVOLUTION FOR **OPERATIONS LIFECYCLE**

Robotic Process Automation (RPA) Platforms are software tools that enable the configuration of "robots" (software bots) to automate repetitive, rule-based, and high-volume tasks typically performed by humans within business processes. These bots interact with digital systems and applications in the same way humans do, mimicking clicks, typing, and data entry across various interfaces without requiring changes to underlying IT infrastructure. By automating mundane and time-consuming operational tasks, RPA Platforms significantly improve process efficiency, reduce human error, accelerate task completion, and free up human employees for higher-value activities. This is crucial for optimizing back-office operations, streamlining customer service processes, and achieving rapid productivity gains across various business functions without extensive system re-engineering.

Without RPA Platforms, organizations rely on manual execution of repetitive, high-volume tasks across various business processes, leading to significant inefficiencies, increased human error rates, and slow processing times. Employees spend valuable time on mundane data entry, form filling, and system navigation, diverting their efforts from more strategic or customer-facing activities. The absence of automated digital task execution hinders scalability, increases operational costs, and makes it challenging to achieve consistent process quality. This ultimately impacts productivity, increases operational risk, and limits an organization's ability to rapidly streamline processes and respond to fluctuating workloads or business demands.

Implementation	Challenges
* Identify repetitive, rule-based processes * Select an RPA platform * Design & develop software bots * Integrate with target applications * Establish governance & change management * Plan for scaling & continuous improvement	* Identifying suitable processes for automation * Managing bot resilience to UI changes * Ensuring security & access control for bots * Scaling bot deployment across the enterprise * Gaining organizational buy-in & change resistance * Measuring clear ROI on automation efforts

Products	Platform Cost	Project Cost	Duration
UiPath	$10-60K	$70K-400K	5-14 m
Automation Anywhere	$10-50K	$60K-350K	5-14 m
Microsoft Power Automate	$2-15K	$10K-70K	1-4 m
Blue Prism	$10-40K	$50K-300K	4-12 m
WorkFusion (Intelligent Automation)	$5-30K	$40K-200K	3-10 m
Redwood Software (Robotic Process Automation)	$5-25K	$30K-180K	3-9 m

Autonomous Vehicle Systems in an operational context refer to self-driving vehicles or machinery used within controlled environments such as warehouses, factories, mines, or logistics yards (e.g., AGVs, AMRs, self-driving forklifts). These systems leverage sensors, AI, and navigation technology to operate independently, perform tasks like material handling, transportation, or inspection without human intervention. By automating routine and often hazardous movement tasks, Autonomous Vehicle Systems significantly improve operational safety, increase efficiency, optimize material flow, and reduce labor costs. This is crucial for optimizing internal logistics, accelerating throughput, and creating safer work environments in industries like manufacturing, warehousing, and mining, driving a new level of operational efficiency and automation.

Without Autonomous Vehicle Systems, organizations rely on human-operated vehicles and manual labor for material handling, transportation, and inspection tasks within their operational facilities. This leads to higher labor costs, increased risk of accidents and injuries, inefficiencies due to human error or fatigue, and slower throughput in repetitive movement tasks. The absence of automated, self-navigating systems hinders the ability to achieve continuous, optimized material flow, limits scalability for increasing operational demands, and exposes workers to unnecessary risks. This ultimately impacts operational efficiency, increases safety incidents, and limits an organization's ability to implement advanced, highly automated operational models.

Implementation	Challenges
* Define operational use cases & environments * Select appropriate autonomous vehicle types * Map facility layout & operational zones * Integrate with WMS/MES for task assignment * Establish safety protocols & human interaction zones * Plan for fleet management & maintenance	* High upfront capital investment * Complex integration with existing infrastructure * Ensuring safety in dynamic environments * Managing navigation & obstacle avoidance * Cybersecurity of autonomous systems * Regulatory compliance & local laws

Products	Platform Cost	Project Cost	Duration
Locus Robotics (AMRs for Warehouses)	$100K-1M+	$200K-3M+	6-18 m
Fetch Robotics (AMRs, now Zebra)	$50K-800K	$100K-2M	6-15 m
Geek+ (Logistics Robots)	$50K-700K	$100K-1.5M	6-15 m
KION Group (AGVs/AMRs)	$50K-1M	$100K-2M+	6-18 m
Waypoint Robotics (AMRs, now Locus)	N/A	N/A	N/A
Seegrid (Autonomous Forklifts)	$80K-1.2M	$150K-2.5M	7-20 m

3D Printing Engines (also known as Additive Manufacturing systems) are advanced industrial machines and associated software that build three-dimensional objects layer by layer from a digital design. These engines can utilize various materials, including plastics, metals, ceramics, and composites. By enabling on-demand, localized, and highly customized production, 3D Printing Engines revolutionize manufacturing processes, allowing for rapid prototyping, complex geometries, and efficient small-batch production. This is crucial for accelerating product development cycles, reducing tooling costs, enabling decentralized manufacturing, and producing customized parts or products with intricate designs, driving innovation in both product design and operational supply chains.

Without 3D Printing Engines, organizations are limited to traditional subtractive manufacturing methods (e.g., machining) or injection molding, which often involve high tooling costs, long lead times for prototypes, and limitations on design complexity. This restricts rapid iteration in product development, makes customized or small-batch production economically unfeasible, and necessitates reliance on centralized manufacturing facilities. The absence of additive manufacturing capabilities hinders the ability to quickly produce complex geometries, create functional prototypes on demand, or implement distributed manufacturing strategies. This ultimately impacts product innovation speed, supply chain resilience, and the cost-effectiveness of producing specialized or low-volume components, reducing operational agility.

Implementation	Challenges
* Identify suitable materials & applications * Select appropriate 3D printing technology * Prepare CAD models for printing * Integrate with design software & PLM * Establish post-processing workflows * Plan for material handling & safety	* High upfront investment in equipment * Limited material compatibility for certain applications * Ensuring part quality & repeatability * Post-processing requirements & expertise * Scaling production from prototype to volume * Managing intellectual property for digital designs

Products	Platform Cost	Project Cost	Duration
Stratasys (PolyJet, FDM)	$50K-1M+	$100K-2M+	6-18 m
EOS (SLM for Metals)	$200K-2M+	$300K-5M+	8-24 m
HP Multi Jet Fusion (MJF)	$100K-500K	$150K-1M	6-15 m
Desktop Metal (Bound Metal Deposition)	$100K-800K	$150K-1.5M	6-15 m
Carbon (Digital Light Synthesis)	$100K-700K	$150K-1.2M	6-15 m
Ultimaker (Desktop FDM)	$5K-50K	$10K-100K	1-3 m

Smart Robotics Platforms refer to advanced robotic systems that incorporate artificial intelligence, machine vision, and sophisticated sensors to perform complex tasks with adaptability and learning capabilities, going beyond traditional, pre-programmed industrial robots. These robots can perceive their environment, learn from experience, and adjust their actions autonomously, making them suitable for variable tasks like picking and placing irregular objects, intricate assembly, or quality inspection. By introducing intelligent automation into operational workflows, Smart Robotics Platforms significantly improve flexibility, enhance precision, increase productivity, and enable collaboration between humans and robots. This is crucial for optimizing manufacturing processes, improving efficiency in logistics, and addressing labor shortages by deploying versatile, intelligent automation solutions.

Without Smart Robotics Platforms, organizations are limited to traditional industrial robots that are highly efficient for repetitive, fixed tasks but lack the flexibility and adaptability for varied or unstructured environments. This restricts automation to predictable workflows, prevents the handling of complex or diverse items, and necessitates significant re-programming for minor task changes. The absence of AI-driven, vision-enabled robotics hinders the ability to automate tasks requiring perception, judgment, or interaction with dynamic environments. This ultimately impacts operational flexibility, limits productivity gains in variable production scenarios, and prevents organizations from deploying advanced automation to address complex challenges or adapt quickly to changing demands.

Implementation	Challenges
* Identify complex, variable tasks for automation * Select appropriate smart robot types (e.g., cobots) * Program AI/vision for task execution * Integrate with MES, WMS, or PLC systems * Establish safety protocols for human-robot interaction * Plan for continuous learning & optimization	* High upfront cost & complex programming * Ensuring safety in collaborative environments * Managing accuracy of AI vision & perception * Integrating with diverse operational systems * Training workforce for human-robot collaboration * Measuring clear ROI on flexible automation

Products	Platform Cost	Project Cost	Duration
Universal Robots (Collaborative Robots)	$30K-100K	$50K-200K	3-9 m
FANUC CRX Series (Collaborative Robots)	$40K-150K	$60K-250K	4-10 m
KUKA LBR iiwa (Collaborative Robots)	$50K-200K	$80K-300K	4-12 m
ABB YuMi (Collaborative Robots)	$50K-200K	$80K-300K	4-12 m
NVIDIA Jetson (AI at the Edge for Robotics)	$0.5K-5K	$5K-50K	1-3 m
RightHand Robotics (Piece Picking)	$100K-500K	$150K-800K	6-18 m

Digital Twin Simulators are advanced software environments that utilize the data and models from digital twins to run "what-if" scenarios, test hypothetical changes, and predict the outcomes of operational adjustments without impacting physical systems. These simulators can model various aspects, from production line performance and supply chain logistics to equipment wear and tear. By providing a safe, virtual sandbox for experimentation and optimization, Digital Twin Simulators enable organizations to identify optimal operating conditions, test new strategies, predict bottlenecks, and train operators more effectively. This is crucial for risk reduction, cost optimization, and accelerating innovation in complex operational environments like manufacturing, smart cities, and energy grids, transforming reactive problem-solving into proactive strategic planning.

Without Digital Twin Simulators, organizations must rely on real-world trials, manual calculations, or less sophisticated modeling to understand the impact of operational changes. This leads to costly disruptions, potential safety risks, and delays in implementing improvements due to the inability to test scenarios in a risk-free environment. The absence of a dynamic, data-driven simulation capability hinders the ability to optimize complex processes, predict the effects of system modifications, or train personnel on new workflows effectively. This ultimately impacts operational efficiency, increases the time and cost of improvements, and limits an organization's capacity for rapid, data-informed innovation and strategic decision-making in highly complex systems.

Implementation	Challenges
* Build robust digital twin models * Integrate with real-time & historical data * Define simulation objectives & scenarios * Select a digital twin simulation platform * Validate simulation results against real-world * Train users on simulation interpretation	* High computational power requirements * Ensuring accuracy & fidelity of simulation models * Managing large volumes of real-time data * Integrating with diverse operational systems * Interpreting complex simulation outcomes * Gaining user trust in simulation predictions

Products	Platform Cost	Project Cost	Duration
Ansys Twin Builder	$10-60K	$80K-500K	6-18 m
Siemens NX (Simulation capabilities)	$15-70K	$100K-600K	6-18 m
Dassault Systèmes 3DEXPERIENCE (SIMULIA)	$20-90K	$120K-700K	6-24 m
Rockwell Automation Emulate3D	$5-30K	$40K-200K	3-10 m
AnyLogic (Simulation Modeling)	$2-15K	$10K-70K	1-4 m
MATLAB/Simulink (MathWorks)	$1-10K	$5K-50K	1-3 m

AI-Driven Dispatch Systems are intelligent platforms that utilize artificial intelligence and machine learning algorithms to optimize the scheduling, routing, and assignment of field service technicians, delivery vehicles, or mobile assets. These systems analyze real-time data (e.g., traffic conditions, technician skills, customer locations, service level agreements) to make dynamic, data-driven decisions that minimize travel time, reduce operational costs, improve service delivery, and maximize resource utilization. By automating and enhancing complex dispatch decisions, AI-Driven Dispatch Systems significantly improve operational efficiency, increase customer satisfaction, and provide greater agility in managing dynamic field operations. This is crucial for industries like logistics, utilities, and field services, where efficient resource deployment directly impacts profitability and service quality.

Without AI-Driven Dispatch Systems, organizations rely on manual dispatching or basic rule-based systems, leading to inefficient routing, suboptimal resource allocation, and delayed service delivery. Dispatchers struggle to account for real-time variables like traffic, last-minute changes, or technician availability, resulting in missed appointments, increased fuel consumption, and customer dissatisfaction. The absence of intelligent, data-driven optimization hinders the ability to maximize technician productivity, minimize operational costs, or adapt quickly to unexpected events. This ultimately impacts service quality, reduces profitability, and limits an organization's capacity to deliver efficient and reliable field operations in a dynamic environment.

Implementation	Challenges
* Define dispatch objectives & constraints * Select an AI-driven dispatch platform * Integrate with CRM, FSM, GPS data * Configure optimization rules & parameters * Train AI models with historical data * Plan for real-time data feeds & monitoring	* Data quality & availability for real-time insights * Complex routing & scheduling algorithms * Integrating with diverse legacy systems * Handling unexpected events & disruptions * User adoption by dispatchers & field teams * Measuring clear ROI on efficiency gains

Products	Platform Cost	Project Cost	Duration
Salesforce Field Service (with Optimization)	$5-30K	$40K-200K	3-10 m
ServiceMax (with Scheduling & Optimization)	$5-25K	$30K-180K	3-9 m
ClickSoftware (now Salesforce FSM)	N/A	N/A	N/A
Route4Me (Route Optimization)	$0.5-5K	$2K-15K	1-3 m
OptimoRoute (Route Optimization)	$0.5-5K	$2K-15K	1-3 m
PTV Optima (Traffic & Logistics)	$5-20K	$20K-100K	2-6 m

Ai6x6™ Transformation Blueprint

Operations 'Interface'

"The user interface is the front door to your system's soul." — Aarron Walter , Founding member of UX team at Mailchimp; known for blending design with operational usability.

Control is no longer centralized—it's distributed through screens, wearables, and voice systems. "Operations Experience" means operational mastery through minimalism. Dashboards don't just report—they predict. Field apps don't just guide—they empower. AR maintenance overlays and operator panels convert complexity into clarity. Transformation maturity surfaces when control feels natural and interfaces speak in urgency. When operations see, act, and learn through experience rather than instruction, performance becomes conversational—not conditional.

"Operations 'Experience" is critical for operations transformation's digital maturity because it emphasizes creating intuitive and accessible user interfaces for operational systems, empowering all users. A digitally mature organization understands that providing frontline workers and managers with user-friendly mobile field service apps, AR maintenance guides, and real-time ops dashboards improves efficiency, reduces errors, and enhances decision-making. This focus on human-machine interfaces and wearable operator panels ensures that complex operational data and controls are easily digestible and actionable, leading to a more engaged, productive, and effective workforce across all operational functions.

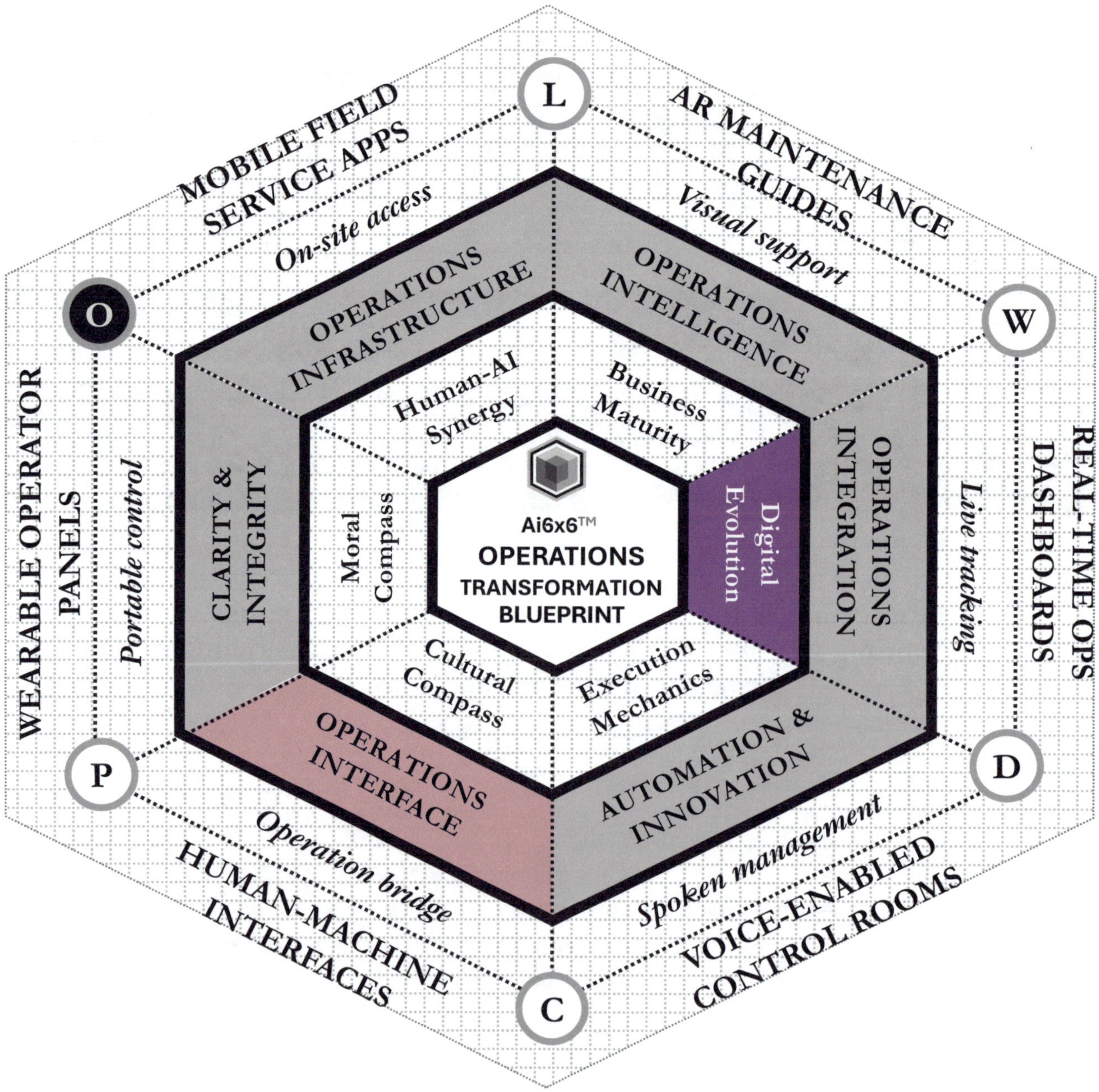

Ai6x6 DIGITAL EVOLUTION FOR **OPERATIONS LIFECYCLE**

Mobile Field Service Apps are specialized applications designed for smartphones and tablets that equip field technicians and service personnel with the tools and information needed to manage their work efficiently while on the go. These apps typically provide access to work orders, customer information, equipment history, diagnostic guides, inventory levels, and allow for real-time updates, data capture (e.g., photos, signatures), and communication with dispatch. By digitizing and mobilizing field operations, Mobile Field Service Apps significantly improve technician productivity, enhance customer service, reduce administrative overhead, and provide real-time visibility into field activities. This is crucial for optimizing service delivery, accelerating response times, and ensuring accurate data collection at the point of service, transforming reactive field work into a streamlined and proactive process.

Without Mobile Field Service Apps, field technicians and service personnel rely on paper-based forms, manual processes, and disconnected communication, leading to inefficiencies, errors, and delays in service delivery. Technicians lack real-time access to critical information, resulting in wasted time, repeat visits, and suboptimal customer experiences. The absence of mobile data capture and real-time updates hinders accurate record-keeping, complicates billing, and delays insight into field operations. This ultimately impacts service quality, increases operational costs due to rework and administrative burden, and limits an organization's ability to scale its field operations and deliver consistent, high-quality service.

Implementation	Challenges
* Define field technician workflows & data needs * Select a mobile field service platform * Configure forms, checklists & workflows * Integrate with CRM, ERP, scheduling systems * Train field technicians on app usage * Plan for offline capabilities & data sync	* Ensuring robust offline functionality & data sync * User adoption by diverse field workforce * Integrating with complex backend systems * Device management & security in the field * Real-time performance on various mobile networks * Customization for unique field processes

Products	Platform Cost	Project Cost	Duration
Salesforce Field Service	$5-30K	$40K-200K	3-10 m
ServiceMax	$5-25K	$30K-180K	3-9 m
Microsoft Dynamics 365 Field Service	$2-15K	$10K-70K	1-4 m
Oracle Field Service	$5-20K	$20K-100K	2-6 m
IFS Field Service Management	$5-25K	$30K-150K	3-9 m
ServiceTitan (for trade businesses)	$1-10K	$5K-50K	1-3 m

AR (Augmented Reality) Maintenance Guides are applications that overlay digital information, instructions, and interactive 3D models onto a real-world physical object or machine through a smartphone, tablet, or smart glasses. These guides provide step-by-step visual instructions, highlight specific components, and offer remote assistance capabilities, allowing technicians to perform complex maintenance, repair, and operational tasks with enhanced accuracy and efficiency. By providing contextual, hands-free guidance and enabling remote expert collaboration, AR Maintenance Guides significantly reduce errors, accelerate training, decrease downtime, and improve first-time fix rates. This is crucial for increasing operational uptime, empowering less experienced technicians, and enhancing safety in demanding industrial and field service environments.

Without AR Maintenance Guides, technicians often rely on paper manuals, static diagrams, or verbal instructions, which can be difficult to interpret, cumbersome to use in the field, and prone to misinterpretation. This leads to increased errors, longer repair times, and a higher reliance on highly experienced (and often scarce) senior technicians. The absence of interactive, visual, and contextual guidance hinders efficient task execution, complicates complex procedures, and slows down knowledge transfer. This ultimately impacts operational efficiency, increases human error rates, and limits an organization's ability to scale specialized maintenance expertise across a distributed workforce, leading to higher operational costs and extended downtime.

Implementation	Challenges
* Identify suitable maintenance tasks for AR * Create 3D models & step-by-step instructions * Select an AR platform/tool * Develop/configure AR application * Train technicians on AR device usage * Plan for content updates & device management	* High upfront cost of AR hardware (glasses) * Creating accurate 3D models & animations * Ensuring precise tracking & overlay in varied environments * Integration with existing technical documentation * User adoption & comfort with AR devices * Managing content updates & versioning

Products	Platform Cost	Project Cost	Duration
PTC Vuforia Expert Capture	$5-30K	$20K-100K	2-6 m
Microsoft Dynamics 365 Guides	$2-15K	$10K-70K	1-4 m
TeamViewer Frontline (AR for industry)	$2-10K	$5K-50K	1-3 m
Resonai (AR for Robotics/Industrial)	$5-25K	$20K-100K	2-6 m
Augmentir (Connected Worker Platform)	$5-20K	$10K-70K	1-4 m
Remote Expert Assist (various platforms)	$1-5K	$5K-30K	1-3 m

3. REAL-TIME OPS DASHBOARDS

Real-Time Operations Dashboards are dynamic, visual interfaces that provide an immediate, up-to-the-minute overview of key operational performance indicators (KPIs), production statuses, equipment health, and other critical metrics. These dashboards aggregate data from various sources (e.g., MES, SCADA, IoT sensors, ERP) and present it through interactive charts, graphs, and alerts. By offering a centralized, continuously updated view of operational realities, Real-Time Ops Dashboards enable faster problem detection, more informed decision-making, and proactive intervention in critical situations. This is essential for optimizing production throughput, minimizing downtime, ensuring quality control, and maintaining a high level of operational awareness across factory floors, logistics centers, and complex industrial environments.

Without Real-Time Ops Dashboards, operations managers and stakeholders rely on static reports, periodic data pulls, or fragmented information sources to gauge performance. This leads to delayed insights, reactive problem-solving, and an inability to identify or address operational inefficiencies as they occur. The absence of a centralized, continuously updated visual representation of critical KPIs hinders rapid decision-making, complicates performance monitoring, and prevents timely intervention in critical situations. This ultimately impacts productivity, increases the risk of costly disruptions, and limits an organization's ability to maintain optimal operational performance and responsiveness in a dynamic environment.

Implementation	Challenges
* Identify critical operational KPIs * Select a real-time dashboarding platform * Integrate with diverse data sources (MES, SCADA, IoT) * Design intuitive visualizations & alerts * Establish data refresh rates & performance * Plan for user training & access control	* Ensuring real-time data freshness & accuracy * Integrating with diverse operational data sources * Handling high data volume & velocity * Designing actionable & uncluttered dashboards * Gaining user adoption across ops teams * Cybersecurity for sensitive operational data

Products	Platform Cost	Project Cost	Duration
AVEVA PI Vision	$5-30K	$40K-200K	3-10 m
Ignition (Inductive Automation)	$5-25K	$30K-180K	3-9 m
PTC ThingWorx (Mashups/Dashboards)	$5-20K	$20K-100K	2-6 m
Grafana (Open Source BI)	N/A	$5K-30K	1-3 m
Tableau / Power BI (with real-time connectors)	$1-10K	$5K-50K	1-3 m
Splunk (Operational Intelligence)	$10-50K	$70K-350K	5-14 m

Voice-Enabled Control Rooms integrate voice recognition and natural language processing (NLP) technologies into central operational control environments, allowing operators to interact with systems, retrieve information, and issue commands using spoken words. This transforms traditional mouse-and-keyboard interfaces into more intuitive, hands-free interactions, particularly beneficial in situations requiring rapid response or when operators' hands are occupied. By enabling swift access to data, faster command execution, and reducing cognitive load, Voice-Enabled Control Rooms enhance operational efficiency, improve response times in critical situations, and increase operator comfort and focus. This is crucial for optimizing control processes, improving safety, and making operational decision-making more agile in complex and high-pressure environments like utilities, logistics, and manufacturing.

Without Voice-Enabled Control Rooms, operators in critical operational environments rely solely on traditional graphical user interfaces (GUIs), requiring constant visual attention and manual input (mouse, keyboard). This can lead to slower reaction times in fast-paced scenarios, increased risk of human error when hands are occupied, and cognitive overload from simultaneously monitoring multiple screens and input devices. The absence of hands-free interaction hinders rapid data retrieval, complicates multi-tasking, and limits the naturalness and speed of human-system interaction. This ultimately impacts operational efficiency, increases response times during critical events, and reduces overall operator effectiveness and safety in complex control environments.

Implementation	Challenges
* Define voice commands & operational contexts * Select a voice AI/NLP platform * Integrate with SCADA, MES, HMI systems * Develop conversational flows & intent recognition * Train voice models for specific vocabulary/accents * Plan for noise cancellation & privacy	* Ensuring high accuracy in speech recognition * Handling background noise & accents * Designing intuitive and forgiving voice commands * Cybersecurity for voice data * User adoption & comfort with voice interaction * Managing complex integrations with control systems

Products	Platform Cost	Project Cost	Duration
Microsoft Azure AI Speech / Bot Framework	$0.5-5K	$5K-30K	1-3 m
Google Cloud AI Platform (Speech-to-Text)	$0.5-5K	$5K-30K	1-3 m
Amazon Alexa Voice Service (AVS)	N/A	$5K-25K	1-3 m
Nuance Communications (Enterprise Solutions)	$5-20K	$20K-100K	2-6 m
IBM Watson Assistant (Voice Integration)	$0.5-5K	$5K-25K	1-3 m
Custom Voice AI Development	N/A	$30K-150K	3-9 m

Human-Machine Interfaces (HMIs) are user interfaces that facilitate communication and control between human operators and industrial machinery, production lines, or process control systems. HMIs typically consist of graphical displays (touchscreens, monitors), pushbuttons, and control devices that present real-time operational data, machine status, alarms, and allow operators to input commands, adjust parameters, and visualize processes. By providing an intuitive and centralized point of interaction, HMIs improve operational control, enhance safety, reduce human error, and increase efficiency on the factory floor and in control rooms. This is critical for optimizing production, monitoring critical parameters, and ensuring the smooth and safe operation of complex industrial equipment and processes.

Without modern Human-Machine Interfaces (HMIs), operators rely on disparate physical controls, analog gauges, and manual data logging to monitor and control machinery. This leads to fragmented information, increased risk of human error, slower reaction times to operational issues, and difficulty in understanding complex process states. The absence of a unified, intuitive graphical interface hinders efficient machine control, complicates troubleshooting, and limits the ability to visualize real-time operational data. This ultimately impacts productivity, increases downtime, poses safety risks, and prevents effective control and optimization of industrial processes, leading to higher operational costs and reduced efficiency.

Implementation	Challenges
* Define operational data & control points * Select an HMI software/hardware platform * Design intuitive screen layouts & navigation * Integrate with PLCs, SCADA, MES systems * Develop alarm management & trending * Plan for operator training & usability testing	* Designing intuitive UIs for complex processes * Ensuring real-time data accuracy & refresh rates * Integrating with diverse industrial control systems * Cybersecurity for HMI systems * Managing version control of HMI projects * User adoption & ergonomic considerations

Products	Platform Cost	Project Cost	Duration
Siemens TIA Portal WinCC (Unified)	$5-30K	$40K-200K	3-10 m
Rockwell Automation FactoryTalk View SE/ME	$5-25K	$30K-180K	3-9 m
AVEVA InTouch HMI	$5-20K	$20K-100K	2-6 m
Ignition (Inductive Automation)	$5-20K	$20K-100K	2-6 m
Wonderware System Platform (AVEVA)	$10-50K	$70K-350K	5-14 m
GE Digital iFIX	$5-20K	$20K-100K	2-6 m

Wearable Operator Panels refer to compact, robust computing devices (e.g., smartwatches, head-mounted displays, arm-mounted tablets) that provide industrial operators with real-time operational data, instructions, and communication capabilities directly on their person. These devices offer hands-free access to critical information, allowing operators to monitor machine status, receive alerts, access digital work instructions, and communicate with supervisors or remote experts while actively performing tasks on the factory floor or in the field. By mobilizing and contextualizing operational data, Wearable Operator Panels enhance productivity, reduce errors, improve safety, and enable more agile decision-making at the point of action, transforming traditional operator workflows into highly efficient and connected experiences.

Without Wearable Operator Panels, industrial operators often rely on fixed workstations, paper manuals, or verbal communication, which can be inefficient, cumbersome, and divert attention from critical tasks. This leads to slower response times, increased errors due to lack of immediate information, and reduced situational awareness. The absence of hands-free, on-person access to real-time data and instructions hinders efficient task execution, complicates troubleshooting, and limits the ability to collaborate effectively in dynamic operational environments. This ultimately impacts productivity, increases operational risks, and limits an organization's capacity to empower its workforce with critical information where and when they need it most.

Implementation	Challenges
* Identify mobile operator tasks for wearables * Select appropriate wearable device types * Develop/configure apps for wearable OS * Integrate with MES, ERP, asset management * Design intuitive UIs for small screens/voice * Plan for device management & charging	* Ensuring device robustness & battery life * User comfort & adoption (ergonomics) * Data security & network connectivity * Designing effective UIs for limited display * Integration with diverse industrial systems * Managing updates & device refresh cycles

Products	Platform Cost	Project Cost	Duration
RealWear HMT (Head-Mounted Tablets)	$2-10K	$10K-70K	1-4 m
Vuzix Smart Glasses	$1-5K	$5K-30K	1-3 m
Google Glass Enterprise Edition	$1-5K	$5K-30K	1-3 m
Microsoft HoloLens (for more complex AR)	$3-15K	$20K-100K	2-6 m
TeamViewer Frontline (Wearable solutions)	$2-10K	$10K-50K	1-3 m
Custom Wearable App Development	N/A	$30K-150K	3-9 m

Ai6x6™ Transformation Blueprint

Clarity & 'Integrity'

"If you think compliance is expensive, try non-compliance." — Paul McNulty , Former U.S. Deputy Attorney General; widely quoted in risk management and enterprise governance.

Compliance is more than constraint—it's operational clarity. ISO certifications, risk assessments, and EHS frameworks must embed into processes before audits arrive. "Clarity & Integrity" means standards are not overhead —they're understructure. Transformation matures when compliance becomes proactive—designing safety, trust, and ethics into every operational moment. Audit trails aren't defensive—they're declarative. Teams no longer fear review—they prepare it in every keystroke and click. When standards drive behavior, operations become not just legal—but literate: fluent in accountability, foresight, and resilience. Compliance stops being a checkbox—it becomes choreography. The organization learns to move like it's always ready—because now, it is.

The "Clarity & Integrity" chapter is fundamental for operations transformation's digital maturity as it underscores the critical importance of embedding quality, safety, and compliance into every operational process. A digitally mature organization establishes robust systems for quality management, audit management platforms, and EHS compliance tools to ensure adherence to regulations and industry standards. This unwavering commitment to operational integrity, supported by ISO certification management and secure audit trails, not only mitigates risks and avoids penalties but also builds trust with stakeholders, ensuring sustainable, responsible, and resilient operations in the digital age.

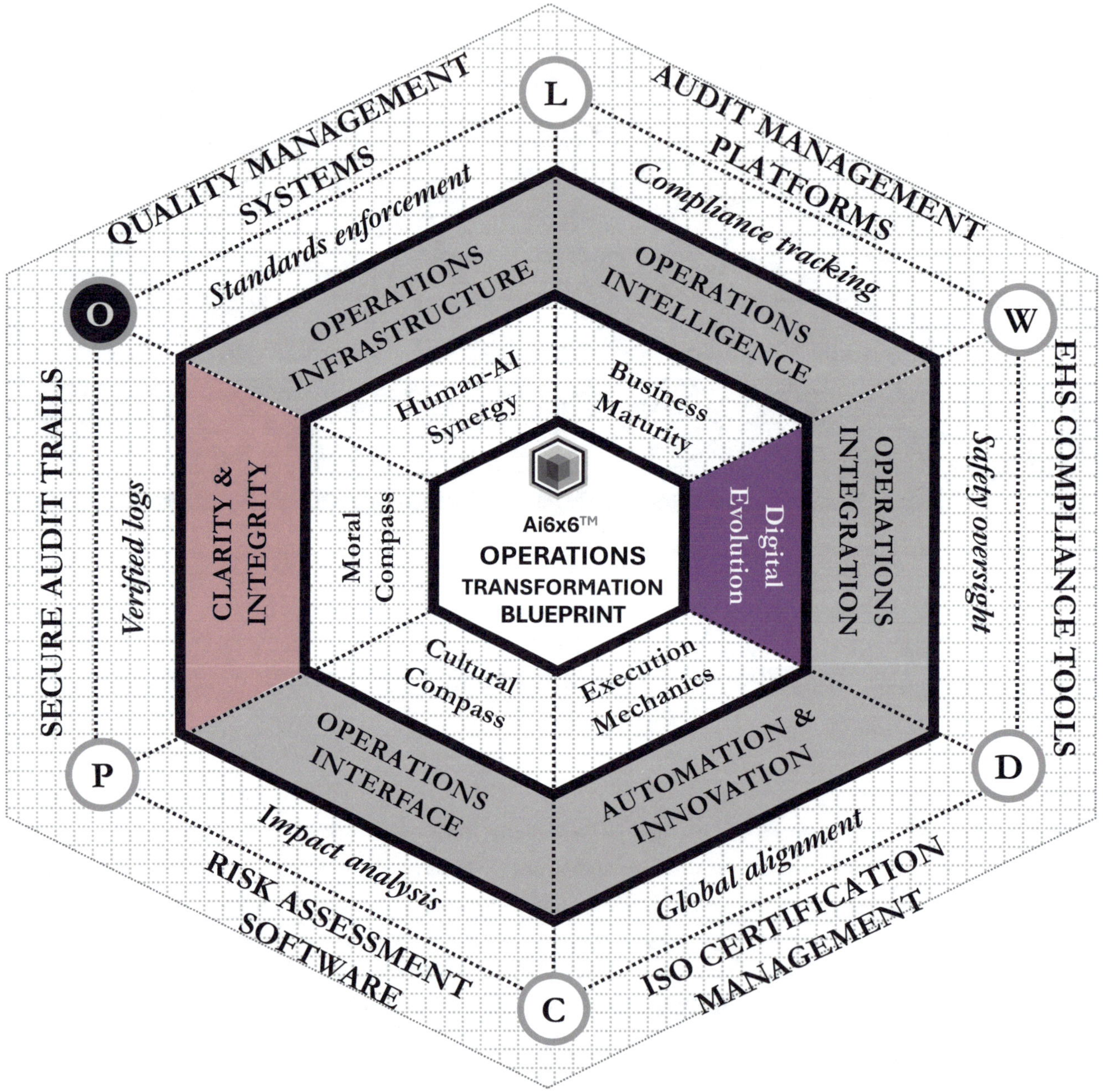

Ai6x6 DIGITAL EVOLUTION FOR **OPERATIONS LIFECYCLE**

1. QUALITY MANAGEMENT SYSTEMS

Quality Management Systems (QMS) are formalized systems that document processes, procedures, and responsibilities for achieving quality policies and objectives within an organization's operations. Often based on international standards like ISO 9001, QMS software solutions provide tools for managing documentation, non-conformances, corrective and preventive actions (CAPA), audits, and training. By centralizing quality-related activities and ensuring adherence to established standards, QMS improves product consistency, reduces defects, enhances customer satisfaction, and facilitates regulatory compliance. This is crucial for maintaining high operational standards, reducing waste, and building a reputation for reliability, making them essential for manufacturers and service providers committed to continuous improvement and customer trust.

Without a robust Quality Management System (QMS), organizations face inconsistent product or service quality, frequent defects, customer complaints, and potential regulatory non-compliance. Relying on fragmented processes, manual documentation, and ad-hoc quality checks makes it difficult to track non-conformances, implement effective corrective actions, or ensure continuous improvement. The absence of a formalized and centralized QMS leads to increased operational costs due to rework and warranty claims, damages brand reputation, and can result in significant legal and financial penalties. This ultimately hinders an organization's ability to compete effectively, achieve customer loyalty, and demonstrate adherence to industry standards, undermining overall operational integrity.

Implementation	Challenges
* Define quality policies & objectives * Select a QMS software solution * Document processes, procedures, and controls * Configure modules (CAPA, document control) * Train employees on QMS usage & principles * Plan for internal & external audits	* Ensuring full adoption across all departments * Managing complex documentation & version control * Integrating with ERP, MES, and other systems * Maintaining compliance with evolving standards * Data accuracy for non-conformance tracking * Demonstrating clear ROI on quality improvements

Products	Platform Cost	Project Cost	Duration
MasterControl QMS	$10-50K	$70K-400K	5-14 m
Sparta Systems TrackWise	$10-60K	$80K-500K	6-16 m
Pilgrim Quality Solutions (now IQVIA)	$5-40K	$50K-300K	4-12 m
ETQ Reliance QMS	$5-30K	$40K-250K	3-10 m
Qualio (Life Sciences QMS)	$2-15K	$10K-70K	1-4 m
QT9 QMS	$1-10K	$5K-50K	1-3 m

Audit Management Platforms are software solutions designed to plan, execute, and report on various types of audits, including internal audits, supplier audits, quality audits, and regulatory compliance audits. These platforms provide tools for scheduling audits, developing checklists, managing findings, tracking corrective actions, and generating comprehensive reports. By centralizing the audit process, Audit Management Platforms streamline compliance efforts, improve transparency, reduce manual overhead, and ensure timely resolution of identified issues. This is crucial for maintaining regulatory compliance, identifying operational weaknesses, mitigating risks, and driving continuous improvement across an organization's processes and supply chain, fostering a culture of accountability and rigorous oversight.

Without dedicated Audit Management Platforms, organizations rely on manual processes, spreadsheets, and disparate documents to manage their audit programs. This leads to inefficient scheduling, inconsistent audit methodologies, difficulty in tracking findings and corrective actions, and a lack of centralized reporting. The absence of a systematic and automated approach to audits hinders the ability to identify compliance gaps proactively, assess risks effectively, or ensure timely remediation of issues. This ultimately increases the risk of regulatory non-compliance, reduces operational transparency, and makes it challenging to demonstrate due diligence, thereby impacting an organization's integrity and its ability to learn from past incidents.

Implementation	Challenges
* Define audit scope & objectives * Select an audit management platform * Customize audit checklists & templates * Configure finding management & CAPA workflows * Train auditors & process owners * Plan for reporting & trend analysis	* Standardizing audit processes across diverse areas * Ensuring comprehensive tracking of findings * Integrating with QMS, ERP, and risk systems * User adoption by auditors & action owners * Data security for sensitive audit information * Demonstrating tangible risk reduction

Products	Platform Cost	Project Cost	Duration
MasterControl Audit Management	$5-30K	$40K-200K	3-10 m
MetricStream GRC (Audit Management)	$10-60K	$80K-500K	6-16 m
AuditBoard (Cloud Audit & Risk)	$5-30K	$30K-180K	3-9 m
Intelex QMS (Audit Management module)	$5-25K	$20K-100K	2-6 m
HighBond (formerly Galvanize)	$5-25K	$20K-100K	2-6 m
LogicManager (Enterprise Risk Mgmt)	$5-30K	$20K-100K	2-6 m

Environmental, Health, and Safety (EHS) Compliance Tools are software solutions designed to help organizations manage and adhere to regulations and best practices related to environmental protection, workplace health, and employee safety. These tools often cover areas like incident management, risk assessments, permits and licenses, waste management, emissions monitoring, and training. By centralizing EHS data, automating compliance workflows, and providing reporting capabilities, EHS Compliance Tools help organizations minimize environmental impact, reduce workplace accidents, ensure worker well-being, and avoid regulatory fines. This is crucial for protecting employees, the community, and the environment, while also maintaining legal standing and corporate social responsibility in all operational facets.

Without dedicated EHS Compliance Tools, organizations often rely on manual tracking, spreadsheets, and fragmented data to manage their environmental, health, and safety obligations. This leads to inefficiencies, increased risk of human error in compliance reporting, and difficulty in identifying and mitigating EHS hazards proactively. The absence of a centralized and automated system for EHS management hinders the ability to track incidents effectively, manage permits, or ensure timely training and compliance with evolving regulations. This ultimately increases the risk of accidents, environmental non-compliance, and substantial fines, impacting brand reputation, operational continuity, and employee well-being.

Implementation	Challenges
* Identify applicable EHS regulations * Select an EHS compliance tool * Configure modules (incident, waste, permits) * Migrate existing EHS data * Train EHS team & employees * Plan for continuous monitoring & reporting	* Complexity of diverse EHS regulations * Data integration from various operational sources * User adoption across different departments * Ensuring data accuracy for regulatory reporting * Managing emergency response plans * Demonstrating tangible safety/environmental gains

Products	Platform Cost	Project Cost	Duration
Sphera EHS	$10-60K	$80K-500K	6-18 m
Cority EHS (formerly Medgate)	$10-50K	$70K-400K	5-14 m
Enablon EHS (Wolters Kluwer)	$10-50K	$60K-350K	5-14 m
Intelex EHSQ	$5-40K	$50K-300K	4-12 m
Gensuite EHS	$5-30K	$40K-200K	3-10 m
VelocityEHS	$2-15K	$10K-70K	1-4 m

ISO Certification Management tools and processes are specifically designed to help organizations achieve and maintain compliance with various International Organization for Standardization (ISO) standards, such as ISO 9001 (Quality), ISO 14001 (Environmental), ISO 45001 (Occupational Health & Safety), or ISO 27001 (Information Security). These solutions provide frameworks for documenting processes, managing audits, tracking non-conformances, and demonstrating adherence to ISO requirements, facilitating the certification process. By systematizing ISO compliance efforts, these tools streamline audits, reduce administrative burden, ensure continuous improvement, and enhance an organization's credibility and market access. This is crucial for demonstrating a commitment to global best practices, improving operational performance, and gaining a competitive advantage in regulated industries and international markets.

Without dedicated ISO Certification Management tools, organizations pursuing or maintaining ISO certifications face a highly manual, complex, and resource-intensive process. This involves extensive paper-based documentation, fragmented tracking of non-conformances, and inefficient preparation for audits, increasing the risk of non-compliance and certification delays. The absence of a structured system for managing ISO requirements hinders effective document control, complicates internal and external audits, and makes it challenging to demonstrate continuous improvement. This ultimately increases the cost and time associated with achieving and maintaining certifications, impacts market access, and limits an organization's ability to leverage ISO standards for true operational excellence and global recognition

Implementation	Challenges
* Identify target ISO standards * Select an ISO certification management tool * Map existing processes to ISO requirements * Document procedures & create controls * Conduct internal audits & prepare for external * Plan for continuous improvement cycles	* Interpreting complex ISO standard requirements * Ensuring all processes are fully documented * Gaining organizational buy-in for new processes * Managing audit findings & corrective actions * Maintaining certification through surveillance audits * Integrating with existing QMS or ERP systems

Products	Platform Cost	Project Cost	Duration
MasterControl QMS (ISO capabilities)	$10-50K	$70K-400K	5-14 m
Intelex EHSQ (with ISO modules)	$5-40K	$50K-300K	4-12 m
Ideagen Q-Pulse	$5-30K	$40K-200K	3-10 m
Qualio (for ISO 13485, 9001 in Life Sciences)	$2-15K	$10K-70K	1-4 m
ComplianceQuest (Salesforce-native QMS)	$5-30K	$30K-180K	3-9 m
SoftExpert Suite (Excellence Management)	$5-25K	$20K-100K	2-6 m

Risk Assessment Software provides structured tools and methodologies to identify, analyze, evaluate, and mitigate potential risks across an organization's operations, projects, and strategic initiatives. These platforms enable users to categorize risks, assess their likelihood and impact, assign ownership, and track mitigation activities. By centralizing risk data and providing clear visibility into the risk landscape, Risk Assessment Software helps organizations proactively identify vulnerabilities, prioritize mitigation efforts, and make informed decisions to protect assets, ensure business continuity, and comply with regulatory requirements. This is critical for safeguarding operational integrity, minimizing financial losses, and fostering a risk-aware culture across the enterprise.

Without dedicated Risk Assessment Software, organizations often rely on ad-hoc methods, spreadsheets, and subjective judgments for identifying and managing risks. This leads to inconsistent risk assessments, missed vulnerabilities, a lack of centralized visibility into the overall risk landscape, and ineffective mitigation strategies. The absence of a structured and systematic approach to risk management hinders the ability to prioritize critical risks, allocate resources efficiently for mitigation, or demonstrate due diligence to stakeholders and regulators. This ultimately increases an organization's exposure to operational disruptions, financial losses, and reputational damage, making it challenging to maintain business continuity and achieve strategic objectives.

Implementation	Challenges
* Define risk categories & assessment methodology * Select a risk assessment software * Identify & document operational risks * Configure risk scoring & reporting * Establish risk mitigation workflows * Plan for continuous monitoring & review	* Ensuring consistent risk identification & scoring * Gaining full participation from risk owners * Integrating with GRC, QMS, and project systems * Handling subjective nature of risk assessment * Maintaining up-to-date risk registers * Demonstrating tangible risk reduction impact

Products	Platform Cost	Project Cost	Duration
MetricStream GRC (Risk Management)	$10-60K	$80K-500K	6-16 m
Archer (RSA Archer) GRC Platform	$10-70K	$100K-600K	6-18 m
LogicManager (Enterprise Risk Mgmt)	$5-30K	$20K-100K	2-6 m
HighBond (Audit, Risk, Compliance)	$5-25K	$20K-100K	2-6 m
ServiceNow GRC (Risk Management)	$10-50K	$70K-400K	5-14 m
Resolver (Risk Management Suite)	$5-30K	$30K-180K	3-9 m

6. SECURE AUDIT TRAILS

Secure Audit Trails are comprehensive, tamper-proof, and time-stamped records of all significant activities, changes, and access events within an organization's operational systems and data. These trails capture who did what, when, and where, providing an irrefutable log of operational events. Implemented through robust logging mechanisms, blockchain technologies, or specialized audit log management solutions, Secure Audit Trails are essential for regulatory compliance, forensic investigations, and maintaining data integrity. By providing verifiable and immutable records, they enable organizations to demonstrate accountability, detect unauthorized activities, reconstruct events accurately, and comply with strict data governance requirements, safeguarding the integrity and trustworthiness of all operational data.

Without Secure Audit Trails, organizations operate with limited visibility into system activities, data changes, and user actions, making it nearly impossible to trace events, detect unauthorized access, or conduct effective forensic investigations. The absence of immutable and verifiable logs leaves organizations vulnerable to undetected data tampering, non-compliance with regulatory requirements (e.g., GDPR, HIPAA, SOX), and difficulty in resolving disputes or demonstrating accountability. This ultimately impacts data integrity, exposes the organization to severe legal and financial penalties, and erodes trust among customers, partners, and regulators, fundamentally undermining the security and reliability of all operational data and processes.

Implementation	Challenges
* Identify critical systems & data for logging * Select an audit trail management solution * Implement robust logging mechanisms * Configure access controls & encryption * Establish log retention policies * Plan for regular monitoring & review	* Managing high volume of log data * Ensuring data integrity & tamper-proofing * Integrating with diverse operational systems * Securing audit trails from unauthorized access * Analyzing logs for anomalies & insights * Complying with diverse regulatory retention laws

Products	Platform Cost	Project Cost	Duration
Splunk Enterprise (Log Management)	$10-60K	$80K-500K	6-18 m
Elastic Stack (ELK - Logging, Analytics)	N/A	$10K-70K	1-4 m
IBM Security QRadar (SIEM)	$10-70K	$100K-600K	6-18 m
Sumo Logic (Cloud SIEM)	$5-40K	$50K-300K	4-12 m
ArcSight (Micro Focus - SIEM)	$10-50K	$70K-400K	5-14 m
LogicMonitor (IT Infrastructure Monitoring)	$2-15K	$10K-70K	1-4 m

SECTION E

BLUEPRINT SERIES

Execution Mechanics

Ai6x6™ Transformation Blueprint

Story 1 — The Manhattan Project Coordination Model (1942–1946): The Manhattan Project united scientists, engineers, and military leaders across dispersed sites under extreme secrecy and urgency. Workstreams were interdependent, requiring precise coordination without full visibility. Structured reporting, controlled interfaces, and disciplined compartmentalization enabled progress despite complexity. Human judgment guided trade-offs while machines and labs executed specialized tasks. The project demonstrated how execution succeeds when governance, clarity, and cadence align. It became a template for managing high-stakes, multi-disciplinary programs.

Story 2 — The Berlin Airlift Logistics System (1948–1949): With land routes blocked, Allied forces had to supply an entire city solely by air. Aircraft schedules, loading patterns, and turnaround times were optimized to the minute. Machines provided lift and speed, but humans orchestrated sequencing, prioritization, and contingency planning. Continuous feedback loops refined operations daily. The airlift proved that disciplined execution can overcome structural constraints. It showed how precision, rhythm, and adaptability create operational breakthroughs.

Story 3 — The Interstate Highway Construction Program (1956–1970s): The United States embarked on building a nationwide highway network requiring coordination across states, contractors, and engineering disciplines. Standardized designs ensured consistency while local teams adapted to terrain and conditions. Machines handled heavy construction, but humans managed sequencing, safety, and quality. The program relied on predictable funding, clear specifications, and phased execution. It demonstrated how large-scale delivery succeeds when structure and flexibility coexist. The highways became a symbol of disciplined national execution.

Story 4 — Toyota's Just-in-Time Logistics (1960s–1980s): As Toyota refined its production system, logistics became a critical execution layer. Parts arrived exactly when needed, minimizing inventory and waste. Machines tracked flow, but humans monitored signals, resolved disruptions, and maintained rhythm. Kanban created transparency across the value chain. The system required precision, trust, and continuous calibration. It showed how execution excellence emerges when every component of the workflow reinforces timing, quality, and flow.

Story 5 — FedEx's Overnight Delivery Network (1973–1990s): FedEx built a hub-and-spoke system capable of moving packages across the country overnight. Aircraft, sorting machines, and routing algorithms provided speed, but humans managed exceptions, weather disruptions, and customer commitments. The network operated on tight cycles that demanded flawless coordination. Data informed decisions while people ensured reliability. The system demonstrated how execution mechanics scale when tech and human oversight operate in synchronized cadence.

Story 6 — The Rise of Agile Software Delivery (2001–2015): Software teams struggled with long cycles, rigid plans, and late-stage failures. Agile introduced iterative delivery, cross-functional collaboration, and continuous feedback. Machines automated testing and deployment, while humans prioritized work, clarified requirements, and adapted to change. Execution shifted from linear phases to dynamic cycles. The approach improved predictability and responsiveness. It became a global standard for modern execution, proving that disciplined iteration outperforms rigid planning.

LESSONS FROM AN OPERATIONS TRANSFORMATION PERSPECTIVE

Operations mature when execution mechanics shift from fragmented activities to synchronized, end-to-end systems. Over time, organizations integrated planning, logistics, production, and delivery into cohesive workflows supported by data and automation. Machines provide speed, consistency, and monitoring, while humans manage exceptions, optimize flow, and uphold standards. Execution maturity emerges when operations maintain rhythm despite variability—absorbing shocks without losing performance. The organization becomes more resilient because processes are designed, measured, and continuously refined. Mature operations deliver stability at scale, enabling strategic goals to be achieved through disciplined, coordinated, and predictable execution.

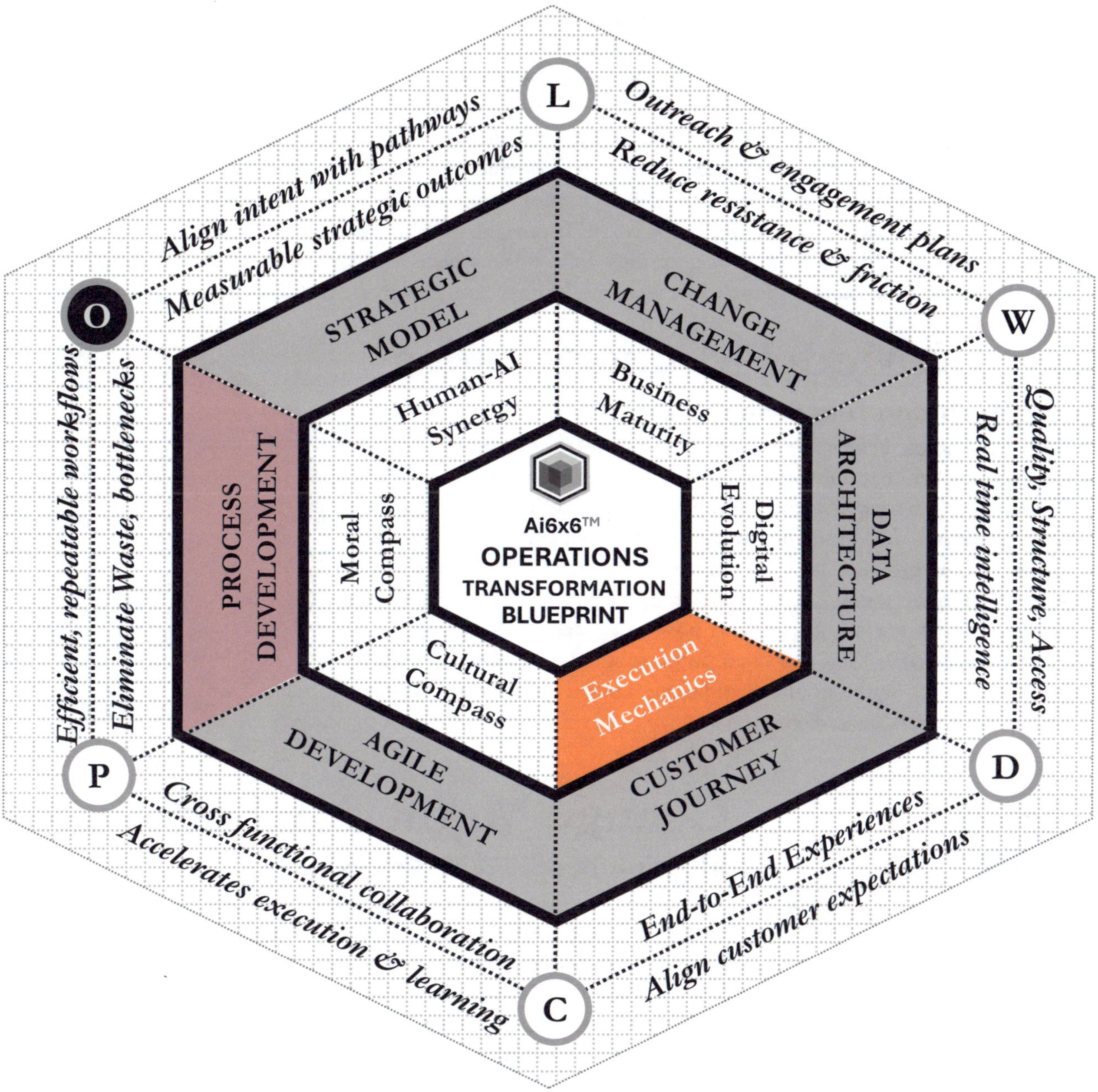

The **Ai6x6 Research Lab** identified this lens as the primary antidote to "operational drift" and the accumulation of waste. We chose it because the human mind requires a disciplined, repeatable system to navigate high-stakes demands like **Six Sigma** and **Lean Management** without losing the rhythm of quality. This lens isn't merely about the mechanics of **RPA Methodology**; it is about the psychological certainty and mental clarity that come from a proven methodology like the **SCOR Model** or the **Digital Twin Method**. It is vital because it provides a mental scaffolding, turning the chaos of a supply chain into a series of surgical, confident optimizations.

This lens allows the operator's heart to remain calm while the organization moves at terminal velocity through **BPM** and **Simulation**. By relying on **Kanban** principles and **Automation**, the leader replaces firefighting with evidence-based conviction. We believe that execution mechanics are the ultimate form of operational mindfulness—the practice of being fully present in the "how" of the flow so that the "what" of reliability is achieved with excellence. This lens ensures that every process has a defined purpose and a coherent structure, providing the steady footing necessary to scale globally with the assurance of a master optimizer.

Execution has always been the crucible where strategy proves its worth. Leaders can articulate bold visions, design elegant frameworks, and outline ambitious roadmaps, but none of it matters if the organization cannot execute with precision, coherence, and resilience. In the age of AI and accelerated change, **Execution Mechanics** becomes the lens that determines whether transformation becomes real or remains theoretical. It forces leaders to confront the uncomfortable truth that most failures are not failures of strategy but failures of execution. The gap between intention and impact is rarely philosophical; it is operational. Execution Mechanics exposes that gap and provides the architecture to close it. This lens reframes execution as a system—an interconnected set of methodologies that convert ambition into momentum. It recognizes that execution is not the final step of strategy but the environment in which strategy lives or dies. In a world where AI accelerates decision cycles, compresses timelines, and amplifies both strengths and weaknesses, leaders must design execution as they design vision.

Strategic Model Methodologies

Turning Ambition Into Structure

Execution Mechanics begins with **strategic model methodologies**, the structured approaches that translate ambition into actionable pathways. These methodologies are not academic exercises; they are the scaffolding that ensures strategy is grounded. Leaders who neglect this dimension find themselves improvising under pressure, relying on intuition rather than disciplined frameworks. Strategic models define the logic of execution—how choices are made, how priorities are sequenced, and how resources are allocated. They ensure that every initiative has a defined purpose, a coherent structure, and a measurable outcome. Without this foundation, execution becomes reactive, fragmented, and vulnerable to disruption.

Change Mgmt. Methodologies

Aligning People With Progress

The next dimension, **change management methodologies**, recognizes that execution is ultimately a human endeavor. No transformation succeeds unless people understand, accept, and commit to the change. Leaders who underestimate this reality discover that resistance is not a barrier but a predictable response to uncertainty. Effective change management acknowledges the emotional, cultural, and psychological dimensions of transformation. It ensures that communication is transparent, expectations are clear, and support is available. It turns change from something imposed into something shared. In this dimension, execution becomes relational—rooted in trust, empathy, and the ability to guide people toward possibility.

Data Architecture Methodologies

Building the Backbone of Execution

Data architecture provide the backbone for modern execution. In a world where decisions must be informed by evidence, organizations cannot afford fragmented data, inconsistent definitions, or unreliable pipelines. Data architecture is not a technical concern; it is a strategic one. It determines whether insights are trustworthy, whether systems can scale, and whether AI can operate responsibly. Leaders who invest in strong data architecture create the conditions for intelligence to flow freely across the enterprise. In this dimension, execution becomes intelligent—powered by clarity, consistency, and evidence.

Customer Journey Methodologies

Anchoring Execution in Value

Customer journey ensure that execution remains anchored in value creation. Too many organizations design processes around internal convenience rather than customer experience. Customer journey thinking forces leaders to see the organization from the outside in, identifying friction, gaps, and opportunities that internal perspectives overlook. It ensures that execution is not only efficient but meaningful. It aligns operations with the expectations, needs, and emotions of the people the organization serves. In this dimension, execution becomes empathetic—shaped by the reality of customers rather than the assumptions of leaders.

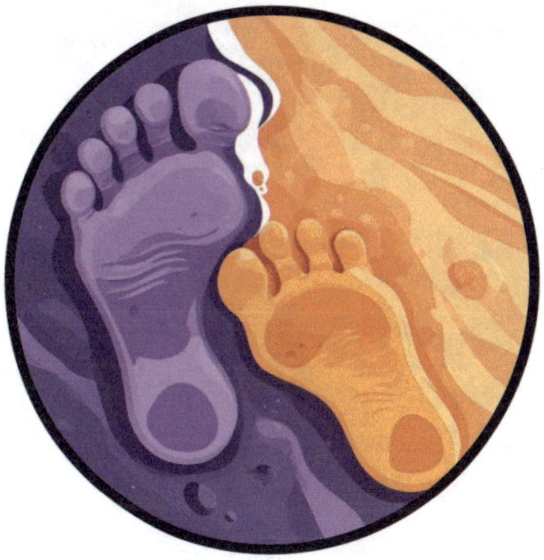

Agile Development Methodologies

Enabling Adaptation at Speed

Agile development introduce the adaptability required in a world where conditions shift rapidly. Traditional execution models assume stability, predictability, and linear progress. Agile assumes the opposite. It embraces iteration, experimentation, and continuous learning. It allows teams to adjust quickly, respond to feedback, and deliver value incrementally. Leaders who adopt agile mindsets create organizations that move with speed without sacrificing quality. In this dimension, execution becomes dynamic—capable of evolving in real time as new information emerges.

Process Improvement Methodologies

Sustaining Excellence Over Time

The final dimension, **process improvement methodologies**, completes the system by ensuring that execution evolves rather than stagnates. They provide the discipline to identify inefficiencies, eliminate waste, and refine workflows. They ensure that execution is not a one-time effort but a continuous practice. Leaders who embrace process improvement build organizations that get better over time, not simply bigger or busier. In this dimension, execution becomes self-correcting—driven by curiosity, discipline, and a commitment to excellence.

Execution Mechanics is the lens that transforms strategy into reality by ensuring that ambition is supported by structure, that change is supported by people, and that progress is supported by systems. It reveals that execution is not the final step of leadership but the continuous arena in which leadership is tested. In a world where AI accelerates both opportunity and exposure, leaders must design execution as deliberately as they design vision, building the operational muscle that allows the organization to move with coherence under pressure. This lens makes clear that execution is not a matter of effort but of engineering—of constructing the models, architectures, and behaviors that convert intention into repeatable performance. Execution becomes a living capability rather than a periodic push, a discipline that strengthens with use rather than erodes under stress.

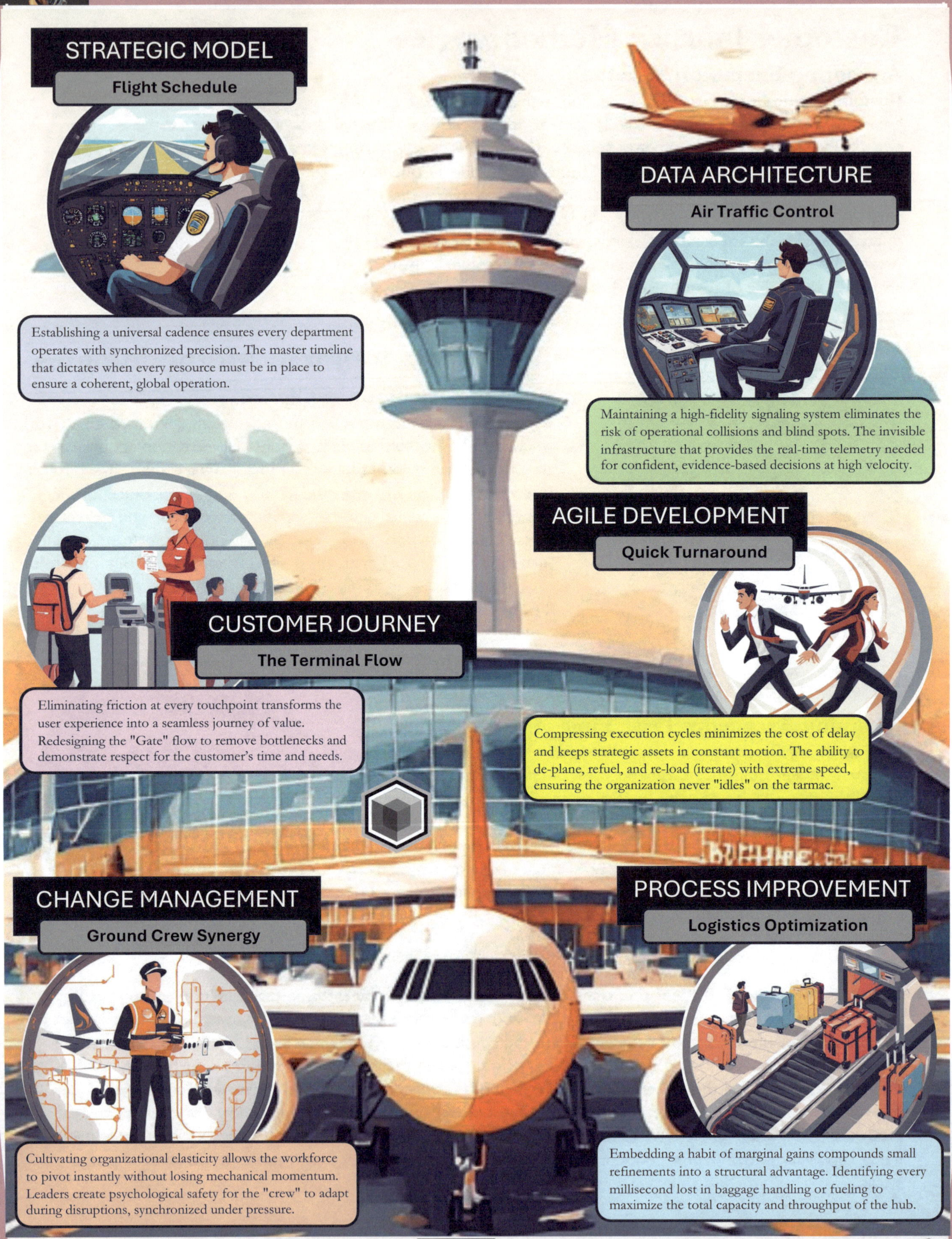

STRATEGIC MODEL
Flight Schedule

Establishing a universal cadence ensures every department operates with synchronized precision. The master timeline that dictates when every resource must be in place to ensure a coherent, global operation.

DATA ARCHITECTURE
Air Traffic Control

Maintaining a high-fidelity signaling system eliminates the risk of operational collisions and blind spots. The invisible infrastructure that provides the real-time telemetry needed for confident, evidence-based decisions at high velocity.

AGILE DEVELOPMENT
Quick Turnaround

Compressing execution cycles minimizes the cost of delay and keeps strategic assets in constant motion. The ability to de-plane, refuel, and re-load (iterate) with extreme speed, ensuring the organization never "idles" on the tarmac.

CUSTOMER JOURNEY
The Terminal Flow

Eliminating friction at every touchpoint transforms the user experience into a seamless journey of value. Redesigning the "Gate" flow to remove bottlenecks and demonstrate respect for the customer's time and needs.

CHANGE MANAGEMENT
Ground Crew Synergy

Cultivating organizational elasticity allows the workforce to pivot instantly without losing mechanical momentum. Leaders create psychological safety for the "crew" to adapt during disruptions, synchronized under pressure.

PROCESS IMPROVEMENT
Logistics Optimization

Embedding a habit of marginal gains compounds small refinements into a structural advantage. Identifying every millisecond lost in baggage handling or fueling to maximize the total capacity and throughput of the hub.

COMMON FRAMEWORK - TOGAF (The Open Group Architecture Framework)

Book 1: Leadership Transformation—Strategic Direction : The frameworks in this pillar ensure that all efforts are strategically aligned and that leaders are focused on the right outcomes for value creation.

1. **McKinsey 7-S Framework:** Aligns the organization's strategy with its structure, systems, staff, and skills to ensure all components support the desired transformation.
2. **Business Model Canvas:** Creates a shared understanding of how value will be created, delivered, and captured, providing a clear map for all teams.
3. **McKinsey Three Horizons of Growth:** Structures innovation efforts across short-term improvements (Horizon 1), scaling existing successes (Horizon 2), and future-proofing the business (Horizon 3), ensuring a sustainable pipeline of value.
4. **Balanced Scorecard:** Translates the strategic vision into measurable performance objectives, connecting operational activities to high-level financial and non-financial goals.
5. **SWOT Analysis:** Provides a structured way to evaluate the strategic context, enabling leaders to leverage strengths, mitigate weaknesses, seize opportunities, and counter threats.
6. **Blue Ocean Strategy:** Drives leaders to look for uncontested market space, shifting the focus from competing for value to creating new value.

Book 2: Workforce Transformation—Adoption and Engagement : This pillar addresses the human element of change, ensuring that employees are prepared, engaged, and able to sustain new ways of working.

1. **Kotter's 8-Step Change Model:** Provides a step-by-step process for leading major change, from creating a sense of urgency to solidifying new approaches in the corporate culture.
2. **ADKAR Model:** Guides change managers in ensuring that individual employees have the Awareness, Desire, Knowledge, Ability, and Reinforcement needed for successful change.
3. **Lewin's Change Management Model:** Conceptualizes change as a three-stage process: "Unfreeze" existing habits, implement the "Change," and "Refreeze" the new processes to make them stick.
4. **Bridges' Transition Model:** Focuses on the psychological journey of individuals, helping them navigate the emotional stages of an ending, a neutral zone, and a new beginning.
5. **Prosci Change Management Methodology:** Provides a structured, research-based methodology for applying change management to any project or initiative.
6. **The Satir Change Model:** Helps leaders understand and anticipate the emotional and performance impacts of change on individuals and teams.

Book 3: Data Transformation—Measurement and Insights - This pillar provides the tools and architectures to manage data as a strategic asset, ensuring that value can be reliably measured and tracked.

1. **Zachman Framework:** Offers a comprehensive, logical structure for classifying and organizing enterprise architecture, ensuring data management is part of the larger business strategy.
2. **DAMA-DMBOK:** Provides a set of best practices for managing data throughout its lifecycle, ensuring the quality and integrity of data used for value measurement.
3. **Data Mesh:** Promotes a decentralized, domain-driven approach to data ownership, empowering teams to create and manage data as a product.
4. **Lambda Architecture (or Kappa Architecture):** Establishes a technical architecture for processing large volumes of data efficiently, supporting real-time value tracking and analytics.
5. **Master Data Management (MDM) Frameworks:** Ensure that an organization's critical business data is consistent, accurate, and provides a "single source of truth" for reliable reporting.
6. **Data Fabric:** Creates a flexible, interconnected data architecture that supports seamless access and integration of data, regardless of its location.

Book 4: Customer Transformation—External Value Delivery: This pillar focuses on defining and delivering value from the customer's perspective, ensuring that transformation efforts translate to measurable customer outcomes.

1. **Customer Journey Mapping:** Visualizes the customer's end-to-end experience, identifying pain points and opportunities for value creation.
2. **Service Blueprinting:** Expands on journey mapping by detailing the front-stage and backstage processes, people, and systems that support a customer experience.
3. **Jobs to be Done (JTBD) Framework:** Helps define "value" from the customer's perspective by focusing on the functional and emotional outcomes they seek.
4. **Design Thinking Methodology:** Provides a human-centered approach to innovation that focuses on empathizing with customers to design meaningful solutions.
5. **CX (Customer Experience) Maturity Models:** Provides a benchmark for tracking and improving customer experience over time, linking strategic initiatives to better customer outcomes.
6. **Persona Development Frameworks:** Creates detailed profiles of target users, ensuring that transformation efforts are grounded in a deep understanding of customer needs.

Book 5: Product Transformation—Internal Value Delivery : This pillar establishes an agile, iterative process for continuously building and improving the products and services that deliver customer value.

1. **Scrum:** Enables cross-functional teams to deliver value incrementally in short sprints, allowing for continuous feedback and adaptation.
2. **Kanban:** Helps manage and visualize workflow, limiting work-in-progress and improving the flow of value to the customer.
3. **Lean Startup:** Applies a "build-measure-learn" feedback loop to product development, ensuring teams are constantly validating their assumptions about value creation.
4. **Scaled Agile Framework (SAFe):** Provides a system for coordinating and aligning multiple agile teams across the enterprise to deliver large-scale solutions.
5. **Feature Flags / Toggles Architecture:** Enables teams to deploy new features and capabilities quickly and safely, allowing for real-time testing of value propositions.
6. **Continuous Integration / Continuous Delivery (CI/CD) Pipeline Architecture:** Automates the software delivery process, allowing teams to deliver new value to customers efficiently and reliably.

Book 6: Operations Transformation—Efficiency and Sustainability
This pillar focuses on optimizing the internal processes and systems that support the entire value chain, ensuring efficiency and sustainability.

1. **Lean Management:** Aims to eliminate waste and optimize the flow of value through the organization's processes, generating cost savings and efficiency gains.
2. **Six Sigma (DMAIC Methodology):** Uses a data-driven approach to improve processes by defining, measuring, analyzing, improving, and controlling them.
3. **BPM (Business Process Management) Lifecycle:** Provides a structured approach to continuously improving business processes, ensuring sustained value delivery over the long term.
4. **SCOR Model (Supply Chain Operations Reference):** Offers a standard framework for managing and improving supply chain efficiency, a key driver of operational value.
5. **Digital Twin Architecture:** Creates virtual models of physical assets or processes, enabling simulation and optimization before real-world changes are made.
6. **Robotic Process Automation (RPA) Frameworks/Methodologies:** Automates repetitive, rule-based tasks, freeing up human resources for higher-value activities.

TOGAF (The Open Group Architecture Framework) is a widely recognized and comprehensive enterprise architecture methodology that provides a structured approach for developing and managing enterprise architecture. It helps organizations design, plan, implement, and govern an enterprise information architecture. At its core is the Architecture Development Method (ADM), a detailed, iterative process for developing various architectural domains: Business, Data, Application, and Technology. For leadership transformation, particularly in achieving digital maturity, TOGAF is indispensable. Digital transformation is not just about implementing new technologies; it requires aligning business strategy with IT capabilities across the entire enterprise. Without a coherent architecture, digital initiatives can become fragmented, leading to integration challenges, redundant systems, security vulnerabilities, and a lack of scalability. TOGAF provides the necessary rigor to ensure that digital investments are strategically aligned, well-integrated, and support the organization's long-term digital vision. It helps leaders understand the current state of their digital landscape, define the desired future state (e.g., cloud-native infrastructure, API-driven integration, AI-enabled applications), and create a roadmap to get there. By fostering a common language and framework for architects and stakeholders, TOGAF enables effective communication and collaboration across complex digital initiatives. It ensures that the digital transformation is not just a series of isolated projects but a cohesive, well-governed evolution of the entire enterprise, a critical component of achieving and sustaining comprehensive digital maturity. It provides the blueprint for how an organization's digital capabilities (applications, data, infrastructure) support its strategic business goals, ensuring that every technological investment contributes to a unified, adaptable, and future-ready digital enterprise. This structured approach helps avoid costly misalignments and ensures that digital initiatives are built on a solid, scalable foundation, allowing the organization to truly harness the power of technology for competitive advantage.

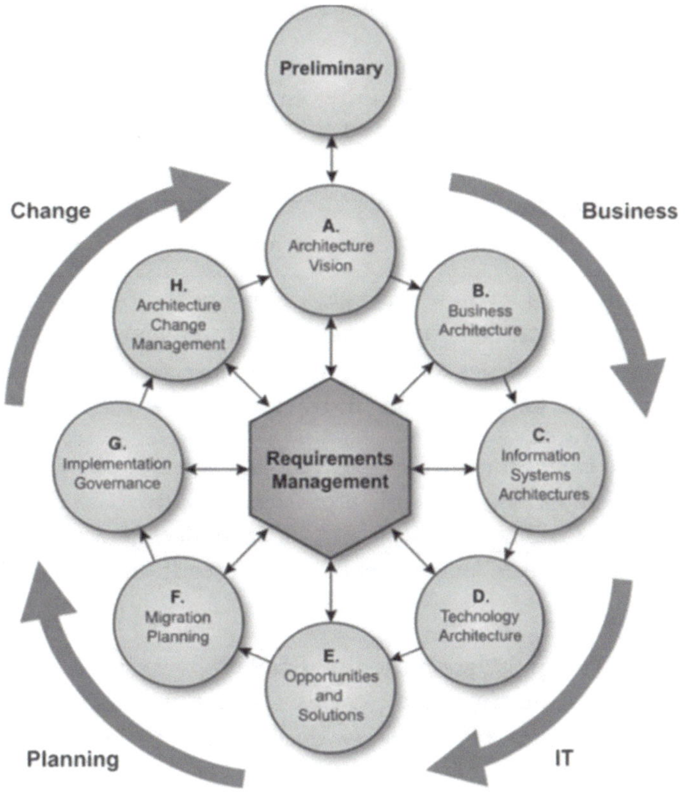

TOGAF Frameworks, Image source: medium.com

1. **Preliminary Phase: Framework and Principles:** Establish the architecture capability for the enterprise. This involves defining the organizational context for architecture work, identifying key stakeholders, establishing the architecture principles (e.g., "cloud-first," "API-driven," "data-driven decision making"), and defining the scope and governance of the architecture effort.

2. **Phase A: Architecture Vision:** Develop a high-level vision for the target architecture that aligns with business strategy and digital transformation goals. This involves identifying key stakeholders, defining the business value of the architecture work, and securing approval for the architecture development cycle.

3. **Phase B: Business Architecture:** Develop a baseline (current state) and target (future state) Business Architecture that describes the organization's processes, functions, and organizational structure. This ensures that the digital transformation efforts are aligned with and support the core business capabilities and strategic objectives, identifying where digital enablement can optimize business value, streamline workflows, and enhance customer interactions through digital means.

4. **Phase C: Information Systems Architectures (Data & Application):** Develop baseline and target Data Architecture (logical and physical data models, data governance, data flows) and Application Architecture (applications, their interactions, and how they support business functions). This phase is critical for designing integrated digital systems, ensuring data flows seamlessly across digital platforms, and supporting new digital applications that drive business value and customer engagement.

5. **Phase D: Technology Architecture:** Develop baseline and target Technology Architecture that describes the logical software and hardware capabilities required to support the Business, Data, and Application components. This includes defining infrastructure (e.g., cloud platforms, edge computing), platforms (e.g., microservices, AI/ML platforms), and security standards necessary for a robust, scalable, and resilient digital enterprise.

6. **Phase E: Opportunities & Solutions:** Identify and evaluate various implementation options and opportunities to deliver the target architecture. This involves assessing gaps between baseline and target architectures, identifying potential solutions (e.g., Commercial Off-The-Shelf (COTS), custom development, cloud migration, SaaS adoption), and defining high-level implementation projects. For digital maturity, this prioritizes digital solutions that offer the greatest business value and strategic alignment.

7. **Phase F: Migration Planning:** Develop a detailed implementation and migration plan. This involves prioritizing projects, estimating costs, defining a timeline, and outlining the sequence of activities required to transition from the baseline to the target architecture. This ensures a structured and managed approach to digital transformation implementation, minimizing disruption and maximizing efficiency.

8. **Phase G: Implementation Governance:** Oversee the implementation of the architecture. This involves establishing governance processes to ensure that projects adhere to the defined architecture, managing architectural contracts, and providing architectural guidance during development and deployment. This continuous oversight is vital for maintaining architectural integrity and consistency throughout the digital transformation.

9. **Phase H: Architecture Change Management:** Establish processes for managing changes to the architecture. This ensures that the architecture remains relevant and responsive to evolving business needs, technological advancements (e.g., new AI capabilities, quantum computing), and market shifts.

10. **Requirements Management:** This phase runs concurrently throughout all ADM phases. It involves managing the architectural requirements, ensuring they are consistent, complete, and aligned with stakeholder needs. For digital maturity, this ensures that evolving digital requirements (e.g., real-time data access, scalability for new digital services) are captured and integrated into the architectural design and implementation.

Ai6x6™ Transformation Blueprint

Process Improvement Frameworks

"If you can't describe what you are doing as a process, you don't know what you're doing."

W. Edwards Deming. The vital need for process definition, standardization, and discipline as the prerequisite for optimization, automation, or successful application of digital technology.

Process Improvement Frameworks (like Lean, Six Sigma, or Theory of Constraints) are the structured methodologies essential for optimizing enterprise operations and maximizing the return on digital investment. These frameworks provide a **disciplined, data-driven approach** (e.g., Six Sigma's DMAIC cycle) to identify, analyze, and eliminate waste, bottlenecks, and variation within core business processes. They are critical for Operations Transformation because they ensure processes are stabilized and efficient *before* being automated or augmented with AI. A robust framework provides the necessary governance and metrics to quantify improvements, ensuring changes are sustained and aligned with strategic business goals. By systematically driving efficiency, these models drastically reduce operational costs, improve quality, and create the standardized, lean processes required for the successful implementation of high-scale automation and the seamless integration of new digital technologies.

The **Process Improvement Framework** is dedicated to enhancing efficiency and sustainability by meticulously optimizing the internal processes and systems that underpin the entire organizational value chain. The core principle of Lean Management is to systematically identify and eliminate all forms of waste, thereby streamlining workflow and generating tangible efficiency gains and cost savings. Complementing this is Six Sigma, which employs a rigorous, data-driven approach, typically using the DMAIC Methodology (Define, Measure, Analyze, Improve, Control), to refine processes and reduce variability. For sustained optimization, the BPM (Business Process Management) Lifecycle provides a structured, continuous approach to managing and improving business processes over the long term. Specific operational areas also utilize specialized frameworks, such as the SCOR Model (Supply Chain Operations Reference), which standardizes and improves the efficiency of supply chain operations. Technological advancements play a key role: Digital Twin Architecture allows for the creation of virtual models for simulation and optimization prior to real-world implementation, while Robotic Process Automation (RPA) Frameworks/Methodologies automate repetitive, rule-based tasks, strategically redirecting human talent toward higher-value activities.

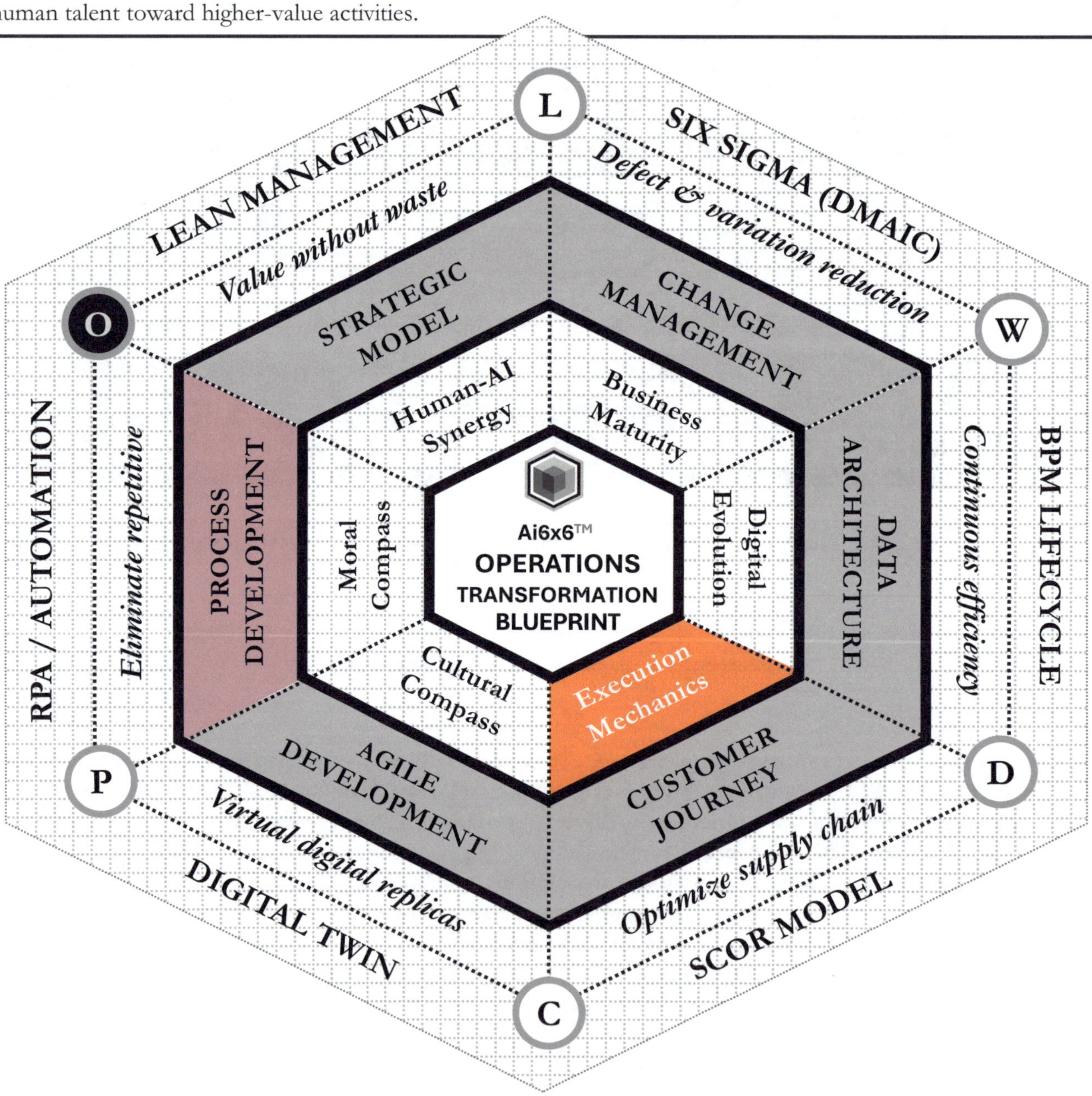

Ai6x6 EXECUTION MECHANICS FRAMEWORK FOR **OPERATIONS LIFECYCLE**

Lean Management is a philosophy and set of practices focused on maximizing customer value while systematically eliminating waste (Muda), inconsistency (Mura), and overburden (Muri) in processes. Originating from Toyota's production system, its core principle is "value without waste."

For operations transformation, Lean Management is foundational because it provides a systematic approach to identifying and removing inefficiencies that hinder operational excellence. Digital maturity in operations is not just about automating processes; it's about automating optimized processes. Applying Lean principles before or during digital transformation ensures that you're not just digitizing waste. It helps organizations streamline workflows, reduce lead times, improve quality, and enhance overall operational efficiency. By mapping value streams, identifying non-value-adding activities, and fostering a culture of continuous improvement (Kaizen), Lean empowers frontline teams to identify and solve problems. This leads to more agile and responsive operations, which can then be further enhanced by digital technologies like RPA, AI, or IoT. Ultimately, Lean Management provides the discipline to continuously optimize digital operational workflows, ensuring that technology investments yield maximum value and that the organization's operations are truly streamlined and adaptable, a hallmark of operational digital maturity.

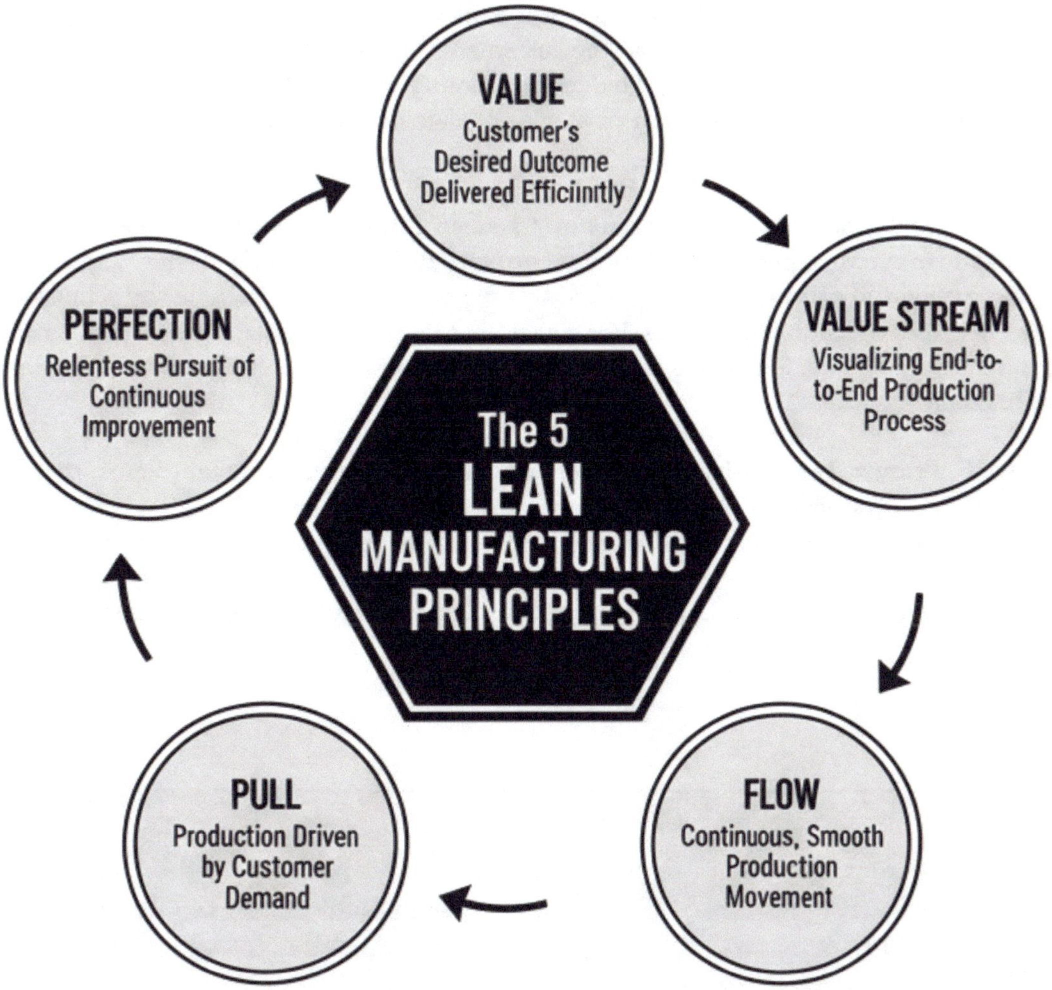

Lean Manufacturing Principles

1. **Identify Value:** Begin by clearly defining what constitutes "value" from the ultimate customer's perspective. This involves understanding their needs, expectations, and what they are willing to pay for. Any activity that does not contribute to this defined value is considered waste. In a digital operations context, this means identifying what truly adds value to the digital service delivery or automated process, eliminating unnecessary digital steps or data processing.

2. **Map the Value Stream:** Visually map all the steps involved in delivering a specific product or service to the customer, from beginning to end. This "value stream map" identifies all activities, information flows (including digital data flows), and resources involved, highlighting both value-adding and non-value-adding activities (waste). For digital operations, this often reveals bottlenecks in digital workflows or redundant data entry across systems.

3. **Create Flow:** Eliminate interruptions, bottlenecks, and waste identified in the value stream to ensure a smooth, continuous flow of work. This involves streamlining processes, reducing handoffs (especially manual ones between digital systems), and optimizing the sequence of activities to minimize delays and improve throughput. In digital operations, this might mean automating process steps or integrating disparate digital systems.

4. **Establish Pull:** Implement a "pull" system where work is only initiated when there is demand from the next step in the process or directly from the customer. This prevents overproduction and ensures that resources are only used when needed, reducing inventory and lead times. For digital operations, this means avoiding unnecessary processing or storage of data until it's actively required, optimizing digital resource utilization.

5. **Seek Perfection (Continuous Improvement / Kaizen):** Foster a culture of continuous improvement (Kaizen) where everyone in the organization is encouraged to systematically identify and eliminate waste, improve quality, and enhance efficiency. This is an ongoing journey of learning and refinement, striving for ever-greater perfection in processes. For digital operations, this translates to continuous optimization of automated workflows, identifying opportunities for further digital enhancement, and refining the use of digital tools.

6. **Empower Teams:** Decentralize decision-making and empower frontline teams to identify and implement improvements within their areas of control. Provide them with the necessary training, tools (including digital analytics tools), and autonomy to analyze their processes, solve problems, and drive continuous improvement initiatives, fostering a sense of ownership and engagement. This empowerment is crucial for achieving digital maturity, as it enables rapid, localized digital process optimization.

> "Simple can be harder than complex: You have to work hard to get your thinking clean to make it simple. But it's worth it in the end because once you get there, you can move mountains."
> — Steve Jobs, Co-founder, Apple

Six Sigma is a rigorous, data-driven methodology for improving process quality by identifying and eliminating the causes of defects and minimizing variation. It aims to achieve "six sigma" levels of quality, meaning only 3.4 defects per million opportunities. The core methodology is DMAIC (Define, Measure, Analyze, Improve, Control).

For operations transformation, Six Sigma is invaluable because it provides a structured, scientific approach to solving complex operational problems and achieving significant quality improvements. Digital maturity in operations requires not just speed and automation, but also high reliability and minimal errors in digital processes. Six Sigma ensures that digital operations are not only efficient but also consistently accurate. It uses statistical analysis to pinpoint root causes of errors in automated workflows, system integrations, or data processing. By systematically measuring performance, analyzing data, and implementing targeted improvements, organizations can significantly reduce defects in their digital operations, leading to higher customer satisfaction, reduced costs, and increased efficiency. This data-driven discipline, coupled with a focus on process consistency and error reduction, is a hallmark of highly mature digital operations that can deliver reliable, high-quality services at scale.

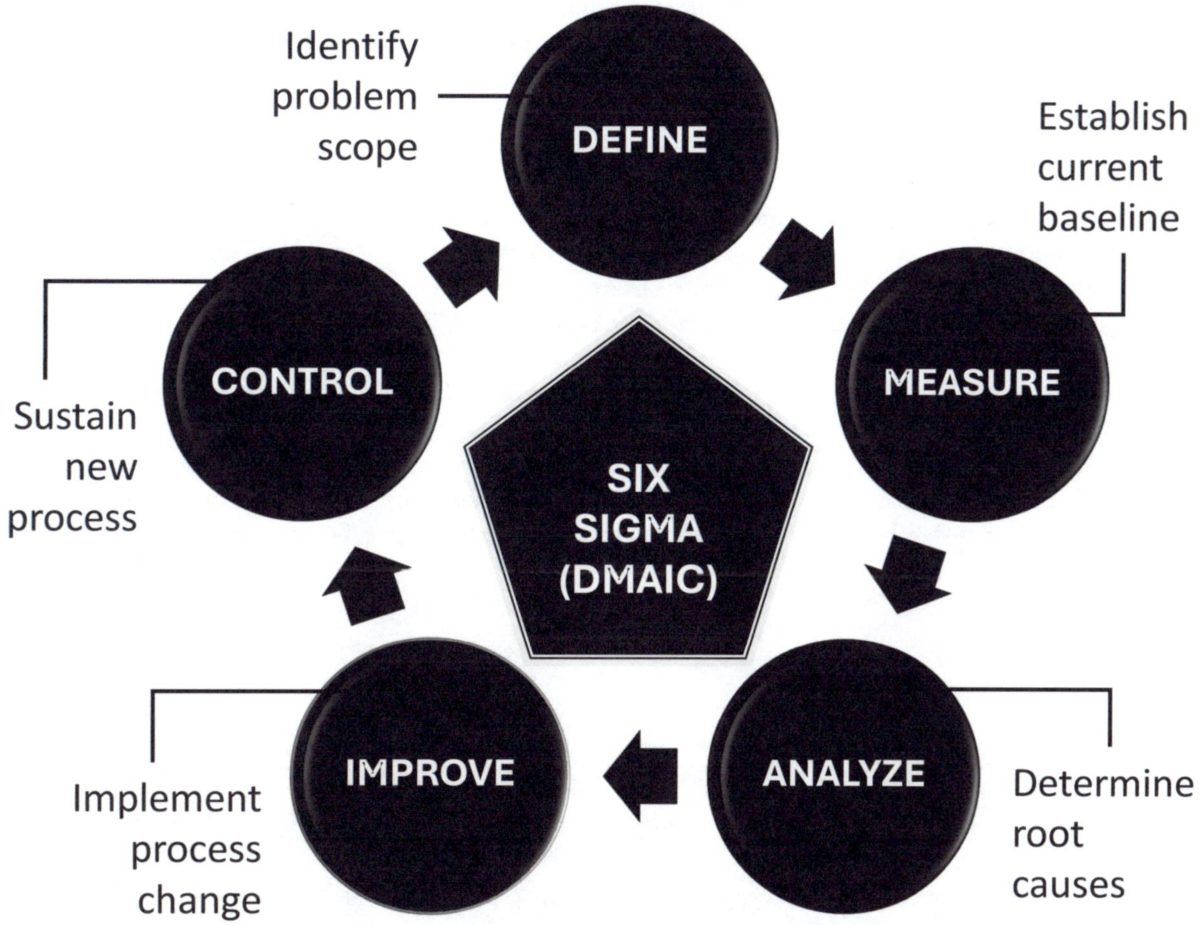

Six Sigma methodology

1. **Define**: Clearly define the problem, the project goals, and the customer (internal or external) requirements for the process. This involves understanding the scope of the process to be improved, its boundaries, and the specific outcomes that need to be achieved. For digital operations, this might involve defining a problem like "high error rate in automated invoice processing" or "slow response time for digital customer inquiries."

2. **Measure**: Collect comprehensive data on the current process performance. This step involves quantifying the problem, establishing baselines for key metrics (e.g., defect rate, cycle time, error frequency), and identifying relevant process inputs and outputs. For digital operations, this means gathering data from digital logs, system performance metrics, and automated process audits to precisely measure current digital process performance.

3. **Analyze**: Analyze the collected data to identify the root causes of defects, inefficiencies, or variation in the process. This phase uses various statistical tools (e.g., regression analysis, hypothesis testing, process mapping) and data visualization techniques to uncover the underlying reasons for poor performance. For digital operations, this might involve analyzing logs to find software bugs, system integration issues, or data quality problems that lead to digital process failures.

4. **Improve**: Develop and implement solutions to eliminate the identified root causes. This often involves process redesign, automation (e.g., implementing RPA for repetitive tasks), new technology implementation (e.g., upgrading a digital platform), or changes to procedures. Solutions are typically piloted and tested on a smaller scale to ensure their effectiveness before full-scale implementation, focusing on digital solutions that directly address the root causes.

5. **Control**: Establish controls and monitoring mechanisms to sustain the improvements over time and prevent regression to the old process. This includes standardizing the new process, updating documentation, training employees on new digital workflows, implementing ongoing performance monitoring (e.g., automated dashboards, real-time alerts), and establishing clear responsibilities for maintaining the improved state. This ensures digital process stability.

6. **Validate Results**: Verify that the implemented improvements have achieved the desired outcomes and delivered the planned benefits. This involves comparing post-implementation data against the baselines established in the "Measure" phase and ensuring that the problem has been effectively resolved and the process is operating at the desired quality level. For digital operations, this means confirming that digital processes are consistently performing with fewer errors and higher efficiency, contributing to overall digital maturity.

> "Be a yardstick of quality. Some people aren't used to an environment where excellence is expected."
> — Steve Jobs, Co-founder, Apple

Business Process Management (BPM) is a discipline that involves continuously analyzing, designing, executing, monitoring, and optimizing business processes. It's not a one-time project but an ongoing lifecycle aimed at achieving organizational efficiency, agility, and alignment with strategic goals.

For operations transformation, BPM is critical because it provides a structured approach to understanding and improving the very fabric of how work gets done. Digital maturity in operations is largely about automating and optimizing processes. BPM ensures that organizations don't just automate existing, inefficient processes, but rather redesign them for maximum effectiveness before applying technology. It helps identify bottlenecks, redundant steps, and opportunities for automation (e.g., using RPA, workflow automation platforms). By modeling processes visually (e.g., with BPMN), it creates a shared understanding across business and IT, facilitating seamless digital implementation. Continuous monitoring of process performance (e.g., cycle times, error rates) allows for data-driven optimization. Whether through incremental improvements or radical re-engineering, BPM ensures that operational processes are continuously refined to leverage digital capabilities, leading to greater efficiency, faster service delivery, and enhanced responsiveness to market changes. This systematic approach to process excellence is a cornerstone of achieving and sustaining operational digital maturity.

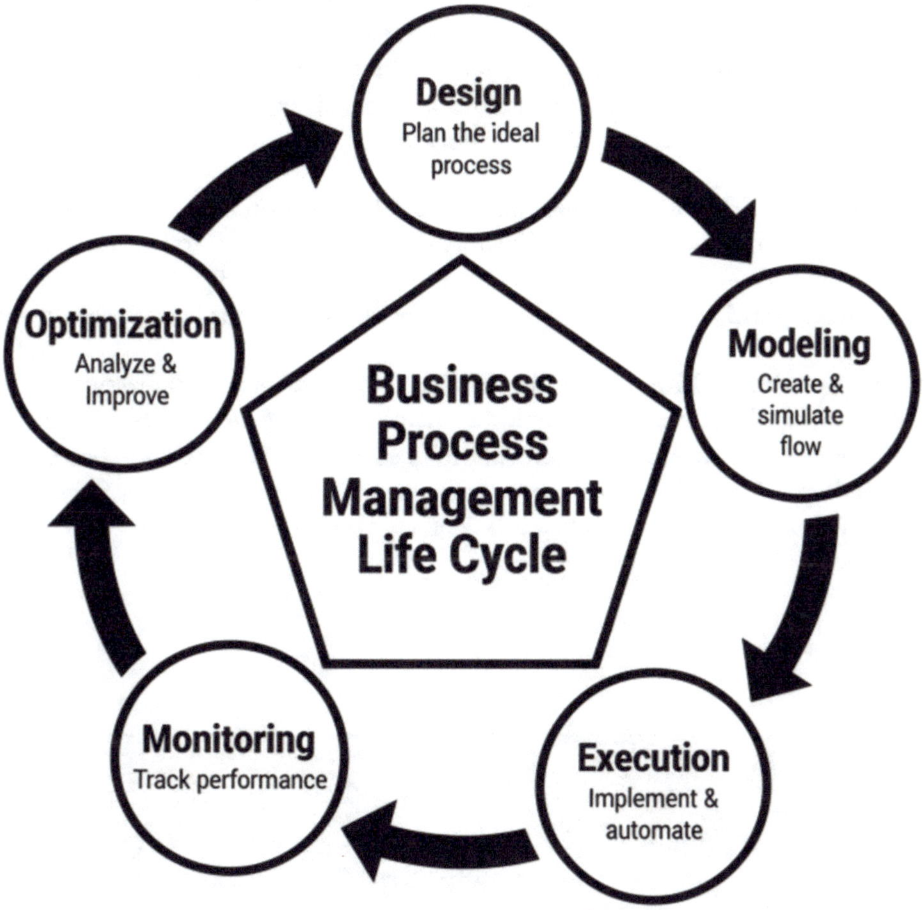

BPM Lifecycle

1. **Process Design:** Model and analyze existing "as-is" processes to understand their current state, identifying inefficiencies, bottlenecks, and areas for improvement. Based on this analysis, design optimized "to-be" processes that are streamlined, more efficient, and better aligned with strategic business goals, often incorporating new technologies or automation. For digital operations, this involves designing processes that leverage digital tools, AI, and automation to maximize efficiency and responsiveness.

2. **Process Modeling:** Document the designed processes visually using standard notation (e.g., Business Process Model and Notation - BPMN). This creates a clear, unambiguous representation of the process flow, roles, and decision points, ensuring shared understanding across all stakeholders (business, IT, operations). In a digitally mature context, these models serve as blueprints for automating workflows and integrating digital systems.

3. **Process Execution:** Implement the designed processes. This can involve configuring Business Process Management (BPM) software to automate workflows, developing new applications, integrating systems (e.g., connecting CRM to ERP), or providing training to employees on new manual workflows and procedures. The goal is to bring the "to-be" process to life, often through digital enablement, ensuring smooth operational flow.

4. **Process Monitoring:** Continuously track the performance of the executed processes against defined Key Performance Indicators (KPIs) and Service Level Agreements (SLAs). This involves collecting real-time data on process cycle times, error rates, resource utilization, and compliance, providing insights into operational effectiveness. For digital operations, automated monitoring tools provide immediate feedback on the health and efficiency of digital workflows.

5. **Process Optimization:** Analyze the monitoring data to identify further opportunities for improvement. This might involve minor tweaks and adjustments (often referred to as Kaizen or continuous improvement) to enhance efficiency, reduce costs, or improve quality. It's an iterative refinement based on performance data, constantly seeking ways to leverage digital tools for better process outcomes.

6. **Process Re-engineering (Optional/As Needed):** If incremental optimization is insufficient to achieve desired results, fundamentally rethink and redesign core processes from scratch. This "clean slate" approach aims for dramatic improvements in cost, quality, service, and speed, often leveraging disruptive technologies (e.g., AI, blockchain) and challenging long-held assumptions about how work should be done. This is a significant step towards achieving breakthrough digital operational maturity.

> "The first rule of any technology used in a business is that automation applied to an efficient operation will magnify the efficiency. The second is that automation applied to an inefficient operation will magnify the inefficiency."
> — Bill Gates, Co-founder, Microsoft

The Supply Chain Operations Reference (SCOR) model is a cross-industry standard framework for analyzing, designing, and optimizing supply chain management processes. It provides a common language and a structured approach to understanding and improving the entire supply chain, from supplier's supplier to customer's customer. SCOR breaks down supply chain operations into five primary management processes: Plan, Source, Make, Deliver, and Return, with an additional "Enable" category for supporting processes.

For operations transformation, particularly in achieving digital maturity, SCOR is invaluable. Modern supply chains are incredibly complex and often fragmented, hindering efficiency and responsiveness. Digital maturity in supply chain operations means leveraging technologies like IoT, AI, blockchain, and advanced analytics for real-time visibility, predictive capabilities, and automated decision-making. SCOR provides the blueprint to identify where these digital capabilities can be most effectively applied. It helps pinpoint bottlenecks in digital information flow, optimize automated logistics, enhance digital collaboration with suppliers, and improve the efficiency of digital manufacturing processes. By standardizing processes and metrics across the supply chain, SCOR enables organizations to measure and improve their digital supply chain performance, leading to greater agility, resilience, and cost-effectiveness. This structured approach to digital supply chain optimization is a clear indicator of operational digital maturity.

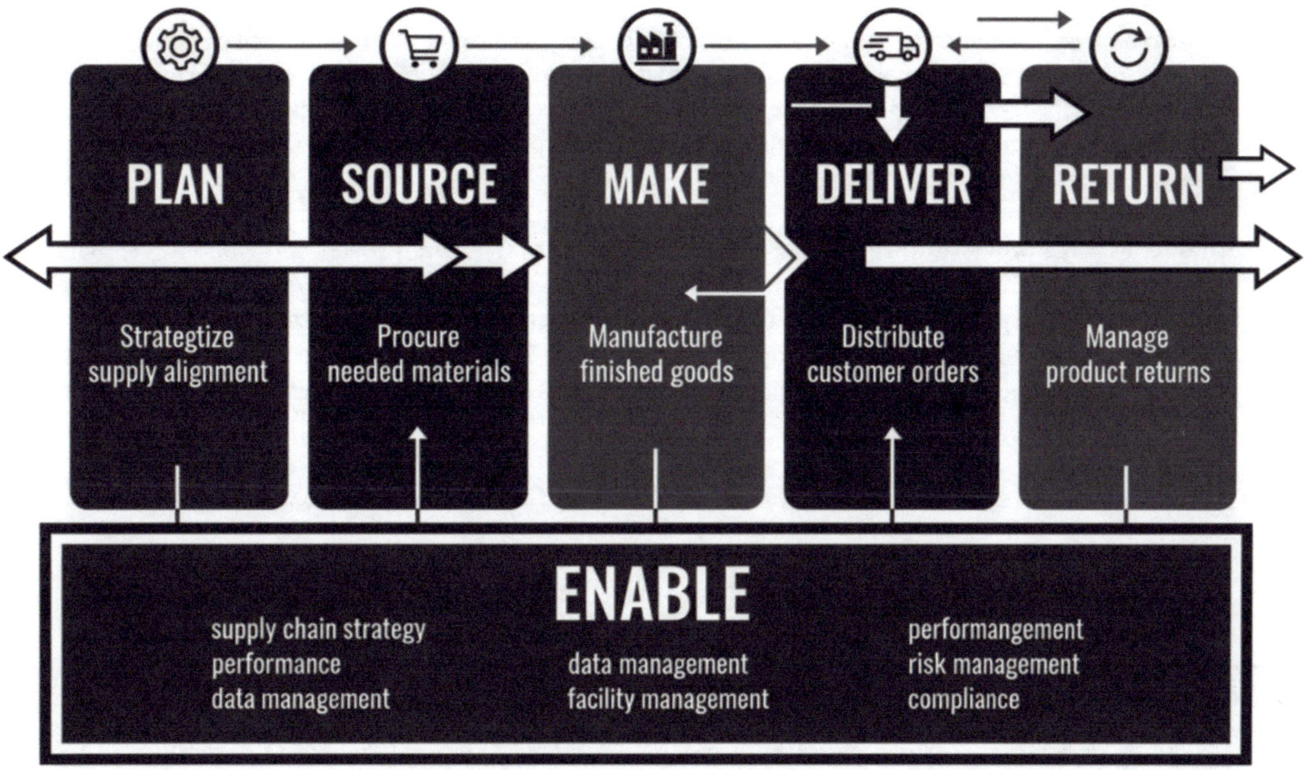

SCOR model

1. **Plan**: Develop strategies for managing all resources that align with demand and supply across the entire supply chain. This includes demand planning, inventory planning, capacity planning, and supply chain network design. In a digitally mature supply chain, this phase heavily relies on advanced analytics, AI-driven forecasting, and real-time data from IoT devices to create highly accurate and responsive plans.

2. **Source**: Procure goods and services to meet planned or actual demand. This covers all activities related to sourcing raw materials, components, and finished products, including supplier selection, contract management, purchasing, quality management, and managing incoming material inventory. For digital maturity, this involves leveraging digital platforms for supplier collaboration, automated procurement, and blockchain for supply chain transparency and traceability.

3. **Make**: Transform products to a finished state to meet planned or actual demand. This includes all manufacturing and production activities, such as production scheduling, actual manufacturing or assembly, quality control, and managing work-in-process inventory. In a digitally mature context, this involves smart factories, automated production lines, and real-time monitoring of manufacturing processes using IoT and AI.

4. **Deliver**: Manage orders, transportation, and distribution of finished goods and services to customers. This covers order management, warehouse management (picking, packing, shipping), transportation management, and installation/delivery to the end customer. Digital maturity in this area means optimizing logistics with AI, real-time tracking, automated warehousing, and last-mile delivery solutions.

5. **Return**: Manage the return of products from customers (e.g., customer returns, defective products, end-of-life products) or from suppliers (e.g., excess materials, defective components). This includes return authorization, inspection, disposition (repair, recycle, dispose), and managing reverse logistics. Digital solutions can streamline this process with automated return portals, predictive analytics for returns, and efficient digital tracking.

6. **Enable**: Manage processes that support and enable the other five core processes (Plan, Source, Make, Deliver, Return). This includes supply chain strategy, performance management, data management, risk management, facility management, and compliance. These are the overarching management processes that ensure the entire supply chain operates effectively, increasingly relying on integrated digital platforms and data analytics for comprehensive oversight and continuous improvement.

> "You've got to have the world's best supply chain. If you don't have it, you're not going to be competitive."
> — Tim Cook, CEO, Apple

Digital Twin Architecture involves creating a virtual replica (a "digital twin") of a physical asset, process, or system. This digital model is continuously updated with real-time data from its physical counterpart (via sensors, IoT devices) and can be used for monitoring, simulation, prediction, and optimization.

For operations transformation, the Digital Twin is a powerful enabler of process improvement, directly contributing to operational digital maturity. Traditional operational management is often reactive, responding to failures after they occur. Digital Twins enable a shift to proactive and even prescriptive management. By simulating "what-if" scenarios in the virtual environment, organizations can test process changes, predict equipment failures (predictive maintenance), and optimize operational parameters without disrupting physical operations. This leads to significant improvements in efficiency, uptime, and cost reduction. For digital maturity, Digital Twin Architecture provides unprecedented visibility into complex operations, allowing for real-time insights into the performance of physical assets and automated processes. It integrates data from various digital sources (IoT, ERP, MES) to create a comprehensive operational view, enabling AI-driven optimization and autonomous control. This ability to continuously monitor, analyze, and optimize operations in a virtual environment before implementing changes in the physical world is a hallmark of highly advanced and digitally mature operational capabilities.

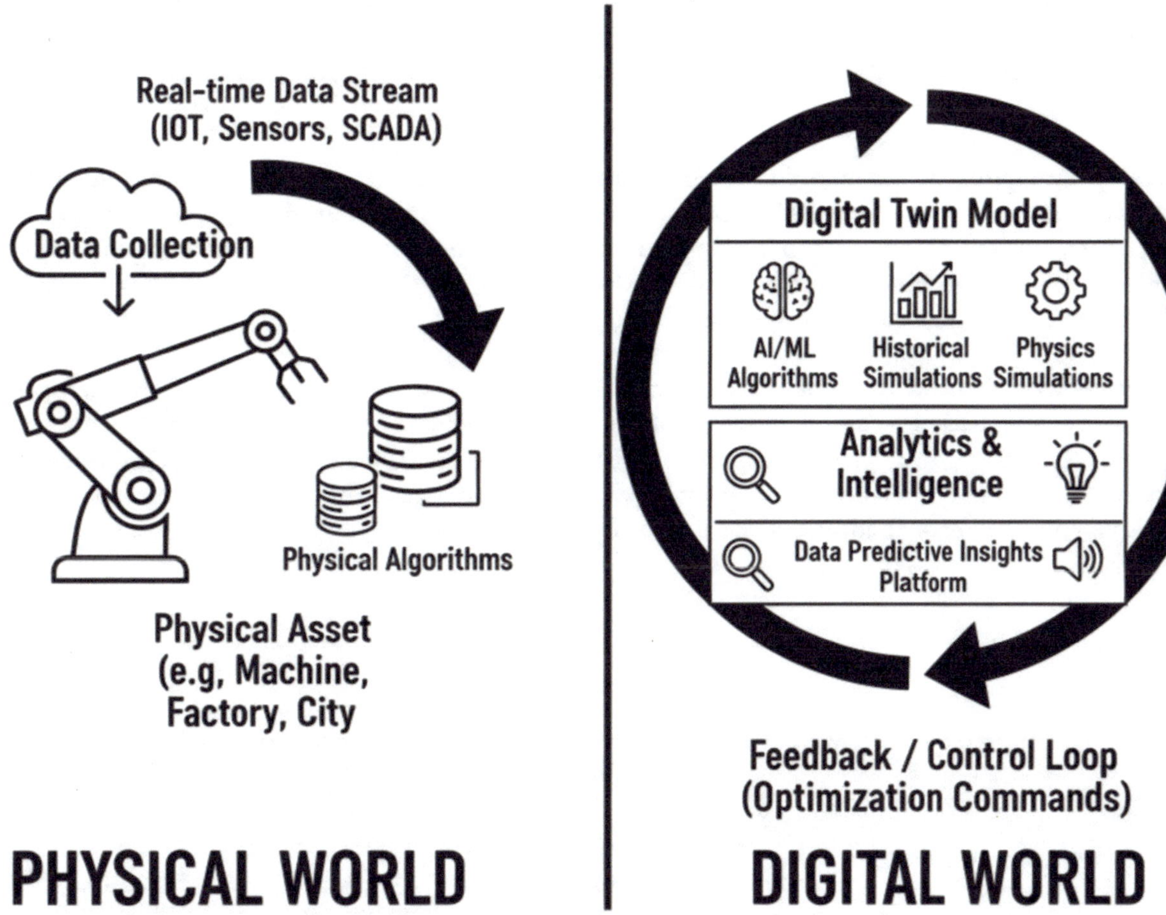

Digital Twin Architecture

1. **Define Scope & Use Case:** Identify the specific physical asset, process, or system that will be "twinned" (e.g., a manufacturing machine, an entire factory, a logistics network, a building's HVAC system, or even a complex business process). Clearly define the business problem or operational improvement it will solve (e.g., predictive maintenance, energy optimization, process simulation, quality control). This step aligns the digital twin initiative with strategic operational goals.

2. **Sensor & Data Collection:** Deploy a network of sensors on the physical asset/process to collect real-time operational data. This data can include temperature, pressure, vibration, energy consumption, throughput rates, environmental conditions, and more. Robust data acquisition systems are essential to feed the digital twin, ensuring a continuous stream of accurate information from the physical world into the digital realm.

3. **Build Digital Model:** Create a virtual, mathematical, or simulation model that accurately represents the physical counterpart's behavior, characteristics, and dynamics. This model can range from simple data visualizations to complex physics-based simulations, depending on the required fidelity and use case. This digital representation is the core of the twin, allowing for analysis and manipulation without affecting the physical asset.

4. **Connect Physical & Digital:** Establish robust and secure data pipelines to continuously stream real-time data from the physical asset to its digital twin. This ensures that the digital model is always synchronized with the physical world, providing an up-to-date and accurate representation. This real-time data flow is critical for dynamic insights and responsive operational decisions in a digitally mature environment.

5. **Develop Analytics & Simulation:** Integrate advanced analytics, AI/ML models, and simulation capabilities into the digital twin. This allows for deriving deep insights from the data, predicting future states (e.g., equipment failure, process bottlenecks), optimizing operational parameters, and testing "what-if" scenarios in the virtual environment without impacting the physical system. This predictive capability is a significant leap towards operational digital maturity.

6. **Integrate with Control Systems:** Enable actionable insights and recommendations from the digital twin to be sent back to control systems for autonomous adjustments or to human operators for informed decision-making. This "closed-loop" capability allows the digital twin to actively influence and optimize the physical operation, moving from reactive to proactive and even prescriptive operational management, a hallmark of advanced digital maturity.

> "The world is moving to simulation first."
> — Jensen Huang, CEO, NVIDIA

Robotic Process Automation (RPA) involves using software robots ("bots") to automate repetitive, rule-based tasks typically performed by humans interacting with digital systems. These bots can mimic human actions, such as clicking, typing, copying, and pasting data across various applications.

For operations transformation, RPA is a powerful tool for process improvement, directly contributing to operational digital maturity. Many organizations still rely on manual, tedious tasks that are prone to human error and inefficiency, even within otherwise digital workflows. RPA addresses this by automating these "swivel-chair" processes, freeing up human workers to focus on higher-value, more complex, and creative tasks. It leads to significant improvements in process speed, accuracy, and cost reduction. While RPA is often seen as a tactical automation solution, its strategic importance for digital maturity lies in its ability to quickly streamline existing digital operational workflows, bridging gaps between disparate legacy systems without requiring complex and costly system integrations. This rapid automation capability allows organizations to achieve quick wins in their digital transformation journey, building momentum and demonstrating tangible benefits. By automating routine tasks, RPA enables operations to scale more efficiently, reduce operational risk, and provide faster, more consistent digital services, pushing the organization further along its path to operational digital maturity.

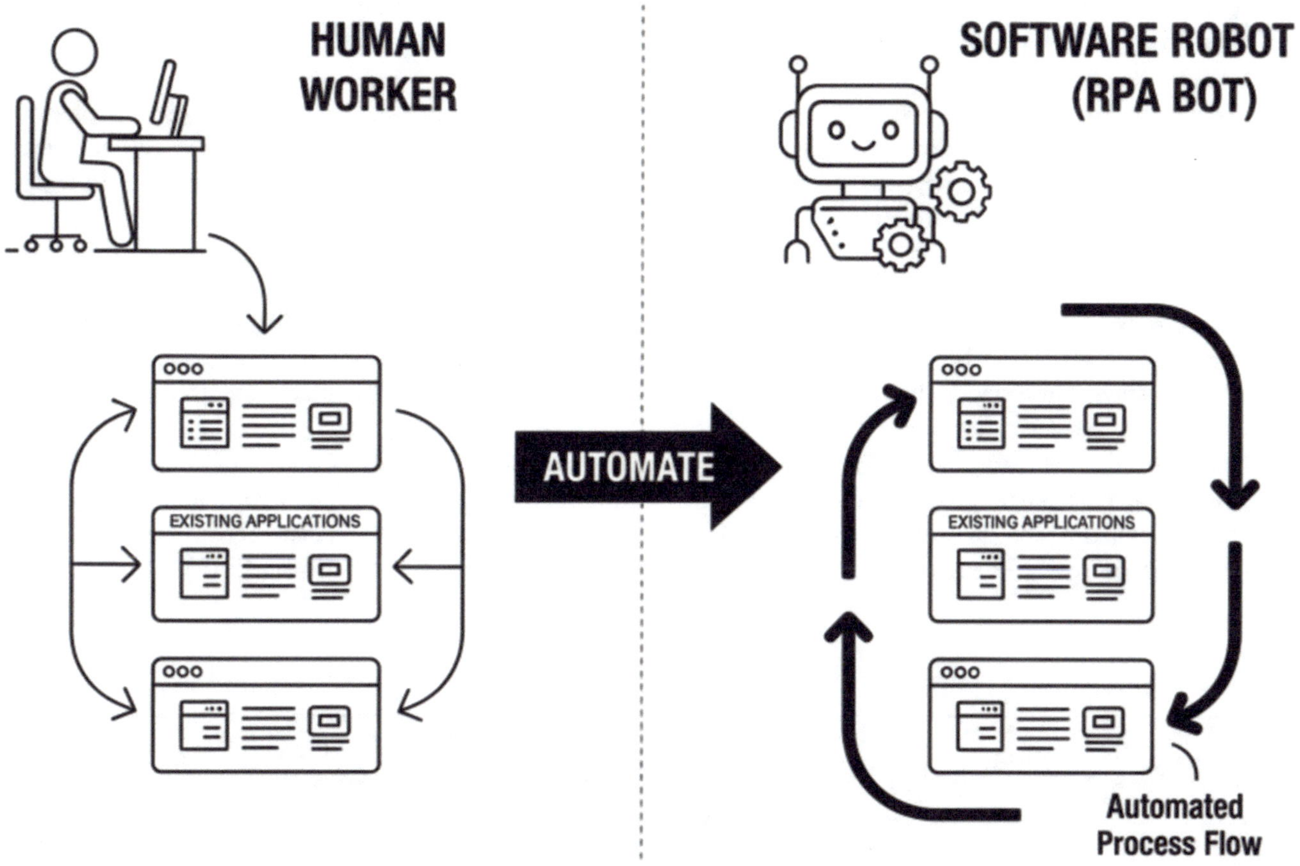

:RPA Lifecycle

1. **Process Identification & Assessment:** Identify business processes that are suitable candidates for RPA. Look for processes that are high volume, repetitive, rule-based, stable, and involve digital inputs (e.g., data entry, report generation, invoice processing, customer onboarding). Assess their potential ROI (Return on Investment) and impact on operational efficiency, ensuring that automation efforts target areas with maximum benefit for digital transformation.

2. **Process Analysis & Documentation:** Map the "as-is" process in granular detail, identifying all steps, decision points, exceptions, and system interactions. Then, design and document the "to-be" automated process, outlining precisely how the RPA bot will interact with applications and data. This detailed documentation is crucial for accurate bot development and ensures clarity and consistency in the automated workflow.

3. **Bot Design & Development:** Design the automation logic and build the RPA bot using specialized RPA software (e.g., UiPath, Automation Anywhere, Blue Prism). This involves configuring the bot to mimic human actions, such as clicking buttons, entering data into fields, copying information between systems, and navigating applications. The goal is to create a robust and efficient digital worker.

4. **Testing:** Thoroughly test the RPA bot in a controlled environment (e.g., a test server, a dedicated sandbox) to ensure accuracy, robustness, and adherence to business rules. Test various scenarios, including exceptions and error handling, to minimize unexpected behavior in production. This rigorous testing is vital to ensure the reliability and integrity of automated digital processes.

5. **Deployment:** Implement the RPA bot into the production environment. This often involves a phased rollout, starting with a pilot to confirm stability and benefits before scaling up across the organization. Proper infrastructure setup, security considerations, and integration with existing IT systems are crucial during deployment to ensure seamless operation.

6. **Monitoring & Maintenance:** Continuously monitor the bot's performance, identify any failures, exceptions, or errors, and provide ongoing maintenance and updates. As underlying systems or business processes change, the bots will need to be adapted to ensure they continue to function correctly and deliver value. This ongoing management is vital for the long-term success of digital automation.

7. **Performance Measurement:** Track key metrics to demonstrate the value of automation. This includes measuring cost savings, reduction in processing time, improvement in accuracy, reduction in manual effort, and overall impact on operational efficiency and employee satisfaction. These metrics provide clear evidence of how RPA contributes to the organization's operational digital maturity and ROI.

> "A lot of people are under the misconception that automation is just about replacing people. No. Automation is fundamentally about making the process better, more efficient, safer."
> — Elon Musk, CEO, SpaceX & Tesla

Ai6x6™ Transformation Blueprint

The Friction-to-Flow Ratio (FFR)

"Efficiency is doing things right; Effectiveness is doing the right things." --- Peter Drucker, the "Father of Management" revolutionized organizational health by focusing on output rather than just activity. For **Operations**, the FFR measures the transition from "Fragile Efficiency" to **Planetary Resilience**. Drucker's distinction reminds operations leaders that "Flow" isn't about moving fast in the wrong direction; it's about removing the systemic friction that prevents a company from doing "the right things" for the planet.

In the industrial era, "Operational Excellence" was achieved through a obsession with Control. We built complex hierarchies of middle management to "check the checker," creating a massive network of human safeguards to prevent error. This model assumed that the primary risk to a system was a lack of oversight. But in the AI era, the machine provides the oversight at a speed and precision no human can match. The new risk is not "Lack of Control"—it is Friction. This is the Identity Chasm of operations: the tension between Manual Oversight and Autonomous Flow. When an operation is technically optimized by AI but remains sluggish because humans insist on manually verifying every step, it is suffering from Control Illusion. The primary metric for this evolution is the Friction-to-Flow Ratio (FFR).

The operational control chasm is the gap between **Manual Oversight** and **Autonomous Flow**. It is the "Control Illusion" where managers believe that "touching the process" makes it safer, even when manual intervention introduces human error and lag. Identifying this chasm is the key to achieving "Frictionless Scale." It allows operators to stop being "cogs in the machine" and start being "architects of the system," moving the focus from the friction of function to the harmony of purpose.

The greatest barrier to operational transformation is the Safety-Through-Slowness Bias. Managers have been conditioned to believe that "Manual Intervention" equals "Lower Risk." They view a frictionless, autonomous process as "Out of Control," even if the data proves it is safer and more efficient. This leads to the "Human Drag" effect, where automated systems are throttled to match the speed of the slowest human reviewer.

To bridge this chasm, the operator must undergo a psychological shift: realizing that their value is no longer in executing the process, but in designing the system that executes itself. This is the move from being a "Manager of Tasks" to being an "Architect of Flow." You must trust that the "Logic" of the machine can handle the "Function," so that you can exercise your Moral Authority over the "Exceptions."

THE RAISON D'ÊTRE:

"TO BELONG"

By mastering the FFR, the operations team fulfills its true purpose: To Belong. In a high-friction environment, people feel like "cogs in a machine"—disposable parts of a mechanical whole. When the "Manual Drag" is removed, the human is elevated to the designer. They no longer belong to the process; they belong to the purpose. We are not removing the human from the loop; we are removing the drudgery that prevents the human from feeling a sense of ownership over the outcome.

DEFINING THE FFR: THE LOGIC OF AUTONOMY

The Friction-to-Flow Ratio is a measurement of Operational Transparency. It asks one simple, objective question: How much of our operational time is spent in manual intervention versus autonomous flow?

In a legacy operation, the FFR is high. For every hour of "Work" done, there is an hour of "Checking" or "Approval." In a high-velocity culture, the FFR is low. The machine handles the standard, high-probability events, while the human only touches the system when it hits a "Moral or Safety Anomaly." The goal is to reach a 1:10 ratio, where human energy is focused exclusively on the 10% of cases that require true Moral Authority.

THE OBJECTIVE: TOWARD FRICTIONLESS SCALE

The goal of the FFR intervention is to reach a state of Equilibrium we call Frictionless Scale. In this state, the organization can grow its output exponentially without growing its headcount proportionally.

Frictionless Scale allows the organization to breathe. By lowering your FFR, you aren't just "cutting costs"—you are "clearing the path." You are allowing the "Logic" of the machine to handle the "Friction," so the "Heart" of the operator can handle the "Flow." This is the only way to build an operation that can handle the infinite complexity of the AI-driven market without breaking the humans who run it.

SAMPLE CASE STUDIES ON CONTROL CHASM

CASE STUDY	THE CHASM REALITY	ACTION & RESULT
1: A GLOBAL LOGISTICS HUB faced massive delays because dispatchers were "Manual Monitorers." Even though AI optimized truck routes in real-time, dispatchers would manually "tweak" them every morning to accommodate familiar roads.	The team was clinging to **Control Illusion**. They believed their manual adjustments made the system "better," when in fact every human touch introduced new delays and broke the machine's optimized flow.	They ran **Shadow Mode Testing**, showing a side-by-side comparison of "Human" vs. "Machine" routes. Once the data proved the machine was 22% faster, they removed manual editing, achieving a 99% on-time rate.
2: A NATIONAL PHARMACY CHAIN struggled with inventory shortages for critical medicines. Store managers were hoarding stock "just in case," causing a massive imbalance where some stores were empty and others were overstocked.	This was **Operational Siloing**. Managers were optimizing for their own store's safety rather than the system's "Flow," creating a "Bullwhip Effect" of artificial scarcity across the entire supply chain.	Leadership implemented **Autonomous Replenishment**. The system began auto-balancing stock between stores without manager approval. Stockouts dropped by 45% while total inventory costs fell by $15M.
3: A MID-SIZED AIRLINE saw its "Turnaround Time" (cleaning and fueling between flights) skyrocket. The bottleneck was a series of 12 manual "Check-offs" required from various ground crews before a plane could be cleared.	The airline was suffering from **Redundant Node Syndrome**. Each manual "click" in the chain added 3 minutes of lag, creating a "Compounding Delay" that resulted in thousands of missed flight connections.	They moved to **Event-Driven Orchestration**, using IoT sensors to signal when a task was done. This removed the need for manual check-offs, cutting turnaround time by 18 minutes per flight and saving millions.
4: A HEAVY EQUIPMENT RENTAL FIRM was losing revenue because 20% of its fleet was always "In Repair." Maintenance was purely reactive; they waited for a machine to smoke before sending a mechanic.	This was **Reactive Rigidity**. The operations team viewed maintenance as a "Cost Center" to be minimized rather than a "Flow Driver," ignoring the high cost of equipment downtime and lost rental days.	They shifted to **Predictive Maintenance**, using AI to dispatch mechanics *before* a failure occurred. Fleet availability rose to 96%, and emergency repair costs were slashed by 50% within the first year.
5: A GLOBAL E-COMMERCE GIANT struggled with "Last-Mile" delivery costs. Their warehouse was highly automated, but the "Hand-off" to local drivers was a chaotic manual process involving paper manifests and verbal instructions.	The firm hit the **Digital-Physical Wall**. They had automated the "Thinking" (sorting) but failed to automate the "Moving" (the hand-off), resulting in a massive bottleneck at the warehouse door.	They implemented **Dynamic Loading Slots**, where AI signaled drivers to specific bays at exact times via a mobile app. This "Frictionless Handoff" reduced driver wait times by 70% and cut delivery costs by 12%.
6: A UTILITY COMPANY found its emergency response times lagging during storms. Crews were dispatched via manual phone calls, a process that became overwhelmed during high-volume events.	The company was trapped in **Legacy Latency**. They were using an "Analog Communication" model for a "Digital Speed" problem, causing a complete system breakdown exactly when the system was needed most.	They launched **Autonomous Dispatching**, where the AI analyzed outage data and sent GPS coordinates directly to the nearest crew's tablet. Response times dropped by 55%, significantly improving public safety.

ASSIGNMENT : THE OPERATIONS CULTURE DIAGNOSTIC AUDIT

The "Identity Chasm" in operations is most visible when a team is "Checking the Checker." You see it in processes that require three signatures for a routine action that the AI has already validated. To bridge this gap, we must diagnose The Control Illusion—the belief that human hands always make a process safer. This audit is designed to expose where "Red Tape" is being disguised as "Risk Management." On the left are the Symptoms of a high-friction operation. On the right are the Strategic Interventions required to recalibrate for Equilibrium.

The Assignment: Identify the "Manual Anchors" in your daily workflow and match them to the flow-centric remedy.

THE SYMPTOMS (THE DRAG)	MATCH	THE INTERVENTIONS (THE REMEDY)
1. The Checker's Paradox: Hiring more people to audit the work of the people who are auditing the AI.		**A. Exception-Only Management:** Shifting human focus exclusively to "Edge Cases" that the AI cannot statistically solve.
2. Manual Override Fever: Staff constantly "tweaking" the AI's optimized routes or schedules based on "gut feeling."		**B. Threshold Governance:** Pre-defining the "Safe Zones" where the machine has absolute autonomy to execute.
3. Approval Congestion: High-speed autonomous insights sitting in an "Inbox" waiting for a human signature.		**C. Negative Consent Flow:** Actions are taken automatically unless a human manually "pauses" the stream within a window.
4. The Shadow Process: Employees maintaining a secret, manual way of doing things because they "don't trust the system."		**D. Radical Simplification:** Auditing every operational step and deleting any that don't add "Moral or Creative Value."
5. Latency Normalization: A culture where "it takes a week to get that done" is accepted, despite the data being ready in seconds.		**E. Flow-Based KPIs:** Measuring managers on "Cycle Time Reduction" and "Autonomy Ratio" rather than "Output."
6. The Safety Shield: Using "Compliance" as an excuse to avoid automating a process that is actually just boring or difficult.		**F. The Trust Audit:** Quantifying the cost (in time and money) of every manual "Check" to reveal the true "Friction Tax."
7. Reporting Bloat: Spending hours creating manual reports to explain data that everyone already sees on the live dashboard.		**G. Automated Transparency:** Ensuring the AI's "Logic" is visible to all, removing the need for "Explainer" meetings.
8. Heroic Intervention: Celebrating the manager who "saved the day" through manual effort, rather than the one who built a self-healing system.		**H. System-Designer Incentives:** Rewarding employees who identify and remove friction from their own roles.
9. Data Silo Friction: Information "stuck" in one department that must be manually "requested" and "formatted" by another.		**I. Unified Operational Fabric:** Creating a single, AI-managed data stream that flows across all departmental lines.
10. The Status-Quo Anchor: "We've always done it this way" being used to block a 10x improvement in cycle time.		**J. Moral Authority Mapping:** Clearly defining the rare moments where human "Heart" must override machine "Logic."

ANSWER MATCH: 1-A, 2-B, 3-C, 4-F, 5-E, 6-D, 7-G, 8-H, 9-I, 10-J

BLUEPRINT SERIES

SECTION F

Cultural Compass

Ai6x6™ Transformation Blueprint

Story 1 — The Post-War Japanese Industrial Ethos (1945–1960): Japan emerged from the war with limited resources but a collective commitment to national rebuilding. Companies embraced a culture of discipline, apprenticeship, and shared responsibility. Workers identified deeply with their organizations, reinforcing loyalty and craftsmanship. Leaders cultivated humility and long-term thinking. This cultural foundation enabled rapid industrial recovery. It demonstrated how shared values can accelerate capability even before systems mature.

Story 2 — NASA's Mission Culture During Apollo (1961–1972): NASA built a culture defined by precision, accountability, and collective purpose. Engineers and astronauts operated with a shared belief that failure was unacceptable, yet learning was essential. Communication was open, direct, and cross-disciplinary. Rituals, reviews, and norms reinforced psychological safety and technical rigor. This culture enabled teams to navigate uncertainty and complexity. It became a global example of mission-driven cohesion.

Story 3 — The Civil Rights Movement's Organizing Ethos (1954–1968): The movement relied on a culture of discipline, nonviolence, and moral clarity. Communities trained in de-escalation, collective action, and strategic communication. Leaders modeled courage and restraint, shaping norms that guided behavior under extreme pressure. Shared values created unity across diverse groups. This cultural coherence amplified impact far beyond organizational structure. It showed how culture can mobilize change at societal scale.

Story 4 — The Nordic Social Trust Model (1970s–1990s): Nordic countries cultivated cultures rooted in transparency, equality, and civic trust. Institutions operated with high accountability and low hierarchy. Citizens expected fairness and contributed to collective welfare. This cultural fabric enabled strong public services, stable governance, and economic resilience. Trust reduced friction in decision-making and collaboration. It demonstrated how cultural norms can become strategic national assets.

Story 5 — The Rise of Silicon Valley's Innovation Culture (1980s–2000s): Silicon Valley developed a culture that celebrated experimentation, risk-taking, and rapid iteration. Informal networks, open information flow, and meritocratic ideals shaped behavior. Failure was reframed as learning, attracting talent willing to push boundaries. Venture capital reinforced speed and ambition. This culture accelerated technological breakthroughs and entrepreneurial growth. It showed how cultural norms can amplify innovation beyond structural advantages.

Story 6 — The Globalization of Corporate Diversity & Inclusion (2000s–2020s): Organizations worldwide recognized that diverse teams outperform homogeneous ones when supported by inclusive cultures. Companies invested in psychological safety, equitable processes, and representation. Employees gained platforms to voice perspectives and challenge norms. Inclusion became a driver of creativity, retention, and ethical integrity. This cultural shift reshaped leadership expectations and organizational identity. It demonstrated how culture evolves to meet societal and competitive demands.

LESSONS FROM AN OPERATIONS TRANSFORMATION PERSPECTIVE

Operational maturity depends on cultural norms that shape discipline, coordination, and resilience. Over time, organizations learned that processes succeed when culture reinforces consistency, accountability, and continuous improvement. Mature cultures encourage teams to surface problems early, collaborate across boundaries, and uphold standards even under pressure. Machines provide structure, but culture determines how people respond to disruptions, ambiguity, and risk. Operations become more reliable when cultural expectations guide behavior without constant supervision. Maturity is achieved when culture and process reinforce each other—creating operations that are stable, adaptive, and grounded in shared responsibility.

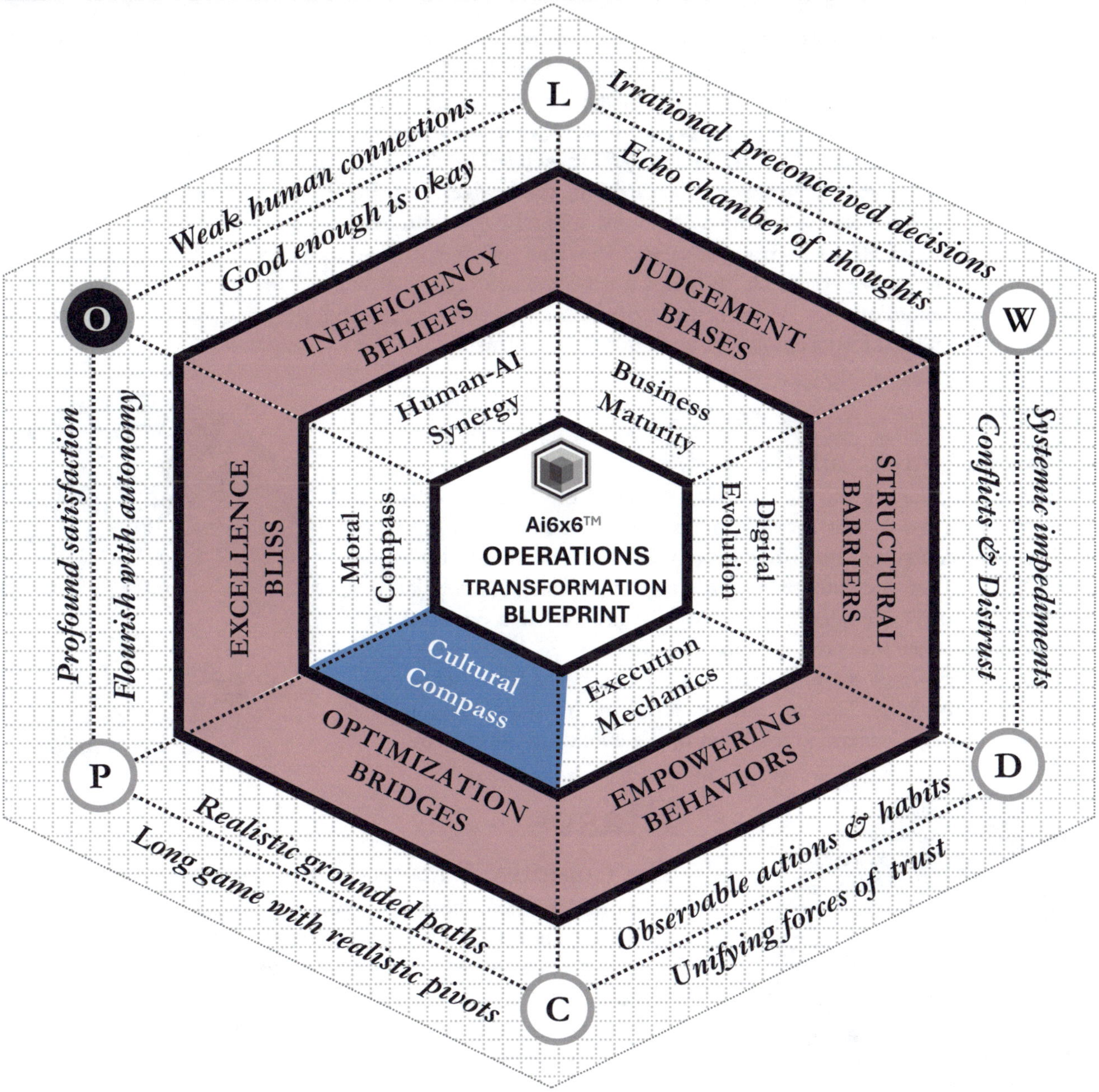

The **Ai6x6 Research Lab** chose this lens because the ultimate vitality of an operation is found in the collective "Bliss" of a workforce that values precision and clarity. We reframed this dimension as the stewardship of **Operations Excellence**, recognizing that culture is the invisible operating system that dictates whether friction is ignored or eliminated. It is vital because it addresses the invisible shadows—the **Metrics Amnesia** and **Interaction Biases**—that live in the minds of the operators. This lens is the bridge between the leader's strategic dashboard and the team's physical reality.

We selected it to ensure leaders build a community centered on excellence rather than just a high-output factory. By dismantling **Barriers** to transformation and building **Operations Excellence Bridges**, the leader creates a "greenhouse" for innovation where **Transformation Behaviors** become the standard. The Lab believes that in an age of automated movement, the human connection to the mission is the only self-sustaining source of operational resilience. This lens forces a look at how beliefs about "Friction" impact the collective trajectory. When culture is treated as a discipline, the organization achieves a state of "Bliss," where the entire team feels fundamentally connected to the mission's success.

Culture has always been the invisible architecture of an organization, shaping how people think, behave, and make decisions long before any formal strategy takes hold. In the age of AI, this architecture becomes even more consequential. Technology may accelerate capability, but culture determines whether that capability becomes meaningful progress or wasted potential. The **Cultural Compass** lens forces leaders to confront the truth that transformation is not sustained by systems alone; it is sustained by the beliefs people hold, the biases they carry, the barriers they navigate, the behaviors they exhibit, the bridges they build, and the moments of profound satisfaction that anchor their commitment. Without cultural alignment, even the most sophisticated AI strategy collapses under the weight of human resistance. This lens reframes culture as a dynamic system rather than a static identity. It recognizes that culture is not a set of values on a wall but a living pattern of interpretation, interaction, and intention. As AI reshapes workflows, decision cycles, and power dynamics, leaders must understand how cultural forces amplify or undermine transformation.

Beliefs

The Narratives Beneath the Surface

Beliefs and assumptions form the first dimension of the Cultural Compass. They are the quiet forces that shape how people interpret change, how they respond to uncertainty, and how they judge the intentions behind leadership decisions. When beliefs are outdated or misaligned, they distort perception and create friction. Leaders who ignore this dimension discover that people do not resist technology—they resist the meaning they attach to it. Addressing beliefs requires leaders to surface the narratives that operate beneath the surface and reshape them with clarity, honesty, and purpose. In this dimension, cultural truth-telling becomes a leadership act rather than a communication tactic.

Biases

The Cognitive Patterns That Shape Judgment

Biases and mental shortcuts represent the second dimension. These cognitive patterns influence how people evaluate information, assign trust, and make decisions. In an AI-enabled environment, biases can distort both human judgment and the interpretation of machine-generated insights. Addressing them requires intentional design, transparent dialogue, and environments where people feel safe challenging their own assumptions. Leaders must create conditions where cognitive unlearning becomes possible. In this dimension, cultural evolution is driven by awareness, humility, and the willingness to question long-held mental models.

Barriers

The Structural Forces That Restrict Change

Barriers and systemic impediments form the third dimension. These are the structural forces—policies, hierarchies, incentives, and legacy processes—that constrain behavior even when people are willing to change. Leaders often misinterpret resistance as emotional when it is structural. Removing barriers is not about demanding more effort; it is about redesigning the environment so that the desired behaviors become the natural ones. In this dimension, organizational de-friction becomes a strategic priority rather than an operational afterthought, enabling people to move with clarity rather than struggle against constraints.

Behaviors

The Visible Expression of Culture

Behaviors and observable actions represent the fourth dimension. Culture is not what people say; it is what they do consistently. In the AI era, behaviors must evolve to reflect curiosity, adaptability, collaboration, and responsible experimentation. Leaders must model these behaviors visibly and repeatedly. They must demonstrate that learning is valued, that questioning is encouraged, and that accountability is shared. In this dimension, behavioral signaling becomes a powerful tool for shaping norms, reinforcing the idea that culture is enacted through daily choices rather than declared through slogans.

Bridges

The Pathways That Connect People and Purpose

Bridges and cultural pathways form the fifth dimension. These are the connections that allow ideas, trust, and collaboration to flow across the organization. Bridges are built through shared language, shared rituals, and shared purpose. They transform isolated teams into aligned communities. Leaders who invest in cultural pathways create environments where people feel connected to something larger than their role. In this dimension, collective identity-building becomes a catalyst for cohesion, enabling the organization to move as a unified whole rather than a collection of disconnected parts.

Bliss

The Emotional Anchor of Commitment

Bliss and profound satisfaction complete the compass. Bliss is not about comfort; it is about meaning. It is the sense of fulfillment that emerges when people feel their work matters, their contributions are recognized, and their growth is supported. Bliss is the emotional anchor that sustains commitment through uncertainty. Leaders who cultivate it create cultures where people do not merely comply—they care. In this dimension, purposeful fulfillment becomes a strategic asset, strengthening resilience and deepening engagement.

The Cultural Compass lens reminds leaders that transformation is not sustained by technology or process alone. It is sustained by the human experience of change. Culture determines whether people interpret transformation as threat or opportunity, whether they respond with resistance or curiosity, and whether they move in alignment or fragmentation. In a world where AI accelerates both progress and pressure, leaders must treat culture as the primary operating system of the enterprise. When leaders align beliefs, challenge biases, remove barriers, shape behaviors, build bridges, and cultivate meaning, they create cultures capable of evolving continuously. They build environments where people feel psychologically safe, intellectually stimulated, and emotionally invested. Culture becomes the force that amplifies strategy rather than undermining it. It becomes the stabilizing anchor in moments of disruption and the accelerating engine in moments of opportunity.

The False Climate — **BELIEFS**

Dismantling "Weather Myths" ensures that unrealistic, universal expectations don't stifle individual growth. Exposing the false belief that all "plants" require the same sun, water, or timing—allowing for specialized care.

The Invasive Weeds — **BIASES**

Identifying and removing hidden mental shortcuts prevents old habits from choking out innovation. Surgically exposing the leadership judgment and systemic biases that create "shadow" cultures and block progress.

The Hard Clay — **BARRIERS**

Breaking through structural resistance allows the organizational roots to expand and stabilize. Removing the deeply embedded obstacles like data, automation and AI anxiety that act as physical walls against digital growth.

Calibrated Care — **BEHAVIORS**

Executing specialized nurturing routines ensures each distinct "species" receives what it needs to thrive. Moving away from "one-size-fits-all" mode to model specific watering, feeding, and light-balancing behaviors required for a diverse ecosystem.

The Trellis System — **BRIDGES**

Constructing pathways for upward growth provides the support needed for teams to reach their full potential. Building the specific leadership and development paths that allow the workforce to climb through the maturity stages.

The Full Harvest — **BLISS**

Securing a state of permanent organizational harmony results in a self-sustaining ecosystem of excellence. Reaching the ultimate destination where the "Cultural Soul" operates in a high-fidelity hum, delivering constant value.

Ai6x6™ Transformation Blueprint

Operations Inefficiency Beliefs

"Manners are the lubricating oil of an organization. It is a law of nature that two moving bodies in contact with each other create friction. This is as true for human beings as it is for inanimate objects." - Peter Drucker

Peter Drucker (1909–2005) was an Austrian-born American management consultant and educator. He is often called the "founder of modern management" for his ideas on corporate governance and the "knowledge worker."

Operations inefficiency beliefs are negative assumptions about how operations should function that create friction and resistance to change. These assumptions create bottlenecks and inefficiencies, such as rigid Compliance Beliefs that prioritize following rules over common sense and efficiency. A team might hold the belief that all problems should be solved in the same way, resisting new approaches that could be more effective. Similarly, a lack of faith in new operational technologies can prevent an organization from modernizing and remaining competitive. To truly transform, an organization must cultivate a new mindset that views operations as a dynamic, flexible system that can be continuously improved, rather than a static set of rules to be followed. This requires a shift from a mindset of friction, as Peter Drucker's quote suggests, to one of seamless flow and continuous improvement.

This chapter secures the Execution Core by focusing on Operations Inefficiency Beliefs, investigating how core convictions drive process Optimization and accelerate Digital Evolution across the enterprise's operational backbone. It explores the crucial psychological contract between human employees and automation systems. This begins with Automation-as-Enabler Beliefs, reframing technology as a tool that enhances human capabilities and frees employees for higher-value work, rather than a replacement. This perspective leads to System-as-Partner Beliefs, viewing operational technology systems as integrated, collaborative elements of the organization. The book then addresses the ingrained convictions necessary for continuous improvement: Efficiency Beliefs (seeing streamlining and waste reduction as a constant pursuit) and rigid Quality-Control Beliefs (ensuring consistent excellence is embedded and non-negotiable in every operational process). Finally, it secures the enterprise with Supply Chain Beliefs (conviction in seamless flow, transparency, and collaboration across the entire value chain) and foundational Resilience Beliefs, ensuring the operational system is robustly designed to withstand and rapidly recover from internal and external disruption.

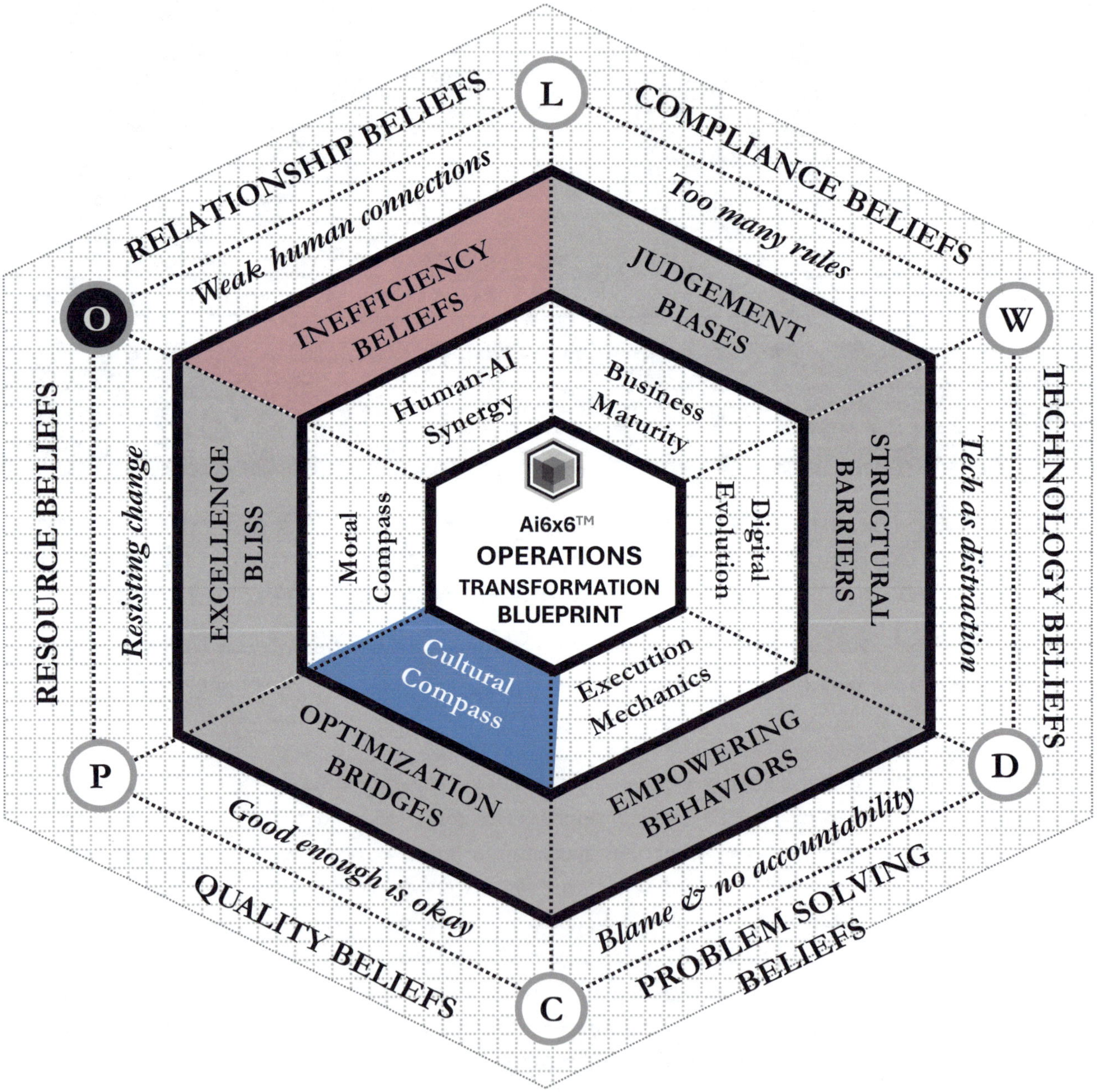

Ai6x6 CULTURAL COMPASS FRAMEWORK FOR **OPERATIONS LIFECYCLE**

If you believe that supplier relationships are purely "transactional" and based only on the lowest price, you are building a fragile, short-term operation. In the AI era, where supply chains are increasingly volatile, your suppliers are actually your first line of defense. Daily, this belief leads to "adversarial" negotiations that destroy trust and prevent the kind of collaborative innovation that saves companies during a crisis. Transformation requires believing that "Supplier Success is Our Success." When you shift this belief, you invest in long-term partnerships that provide priority access during shortages and co-investment in new technologies. This ensures your operations are supported by a loyal network that provides a competitive edge through mutual growth and shared, real-time intelligence.

DO'S FOR THE LEADER	DON'TS FOR THE LEADER
Build strong supplier relationships.	Treat suppliers as a single transaction.
Communicate openly with partners.	Keep your partners in the dark.
Work together to solve problems.	Blame suppliers for all issues.
Have fair, long-term contracts.	Use short-term, aggressive contracts.
Share information for better planning.	Hide your plans from your partners.
Celebrate shared successes.	Only point out partner failures.

SITUATION	SUGGESTED ACTION
Teams assume strong relationships alone guarantee supplier performance without structured accountability mechanisms	Balance trust with metrics, enforce expectations, and maintain transparent performance reviews
Leaders overvalue personal rapport and overlook objective indicators of supplier reliability or risk	Prioritize data, assess patterns, and integrate evidence into supplier evaluations
Procurement teams hesitate to challenge suppliers because they fear damaging long standing relationships	Establish negotiation frameworks, reinforce professionalism, and align discussions with operational goals
Organizations rely on informal agreements instead of formalizing commitments through structured contracts	Document terms, clarify responsibilities, and ensure alignment through enforceable agreements
Teams overlook emerging supplier issues because relationships mask early warning signs	Monitor signals, escalate concerns, and maintain balanced oversight regardless of rapport
Executives assume relationship strength reduces the need for diversified supplier portfolios	Review concentration risk, evaluate alternatives, and maintain resilience through strategic diversification

Believing that compliance is a "necessary evil" or a "box-ticking exercise" leads to an operation that is constantly at the precipice of regulatory failure. In finance or media, if your team sees rules as "obstacles to speed," they will inevitably find ways to bypass them. Daily, this manifests as "compliance shortcuts" that eventually lead to massive fines, legal battles, or permanent brand damage. Transformation requires believing that "Compliance is the Ultimate Quality Standard." When you shift this belief, you integrate regulatory requirements directly into your automated AI workflows. This ensures that your operations are "compliant by design," making audits effortless and turning regulatory adherence into a mark of operational excellence that builds deep customer trust and long-term market stability.

DO'S FOR THE LEADER	DON'TS FOR THE LEADER
Be strict about following all rules.	Cut corners to save time or money.
Make compliance part of the process.	Treat compliance as a checklist item.
Train staff on all regulations.	Assume everyone knows the rules.
Update processes for new laws.	Fail to adapt to changing regulations.
Audit for compliance regularly.	Only check for compliance once a year.
Be transparent with regulators.	Hide or obscure compliance issues.

SITUATION	SUGGESTED ACTION
Teams treat compliance as optional when operational pressure increases or deadlines tighten significantly	Reinforce standards, clarify consequences, and integrate compliance into performance expectations
Leaders assume strict compliance slows operations without evaluating long-term risk reduction benefits	Review tradeoffs, quantify exposure, and align compliance with strategic risk management
Operators bypass controls believing shortcuts improve efficiency without meaningful consequences	Strengthen oversight, update training, and reinforce adherence through consistent enforcement
Teams misinterpret regulations because guidance is outdated or inconsistently communicated across functions	Update documentation, centralize interpretations, and ensure clarity through structured communication
Organizations over-comply in some areas while neglecting critical requirements due to misunderstanding	Conduct compliance mapping, assess gaps, and rebalance efforts based on risk and regulatory priority
Executives underestimate compliance complexity because they rely on outdated assumptions about regulatory environments	Review changes, engage experts, and integrate updated requirements into operational planning

The assumption that "all new technology is a silver bullet" or, conversely, that "new tech is too risky" creates massive strategic friction. If you believe technology replaces the need for process discipline, you will simply automate a mess. Daily, this looks like expensive AI tools being forced onto broken, manual workflows, resulting in faster-produced errors and frustrated staff. Transformation requires believing that "Technology Amplifies Process." You must shift your belief to see tech as the enabler of a well-defined human strategy. When you believe technology must be fit-for-purpose rather than just "new," you choose tools that solve specific operational bottlenecks, ensuring that your digital investments actually deliver the intended ROI and improve the lives of your operators.

DO'S FOR THE LEADER	DON'TS FOR THE LEADER
Adopt new tech to improve processes.	Stick with outdated, inefficient tech.
Invest in tech that solves problems.	Buy new tech just for the sake of it.
Train employees on new technology.	Throw new tech at people with no training.
Integrate new tech with old systems.	Create a mess of unconnected systems.
Test new tech before rolling it out.	Blindly implement new technology.
Use tech to automate repetitive tasks.	Have people do tasks a machine could do.

SITUATION	SUGGESTED ACTION
Teams delay adoption believing new technologies introduce unnecessary operational instability or unpredictable behavior	Pilot in controlled environments, validate performance, and scale only when operational readiness is proven
Leaders push rapid deployment assuming technology alone will resolve entrenched operational inefficiencies	Diagnose root causes, align tools with redesigned workflows, and ensure adoption supports measurable improvements
Operators resist automation because they assume it cannot handle real-world operational edge cases	Demonstrate capability, test extremes, and refine systems using frontline operational feedback
Teams underestimate integration complexity when connecting new tools to aging operational infrastructure	Assess compatibility, run integration tests, and sequence deployment based on validated feasibility
Organizations adopt tools based on market trends rather than operational fit or strategic necessity	Define evaluation criteria, assess alignment, and prioritize technologies that strengthen core operational capabilities
Executives assume technology investments guarantee transformation without sustained process discipline	Reinforce governance, track outcomes, and integrate technology into continuous improvement cycles

If you believe the best way to solve an operational problem is through "top-down directives" from people far from the action, you will stifle the frontline innovation needed to survive. Daily, this belief leads to solutions that look perfect in a boardroom PowerPoint but fail miserably in the warehouse or on the retail floor. Transformation requires believing that "The Best Solutions are Found at the Gemba"—the place where the work actually happens. When you shift this belief, you empower frontline workers with AI tools and data to solve their own local frictions. This ensures that problem-solving is democratic, fast, and grounded in reality, leading to a more responsive, agile, and respected operational culture.

DO'S FOR THE LEADER	DON'TS FOR THE LEADER
Investigate the root cause of issues.	Only treat the symptoms of a problem.
Use data to find and solve problems.	Solve problems based on gut feelings.
Encourage problem-solving at all levels.	Expect a single person to solve problems.
Create a blameless problem-solving culture.	Blame someone whenever a problem occurs.
Document solutions to prevent repeats.	Let the same problems happen again.
Learn from every failure.	Let failures just fade away.

SITUATION	SUGGESTED ACTION
Teams jump to solutions without understanding operational context or constraints shaping the issue	Slow the process, gather evidence, and validate assumptions before committing to corrective actions
Leaders rely on intuition rather than structured analysis when addressing recurring operational failures	Use frameworks, review data, and ensure decisions reflect objective insights rather than personal preference
Operators assume visible symptoms represent the true cause of operational breakdowns	Conduct root-cause analysis, test hypotheses, and validate findings through cross-functional review
Teams prioritize speed over accuracy when responding to operational disruptions	Balance urgency with rigor, assign roles, and ensure decisions are informed by reliable information
Organizations treat each issue as isolated rather than identifying systemic operational patterns	Consolidate incidents, analyze trends, and implement structural improvements addressing shared causes
Executives assume complex problems require complex solutions rather than targeted operational adjustments	Prioritize simplicity, test incremental changes, and scale improvements based on validated impact

Assuming that quality is "someone else's job," like the QA department, creates a siloed barrier that kills operational flow. If your manufacturing or finance team believes their job is just "output" while someone else "checks the work," errors will skyrocket. Daily, this manifests as high "rework" rates and toxic finger-pointing when mistakes reach the end customer. Transformation requires believing that "Quality is an Input, Not an Output." You must shift to a "Total Quality Management" mindset where every AI node and every human operator is responsible for excellence at their station. This belief ensures that errors are caught at the source, dramatically reducing waste and building a market reputation for "unshakeable reliability" that competitors cannot match.

DO'S FOR THE LEADER	DON'TS FOR THE LEADER
Build quality into every step.	Only check for quality at the end.
Set clear quality standards.	Have vague or non-existent quality rules.
Empower staff to stop bad quality.	Push people to release a bad product.
Reward quality over quantity.	Reward high volume over good quality.
Use metrics to track quality.	Don't measure quality in any way.
Learn from every quality issue.	Hide or ignore all quality problems.

SITUATION	SUGGESTED ACTION
Teams equate high output volume with quality, ignoring downstream rework and operational instability	Define quality thresholds, track defect patterns, and enforce standards that prioritize reliability over raw throughput
Leaders assume quality issues originate only on the frontline rather than across interconnected workflows	Map failure points, review handoffs, and assign shared accountability across operational functions
Operators believe minor deviations are harmless because immediate impact is not visible during execution	Quantify downstream effects, expose hidden costs, and reinforce discipline through targeted audits
Teams prioritize visual completeness over operational resilience when evaluating deliverables	Stress-test outputs, validate performance, and ensure quality criteria reflect real operating conditions
Organizations treat compliance as a proxy for quality without measuring actual customer or system outcomes	Compare compliance results with operational performance, and recalibrate quality metrics accordingly
Executives assume quality improvements require major investments rather than disciplined operational refinement	Identify friction points, implement targeted fixes, and scale improvements based on validated impact

The belief that resources should be "static and specialized" leads to a rigid operation that cannot adapt to rapid market shifts. If you believe a retail store or a factory should only do "one thing," you will be left behind when consumer demand changes overnight. Daily, this looks like "idle capacity" in one area while another is overwhelmed and failing to meet targets. Transformation requires believing in "Dynamic Resource Elasticity." When you shift this belief, you use AI to cross-train staff and create modular, flexible production lines. This ensures your resources—both human and machine—can shift to where the value is highest, creating a fluid operation that can "shape-shift" to meet the demands of the moment with total efficiency.

DO'S FOR THE LEADER	DON'TS FOR THE LEADER
Be flexible with resource allocation.	Have rigid, inflexible resource plans.
Cross-train employees for flexibility.	Let people get stuck in a single role.
Manage resources efficiently.	Waste resources on low-value tasks.
Use tech to optimize resource use.	Manage resources manually.
Plan for sudden resource changes.	Be unprepared for shifts in demand.
Forecast future resource needs.	Only react to a resource crisis.

SITUATION	SUGGESTED ACTION
Teams assume adding more people automatically resolves operational bottlenecks regardless of skill alignment	Assess capability gaps, match resources to constraints, and deploy staffing based on operational need
Leaders believe resource flexibility is optional rather than essential for handling volatile operational demand	Build cross-training programs, diversify skills, and maintain adaptable capacity across teams
Operators resist role flexibility because they fear unclear expectations or increased workload	Clarify responsibilities, provide structured support, and reinforce flexibility as a shared operational advantage
Teams rely on fixed staffing models even when demand patterns shift unpredictably	Implement dynamic scheduling, monitor real-time load, and adjust allocation based on operational signals
Organizations assume specialized roles cannot adapt to broader responsibilities across workflows	Identify transferable skills, provide targeted training, and expand capabilities through structured development
Executives underestimate the value of adaptable resources when planning for disruptions or rapid scaling	Integrate flexibility into workforce planning, assess readiness, and invest in scalable operational capacity

Ai6x6™ Transformation Blueprint

Operations Judgment Biases

"The single biggest problem in communication is the illusion that it has taken place." - George Bernard Shaw

George Bernard Shaw (1856–1950) was an Irish playwright, critic, and political activist. He is considered one of the greatest playwrights in the English language and is known for his witty social commentary and advocacy for social reform.

Operations judgment biases are cognitive shortcuts that can compromise operational effectiveness and lead to errors. A common pitfall is Algorithmic Aversion Bias, where human operators mistrust AI-driven processes, preferring to rely on their own judgment even when the system is more accurate. This prevents an organization from leveraging the benefits of automation and data-driven insights. Another critical issue is the Moral Crumple Zone Bias, where human operators are unfairly blamed for a system's failure, even when the system itself is poorly designed. Over-reliance Bias is where an organization trusts a system without questioning its output, leading to potential errors and a lack of critical vigilance. Overcoming these biases requires a commitment to a rigorous, evidence-based approach to system management, where the relationship between humans and technology is one of partnership and mutual trust.

This chapter analyzes Operations Judgement Biases, investigating how human cognitive shortcuts interact with operational systems and automation, which disrupts Optimization and slows Digital Evolution. This domain critically explores the crucial psychological contract between human operators and sophisticated automation systems. The journey begins with managing Algorithmic Aversion—the unwarranted distrust humans feel toward algorithms, leading to underutilization of powerful tools. This is often complicated by Anthropomorphism Bias, where operators attribute human traits to AI, leading to inaccurate expectations of system behavior. The chapter addresses the risk of Transparency Illusion—the false belief that data visibility in dashboards equates to actual system understanding—which hinders true process mastery. Crucially, it examines the organizational failure mode known as the Moral Crumple Zone, where humans are unfairly blamed for failures caused by opaque systemic flaws, eroding trust. Finally, it explores the dangers of Delegation Bias (inaccurate assessment for task delegation to AI) and the safety threat of Over-reliance Bias, where over-trusting automation reduces human vigilance and manual skill, leaving the system brittle in the face of unexpected disruption.

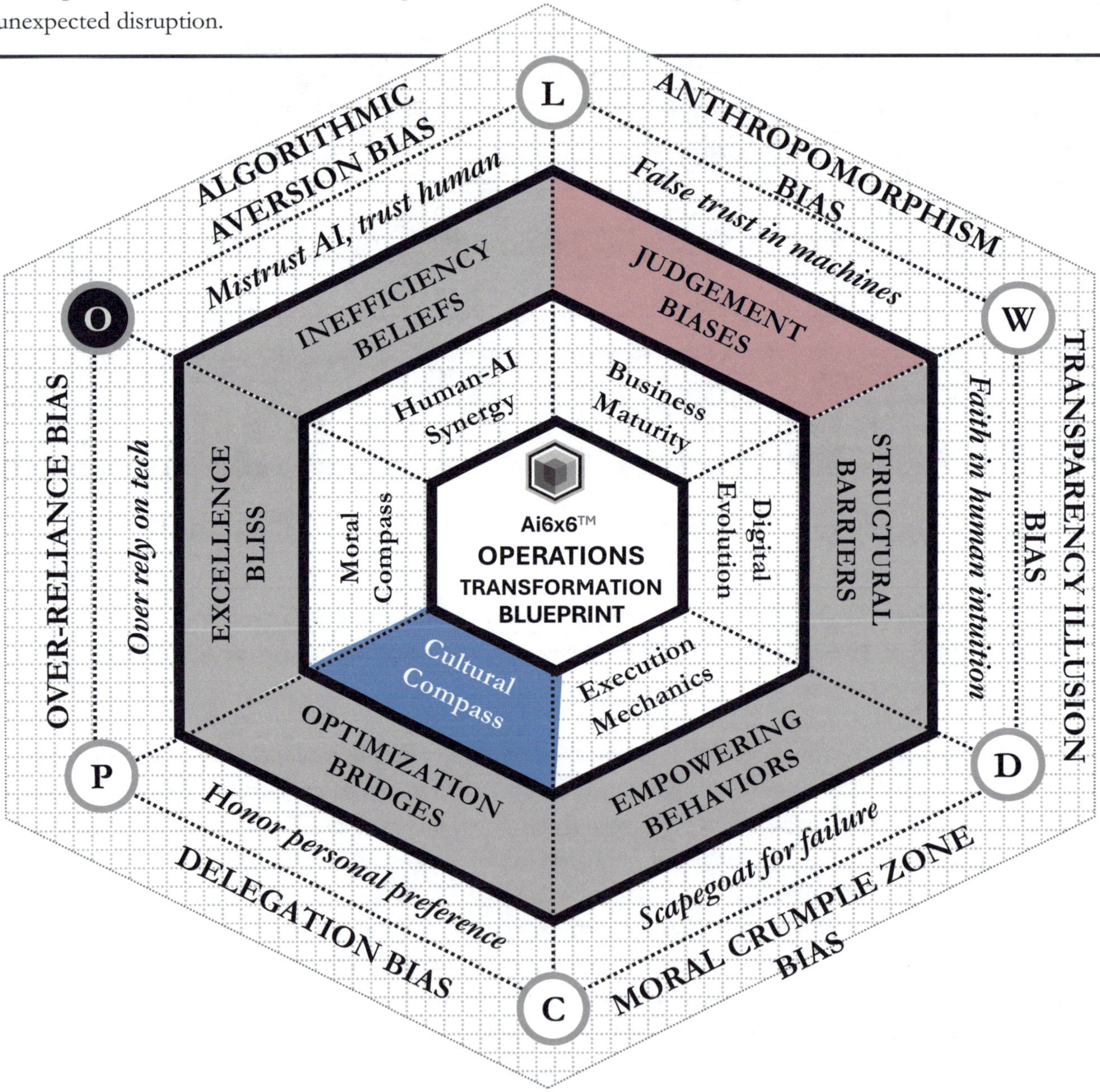

Ai6x6 CULTURAL COMPASS FRAMEWORK FOR OPERATIONS LIFECYCLE

This bias occurs when human operators mistrust AI-driven processes, even when the data proves the AI is more accurate. In finance or supply chain forecasting, an operator might "manually override" a precise AI prediction because of a "gut feeling," often leading to massive overstocking or missed revenue. Daily, this results in the chronic under-utilization of expensive AI investments and a return to "guess-based" management. Transformation requires building "Explainable AI" that builds human confidence through transparency. By recognizing this bias, you can implement training that shows the AI's reasoning. This ensures that humans use AI as a powerful "co-pilot," combining data-driven precision with human intuition only where it adds true, documented value to the operational process.

DO'S FOR THE LEADER	DON'TS FOR THE LEADER
Use AI as a tool to enhance humans.	Mistrust AI-driven processes.
Explain how the AI works.	Use AI as a black box system.
Train employees to work with AI.	Assume employees don't need training.
Use AI for complex data analysis.	Force humans to do tasks AI could do.
Validate AI output with human oversight.	Over-rely on AI output without checking it.
Use AI to increase efficiency.	Let fear of AI prevent efficiency.

SITUATION	SUGGESTED ACTION
Teams override AI outputs because they assume intuition is safer during operational uncertainty	Compare outcomes, track error rates, and require evidence before rejecting machine-generated recommendations under real operating conditions
Operators distrust alerts after one false positive significantly disrupts workflow confidence and stability	Calibrate thresholds, validate signals, and retrain teams to interpret alerts within operational context and system behavior
Leaders delay AI adoption fearing loss of control over critical operational decisions	Define guardrails, maintain oversight, and deploy AI where decision latency creates measurable operational risk
Teams ignore predictive models believing historical instincts outperform algorithmic forecasts consistently	Run parallel tests, measure accuracy, and integrate model performance into planning cycles and operational reviews
Operators resist automation assuming AI cannot handle real-world edge conditions reliably	Expose model limits, test extreme scenarios, and refine systems using frontline operational feedback and observations
Executives dismiss AI outputs when results contradict long-held assumptions about system behavior	Investigate discrepancies, review evidence, and update mental models based on validated operational data and patterns

2. ANTHROPOMORPHISM BIAS | The Ghost in the Machine

Attributing human traits—like "intelligence," "loyalty," or "intent"—to machines can lead to dangerous operational errors and safety lapses. If a warehouse manager "trusts" a robot to "know what it's doing," they may stop performing necessary safety checks or quality audits. Daily, this bias leads to a "false sense of security" that ignores the fundamental technical limitations of any automated system. Transformation requires maintaining a "Cold-Eyed Machine" perspective. You must ensure operators understand exactly what an algorithm cannot do. By recognizing this bias, you maintain the "critical distance" needed for safe operations, ensuring that humans manage systems as tools of logic, not as surrogate colleagues with "common sense" they don't actually possess, preventing catastrophic over-reliance.

DO'S FOR THE LEADER	DON'TS FOR THE LEADER
View systems as tools to be used.	Attribute human traits to systems or machines.
Understand a system's limitations.	Assume a system has feelings or goals.
Use clear, technical language.	Use emotional or vague language.
Be rational about a system's output.	Be emotional about a system's output.
Learn how the system was built.	Assume the system thinks like a human.
Test the system for what it is.	Test the system for what it is not.

SITUATION	SUGGESTED ACTION
Teams assume systems act intentionally instead of recognizing deterministic behavior under constraints	Review logic paths, expose limitations, and reinforce system behavior as rule-driven rather than emotional or adaptive
Operators trust outputs believing the system understands operational context intuitively and accurately	Train teams on model boundaries, highlight blind spots, and require verification for high-impact operational decisions
Leaders attribute failures to system moodiness rather than configuration drift or data degradation	Inspect logs, validate inputs, and correct underlying technical conditions driving inconsistent performance
Teams expect systems to adapt automatically without explicit retraining or configuration updates	Establish retraining schedules, monitor drift, and maintain operational ownership of model upkeep and accuracy
Operators hesitate to override systems believing intervention disrupts automated processes unnecessarily	Clarify override protocols, reinforce authority, and ensure operators understand system neutrality and constraints
Executives assume AI can learn anything without structured data pipelines or governance	Define data requirements, enforce quality standards, and align expectations with actual system capabilities and limits

3. TRANSPARENCY ILLUSION BIAS | The Illusion of Control

This is the false belief that an operational system is "transparent" just because a user has access to a massive dashboard of raw data. In reality, too much data creates "noise," not "clarity." Daily, this manifests as "dashboard fatigue," where managers miss critical failure signals because they are buried in thousands of irrelevant metrics. Transformation requires "Curated, Actionable Transparency." You must use AI to highlight only the "insights that matter" for immediate decision-making. By recognizing this bias, you shift from "data access" to "information utility." You ensure that transparency actually leads to better outcomes, providing operators with a "clear view" of the process rather than just an overwhelming and useless "data dump" that slows down response times.

DO'S FOR THE LEADER	DON'TS FOR THE LEADER
Make system info truly understandable.	Believe a system is transparent just because it has some data.
Provide clear context for data.	Provide raw data with no explanation.
Create a single source of truth.	Let each team have a different view of data.
Explain how the system works.	Let users assume they know how it works.
Use visualization to explain data.	Use only text and numbers.
Get feedback on transparency.	Assume your system is transparent.

SITUATION	SUGGESTED ACTION
Teams assume dashboards show complete truth without understanding hidden dependencies or blind spots	Map data flows, expose gaps, and validate whether displayed metrics reflect actual system conditions accurately
Operators trust logs without recognizing missing context or suppressed error signals	Review log completeness, audit pipelines, and ensure visibility into upstream and downstream operational events
Leaders believe transparency exists because they can view high-level summaries and charts	Demand deeper diagnostics, inspect feature behavior, and require explanations for critical decision paths and anomalies
Teams misinterpret partial data as complete insight during incident response activities	Cross-reference sources, validate assumptions, and escalate when inconsistencies appear across systems
Operators assume system health is stable because monitoring tools show green indicators consistently	Verify thresholds, test failure modes, and ensure monitoring reflects real operational stress and load
Executives rely on simplified reports without understanding underlying data quality issues	Review lineage, assess accuracy, and validate metrics before using them for strategic operational decisions

4. MORAL CRUMPLE ZONE BIAS | The Scapegoat

This bias occurs when human operators are blamed for a system's failure, even when the failure was actually caused by a flawed AI or process design. In manufacturing or finance, if a "pilot error" is blamed for a system crash that was inevitable given a confusing UI, you are in the moral crumple zone. Daily, this leads to a "culture of fear" and high stress for operators, who feel set up to fail. Transformation requires "Systemic Accountability." By recognizing this bias, you shift the focus from "who failed" to "how the system failed the human." You use AI to identify design flaws that burden humans, ensuring a fairer, more resilient operational environment where the system is hardened against human error.

DO'S FOR THE LEADER	DON'TS FOR THE LEADER
Create a clear accountability framework.	Blame a human operator for a system's failure.
Blame the system, not the person.	Hunt for a scapegoat after a failure.
Investigate the root cause of failures.	Only punish the person closest to the error.
Improve the system to prevent repeats.	Punish a person and do nothing else.
Create a blameless post-mortem.	Punish people for making mistakes.
Focus on the process, not the person.	Focus on the person, not the process.

SITUATION	SUGGESTED ACTION
Operators are blamed for incidents triggered by flawed automation or incomplete safeguards	Trace failure origins, review design decisions, and assign accountability to responsible system owners and architects
Teams penalize staff when AI recommendations create operational errors or disruptions	Audit model behavior, assess training data, and correct algorithmic flaws before assigning responsibility
Leaders assume human error caused failures without reviewing system constraints or interface design	Analyze workflows, inspect usability issues, and identify design factors contributing to recurring mistakes
Operators hesitate to report issues fearing blame for system shortcomings or failures	Establish blameless reporting, investigate system factors, and reinforce psychological safety across teams
Teams escalate incidents as human failures even when automation produced misleading signals	Validate alert accuracy, correct thresholds, and update monitoring logic to prevent misclassification
Executives overlook systemic design flaws because accountability defaults to individuals	Conduct systemic reviews, document design gaps, and align accountability with actual failure sources

5. DELEGATION BIAS | The Human Factor

Delegating tasks based on "personal preference" or "long-standing habit" rather than objective criteria is a major source of operational inefficiency. In media or retail, a manager might delegate a complex task to a "favorite" staff member or an "old" automated script simply because it's comfortable. Daily, this leads to "skill-task mismatch" and sub-optimal output that frustrates the team. Transformation requires "Algorithmic Task Allocation." By recognizing this bias, you use AI to match the "right task" to the "best resource" (human or machine) based on real-time capacity and proven skill data. This ensures your operations run at "peak efficiency," with every resource being utilized for its highest and best use, regardless of outdated personal preferences.

DO'S FOR THE LEADER	DON'TS FOR THE LEADER
Delegate tasks based on objective criteria.	Delegate tasks based on personal preference.
Use data to inform who does what.	Give tasks to a favorite or a friend.
Be fair and transparent in delegation.	Be unfair or secretive in delegation.
Delegate to empower people.	Delegate to avoid doing a task yourself.
Use clear rules for delegation.	Have no clear rules for who does what.
Match tasks with skills.	Give tasks to people who are not skilled.

SITUATION	SUGGESTED ACTION
Teams assign critical tasks to familiar individuals rather than those with proven operational capability	Define criteria, assess skills, and allocate responsibilities based on demonstrated performance rather than comfort
Leaders delegate high-risk work to whoever is available instead of evaluating operational readiness	Review workload, validate expertise, and match assignments to capability and situational complexity
Operators receive tasks misaligned with their strengths because delegation follows habit rather than need	Map competencies, rebalance assignments, and ensure work aligns with operational skill distribution
Teams avoid delegating to newer staff due to misplaced assumptions about inexperience	Provide structured onboarding, test capability, and integrate new operators into meaningful operational responsibilities
Organizations rely on informal delegation patterns that ignore changing operational demands	Establish delegation rules, monitor alignment, and adjust assignments based on real-time operational signals
Executives assume delegation quality improves naturally without governance or performance feedback	Implement oversight, track outcomes, and refine delegation practices through continuous operational review

This bias happens when humans rely so heavily on an automated system that they lose the ability to perform the task manually or "question" the system's output. In finance, if an analyst stops checking the "sanity" of an AI's risk report, a catastrophic error could go unnoticed for months. Daily, this leads to "skill atrophy" and a "passive monitoring" mindset that is easily bypassed by errors. Transformation requires "Active Vigilance Protocols." By recognizing this bias, you implement "random spot-checks" and "manual refresh" sessions. This ensures that your human operators remain "engaged and capable," acting as the final, intelligent safeguard against the "hallucinations" or edge-case failures of even the most advanced and expensive AI systems.

DO'S FOR THE LEADER	DON'TS FOR THE LEADER
Use a system as a tool, not a crutch.	Rely too heavily on a system or process.
Encourage critical vigilance.	Blindly accept a system's output.
Set clear limits on a system's use.	Let a system be used for everything.
Train employees to question the system.	Assume the system is always correct.
Create a human-in-the-loop process.	Let the system run with no human oversight.
Get feedback on system usage.	Assume a system is being used correctly.

SITUATION	SUGGESTED ACTION
Teams trust automated outputs even when operational conditions clearly contradict system recommendations	Cross-check signals, validate assumptions, and escalate discrepancies for immediate investigation
Operators follow workflows mechanically without verifying whether steps still reflect current operational realities	Review procedures, retire outdated steps, and ensure workflows evolve with system behavior
Leaders assume dashboards reflect full operational truth without validating data completeness or accuracy	Audit data pipelines, inspect lineage, and confirm metrics represent actual system conditions
Teams ignore anomalies because they assume the system would alert them if something were wrong	Test alert thresholds, simulate failures, and ensure monitoring captures meaningful deviations
Operators rely on historical system performance without considering model drift or configuration changes	Track drift indicators, review updates, and recalibrate systems regularly
Executives assume automation eliminates the need for human oversight during critical operations	Maintain supervisory roles, define intervention points, and enforce human-in-the-loop governance

Ai6x6™ Transformation Blueprint

Operations Structural Barriers

"Measurement is the first step that leads to control and eventually to improvement. If you can't measure something, you can't understand it. If you can't understand it, you can't control it. If you can't control it, you can't improve it." - H. James Harrington

H. James Harrington (1929–2024) was a business expert and consultant. He is a leading figure in quality management and is known for his work in business process improvement, earning numerous awards for his contributions.

Operations structural barriers are cultural and procedural impediments that prevent learning and improvement. These barriers lead to stagnation and inefficiency, such as Process Amnesia where an organization fails to learn from past processes and their outcomes, leading to the same mistakes being repeated. A Knowledge Amnesia Barrier is the loss of institutional knowledge due to a lack of documentation or high employee turnover, crippling the organization's ability to maintain its expertise. A Supplier Amnesia Barrier is the failure to remember past issues or successes with suppliers. Overcoming these barriers requires a focused effort to redesign systems, document processes, and instill a culture of continuous learning and proactive problem-solving. As H. James Harrington's quote suggests, if you can't measure something, you can't improve it. This is about building a culture where data is a tool for empowerment, not a source of fear.

The final framework secures the Execution domain by meticulously analyzing Operations Structural Barriers, investigating the structural and technical obstacles that interact with processes and automation, which disrupts Optimization and slows Digital Evolution. The primary friction source is Legacy Processes, workflows and routines that are outdated, manual-intensive, and designed for an earlier technological era. This is compounded by Lack of Integration, where mission-critical systems and applications cannot communicate seamlessly, forcing data transfer and manual reconciliation. The result is often Manual Bottlenecks, points in the process that require human intervention, halting the flow of goods or information and preventing true scalability. The component addresses the rigidity of Inflexible Systems, technology that cannot adapt to changing business rules or market demands without costly, lengthy custom code changes. It details the necessity of Training, since a Lack of Training means employees don't know how to use new systems efficiently, reverting to old, inefficient methods. Finally, it highlights the vulnerability of Disrupted Supply Chains, where breaks in the flow of goods and information—whether physical or digital—jeopardize customer delivery and core business stability.

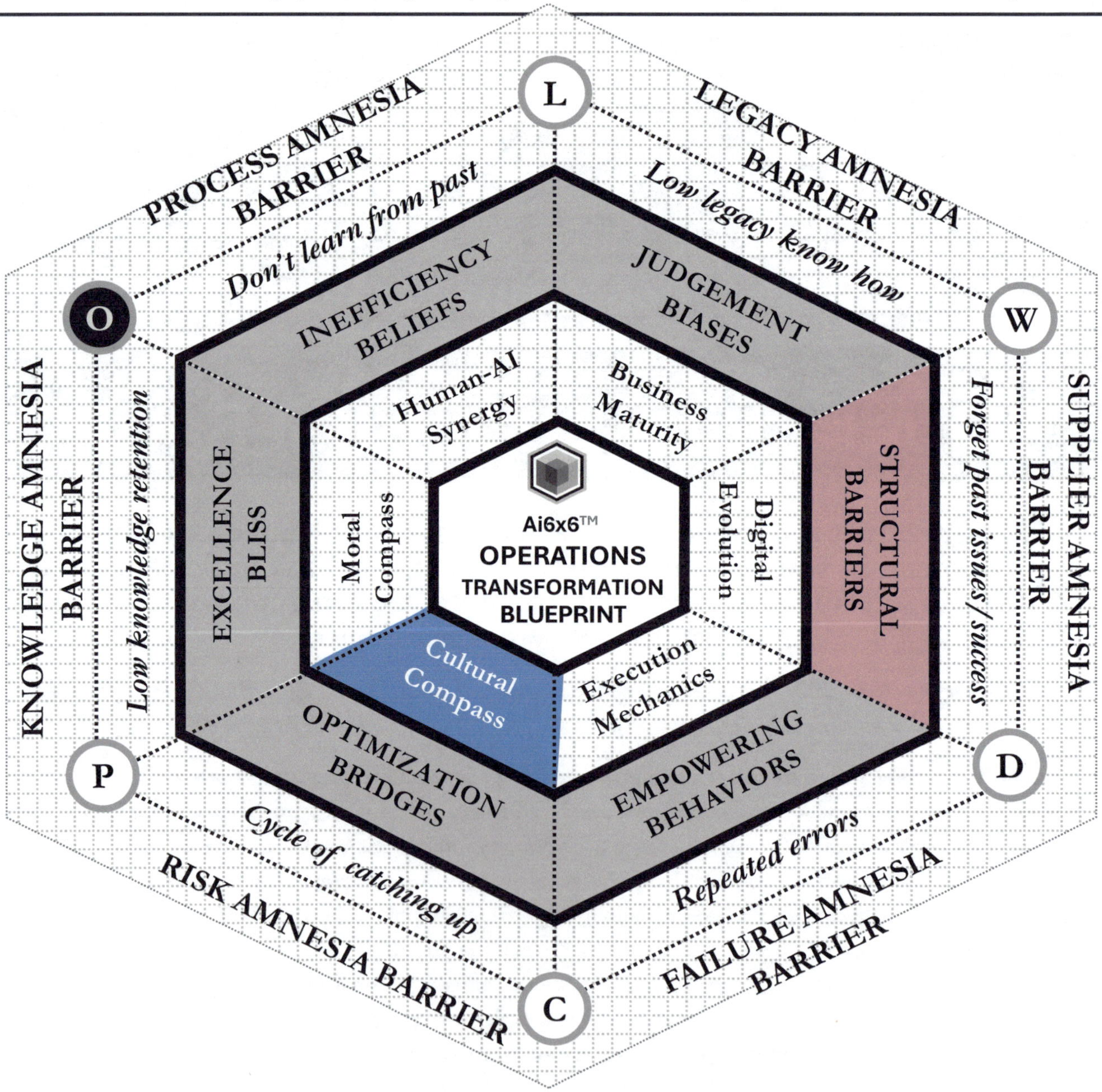

Ai6x6 CULTURAL COMPASS FRAMEWORK FOR OPERATIONS LIFECYCLE

1. PROCESS AMNESIA BARRIER | The Past is Prologue

This barrier occurs when an organization fails to internalize the logic of past operational workflows, leading to the expensive repetition of inefficient habits. In complex sectors like manufacturing or supply chain, this often manifests as "solving" the same bottleneck every peak season without ever formalizing the fix in the system's architecture. Daily, this looks like "groundhog day" meetings where veteran staff and new hires debate issues that were supposedly resolved eighteen months prior. Transformation requires building a "digital memory" through AI that captures and tags process telemetry in real-time. When you break this barrier, you ensure that every operational cycle is measurably smarter than the last. You move from a state of constant, exhausting "reinvention" to one of continuous, documented refinement that compounds.

DO'S FOR THE LEADER	DON'TS FOR THE LEADER
Document past processes and outcomes.	Let old processes be forgotten.
Learn from old processes.	Fail to learn from past processes.
Update processes based on past data.	Repeat the same process mistakes.
Get feedback on old processes.	Assume old processes were perfect.
Make it easy to find old processes.	Have no documentation for old processes.
Use a single source of truth.	Let each team have a different process.

SITUATION	SUGGESTED ACTION
Teams repeat inefficient workflows because earlier process reviews were never documented or shared effectively	Establish structured retrospectives, capture insights, and ensure operational learnings remain accessible across teams to prevent recurring inefficiencies
Leaders make decisions without referencing historical performance data from similar operational cycles or initiatives	Centralize metrics, require historical baselines, and integrate past outcomes into decision frameworks guiding operational planning and prioritization
Operators revert to outdated routines because improvements from earlier initiatives were not institutionalized	Standardize updated procedures, reinforce adoption, and embed new practices into training and operational governance to ensure consistency
Teams overlook recurring bottlenecks because previous root-cause findings were not preserved or communicated	Create searchable repositories, mandate documentation, and integrate findings into continuous improvement cycles across operational units
Executives underestimate process risks because earlier warning signs were forgotten or not recorded	Implement risk logs, review patterns, and incorporate historical signals into operational forecasting and scenario planning
Organizations lose operational discipline when institutional memory fades after leadership or team transitions	Codify processes, maintain continuity plans, and ensure knowledge transfer during organizational changes and restructuring

2. LEGACY AMNESIA BARRIER | The Forgotten Foundations

Legacy Amnesia is the dangerous lack of understanding regarding why old systems were built with specific, seemingly odd constraints. In finance or retail, teams often scrap "clunky" old systems only to find they have accidentally deleted critical regulatory compliance or niche edge-case logic that was hard-coded for a reason. Daily, this results in "migration disasters" where the shiny new AI tool fails because it ignored a fundamental business rule buried in thirty years of legacy COBOL or spreadsheet macros. Transformation requires a "system archaeology" mindset where you use AI to document and interpret legacy logic before replacement. By breaking this barrier, you ensure that modernization preserves hard-won institutional wisdom while shedding technical debt, creating a bridge between past reliability and future agility.

DO'S FOR THE LEADER	DON'TS FOR THE LEADER
Understand why old systems were built.	Just tear down old systems without understanding them.
Document legacy system history.	Let the history of old systems be forgotten.
Use data to understand a legacy system.	Assume an old system is completely useless.
Learn from old systems.	Repeat the same old mistakes with new tech.
Integrate new tech with old systems.	Create a mess of unconnected systems.
Preserve what works in old systems.	Throw out the good parts with the bad.

SITUATION	SUGGESTED ACTION
Teams criticize legacy systems without understanding constraints that shaped original architectural decisions	Document historical context, interview architects, and evaluate legacy choices before planning modernization or replacement
Executives underestimate operational dependencies embedded in legacy platforms supporting critical workflows	Map dependencies, assess impact, and prioritize modernization based on operational risk and business continuity requirements
Engineers propose replacements without recognizing regulatory or compliance requirements embedded in legacy logic	Review compliance history, validate constraints, and ensure new systems preserve required safeguards and operational controls
Teams misjudge migration complexity because legacy integrations are poorly understood or undocumented	Conduct technical archaeology, rebuild documentation, and sequence modernization based on validated system knowledge and dependencies
Leaders push rapid decommissioning without understanding legacy system value to frontline operations	Engage operators, assess usage patterns, and align modernization timelines with operational realities and customer commitments
Organizations repeat past mistakes because legacy system lessons were never captured or analyzed	Create historical reviews, preserve architectural rationale, and integrate learnings into modernization governance and decision frameworks

3. SUPPLIER AMNESIA BARRIER | The Supplier's Shadow

This is the failure to maintain a cohesive, cross-departmental history of supplier performance, leading to inconsistent procurement and recurring supply chain disruptions. If your retail or media operation "forgets" a vendor's past delivery failures during a critical campaign, you are hitting this barrier head-on. Daily, this manifests as high-risk contracts being awarded to unreliable partners simply because the procurement team has "reset" its memory due to staff turnover or siloed data. Transformation requires a unified "Supplier Intelligence" platform that tracks every interaction. By breaking this barrier, you move from transactional, price-based purchasing to strategic, relationship-based sourcing. You use deep performance data to predict future reliability, ensuring your supply chain is resilient and backed by partners who have proven their operational worth.

DO'S FOR THE LEADER	DON'TS FOR THE LEADER
Document past issues with suppliers.	Forget past issues or successes with suppliers.
Remember what worked and what didn't.	Let history repeat itself with a supplier.
Communicate past issues to new staff.	Let new staff make the same mistakes.
Use supplier data to make choices.	Rely on memory or gut feelings.
Be fair to suppliers, but remember.	Be too forgiving or too strict.
Learn from every supplier interaction.	Treat every interaction as a blank slate.

SITUATION	SUGGESTED ACTION
Teams reengage suppliers without recalling previous performance issues affecting delivery timelines or quality	Maintain supplier scorecards, review historical performance, and incorporate insights into selection and negotiation decisions
Executives overlook suppliers who previously delivered exceptional value during operational disruptions	Track positive outcomes, reward reliability, and strengthen partnerships with proven contributors across operational cycles
Procurement repeats negotiation mistakes because prior contract learnings were not documented	Capture negotiation outcomes, maintain playbooks, and apply lessons to future supplier engagements and contracting strategies
Teams underestimate supplier risk because earlier incidents were forgotten or minimized	Record incidents, analyze patterns, and integrate supplier risk scoring into operational planning and vendor management
Organizations fail to leverage supplier innovations because past collaboration successes were not preserved	Document innovation wins, formalize collaboration models, and encourage joint development initiatives with high-performing suppliers
Leaders rotate suppliers frequently without understanding historical cost, quality, and reliability trends	Consolidate supplier intelligence, review longitudinal data, and base decisions on multi year performance and strategic fit

Failure Amnesia is the refusal to internalize the painful lessons from past operational breakdowns, leading to a "patch and pray" culture. In heavy manufacturing, a machine failure that isn't root-caused will inevitably happen again, often with higher costs. Daily, this looks like a reactive environment where symptoms are treated with "quick fixes" while the underlying structural causes are ignored. Transformation requires a "Blameless Post-Mortem" behavior supported by high-fidelity AI data logs. When you break this barrier, you turn every error into a permanent process improvement. You shift the focus from "who to blame" to "how to harden the system." This ensures that the organization's "immune system" grows stronger with every challenge, eventually reaching a state of near-zero recurring operational errors.

DO'S FOR THE LEADER	DON'TS FOR THE LEADER
Document all failures in a blameless way.	Fail to learn from past mistakes.
Use a blameless post-mortem process.	Hunt for a scapegoat after a failure.
Learn from past mistakes.	Let the same errors happen again.
Share what you learned from failures.	Hide or cover up past failures.
Create a safe space for mistakes.	Punish people for making mistakes.
Improve processes based on failures.	Assume failure is just a fact of life.

SITUATION	SUGGESTED ACTION
Teams repeat operational failures because previous incidents were not analyzed with sufficient rigor	Conduct structured post-mortems, document findings, and integrate corrective actions into operational standards and governance
Executives underestimate systemic weaknesses because earlier failures were treated as isolated events	Identify patterns, assess systemic risks, and implement cross-functional remediation plans addressing root causes
Operators repeat errors because training materials were never updated after prior incidents	Revise training, reinforce updated procedures, and validate adoption through operational audits and performance checks
Teams ignore early warning signs because past signals were dismissed or forgotten	Track anomalies, escalate patterns, and integrate predictive monitoring into operational workflows and dashboards
Organizations fail to improve because failures are not shared across business units	Create cross-unit learning forums, publish insights, and institutionalize shared improvement practices across operations
Leaders repeat strategic missteps because historical decision outcomes were never reviewed objectively	Establish decision reviews, analyze impact, and incorporate learnings into strategic planning and operational prioritization

This barrier involves a collective forgetting of past market shocks or operational risks, leading to a lack of preparation for future "Black Swan" events. In finance, this is seen when teams ignore the lessons of past volatility during a long period of market stability. Daily, this leads to lean operations that are technically efficient but "brittle"—shattering the moment a global shipping lane closes or a cyber-attack occurs. Transformation requires "Predictive Risk Modeling" that refuses to let the past fade. By breaking this barrier, you build "resiliency buffers" back into your core operations. You use AI to simulate past crises against current data, ensuring your supply chain and financial processes are ready for the unexpected, transforming passive stability into true, proactive anti-fragility.

DO'S FOR THE LEADER	DON'TS FOR THE LEADER
Document past risks and how they were handled.	Fail to remember and prepare for past risks.
Create a risk log for all projects.	Have no system for tracking risks.
Plan for potential risks.	Be unprepared for shifts in demand.
Use past data to predict future risks.	Assume a risk won't happen again.
Communicate risks to all teams.	Keep risks a secret from your teams.
Learn from every risk that occurs.	Let the same risks happen again.

SITUATION	SUGGESTED ACTION
Teams underestimate operational threats because earlier risk events were not documented or reviewed thoroughly	Maintain risk histories, analyze patterns, and integrate lessons into forecasting and mitigation planning across operations
Leaders repeat strategic missteps because prior risk signals were ignored during earlier decision cycles	Establish escalation protocols, require evidence reviews, and incorporate historical indicators into executive decision frameworks
Operators overlook early warning signs because past anomalies were dismissed as isolated incidents	Track deviations, investigate patterns, and embed predictive monitoring into frontline workflows
Teams fail to prepare for disruptions because earlier contingency plans were never updated or tested	Refresh scenarios, validate assumptions, and conduct regular drills to ensure readiness
Organizations misjudge supplier exposure because previous disruptions were forgotten or minimized	Maintain supplier risk profiles, review incident histories, and integrate findings into procurement strategies
Executives assume stability because past volatility was not captured in operational reporting	Build volatility dashboards, review long-term trends, and incorporate variability into planning cycles

6. KNOWLEDGE AMNESIA BARRIER | The Knowledge Drain

The loss of institutional knowledge due to employee turnover or a lack of documentation is a silent killer of operations. When a "star operator" in a media firm or bank leaves, they often take the "secret sauce" of the process with them, leaving the rest of the team guessing. Daily, this manifests as a sudden, inexplicable drop in quality or speed when key personnel move on. Transformation requires "Knowledge Orchestration" where AI captures expert heuristics into "Living Standard Operating Procedures." By breaking this barrier, you ensure that excellence is a property of the system, not just the individual. This allows your operations to scale without being held hostage by the availability of specific experts, ensuring continuity regardless of who is in the seat.

DO'S FOR THE LEADER	DON'TS FOR THE LEADER
Document all institutional knowledge.	Let knowledge be lost due to turnover.
Create a central knowledge base.	Let each team have its own knowledge base.
Cross-train employees for flexibility.	Let people get stuck in a single role.
Make it easy to find information.	Have no documentation or a bad system.
Use tech to store and share knowledge.	Rely on human memory for everything.
Onboard new employees with a clear process.	Have no onboarding process.

SITUATION	SUGGESTED ACTION
Teams lose critical expertise when experienced employees depart without structured knowledge transfer	Implement transition plans, document processes, and ensure continuity through cross-training
Leaders underestimate onboarding complexity because institutional knowledge was never formalized or centralized	Build knowledge repositories, standardize documentation, and maintain accessible operational references
Operators repeat mistakes because procedural insights were never captured after earlier incidents	Record learnings, update playbooks, and integrate improvements into training materials
Teams duplicate work because prior research and decisions were not preserved or shared effectively	Consolidate archives, maintain searchable systems, and enforce documentation standards across functions
Organizations rely on tribal knowledge that disappears when teams reorganize or shift priorities	Establish governance, maintain ownership, and ensure knowledge stewardship across operational domains
Executives misjudge capability gaps because historical expertise levels were never tracked or assessed	Conduct capability mapping, review trends, and align workforce planning with documented skill requirements

Ai6x6™ Transformation Blueprint

Operations Empowering Behaviors

"The art of simplicity is a puzzle of complexity." - Douglas Horton

Douglas Horton (1891–1968) was a prominent American Protestant clergyman and academic leader. He is best known for his significant work in ecumenical relations and uniting different Protestant denominations.

Operations empowering behaviors are positive and proactive actions that reorient the entire organization around operational excellence. Improvement Behavior is the ongoing practice of making incremental changes to improve processes, which is a direct antidote to process amnesia. Modernization Behavior is the practice of consistently updating old systems to be more efficient, which stands in stark contrast to the stagnation caused by technological apprehension. Partnership Behavior is the practice of building strong, collaborative relationships with suppliers and partners, moving beyond a purely transactional approach. As Douglas Horton's quote suggests, the art of simplicity is a puzzle of complexity. These behaviors, when adopted across the entire organization, create a virtuous cycle where efficiency and quality are constantly being improved, leading to a more resilient and effective operational system.

This framework secures the Execution domain by meticulously analyzing Operations Empowering Behaviors, investigating the human actions that interact with systems and automation to drive Optimization and accelerate Digital Evolution. A critical action is Adaptive Behavior, requiring personnel to quickly adjust to new automated workflows and updated processes as they are deployed and modified. Transformation success depends on Trusting Behavior, relying appropriately on automated systems for efficiency while maintaining necessary human oversight and vigilance. Teams must exhibit Debugging Behavior, actively identifying, logging, and fixing system errors and failures to maintain uptime and performance integrity. Efficiency is driven by Optimizing Behavior, where employees continuously look for ways to improve process efficiency and reduce waste within the operational systems. This requires Training Behavior, the commitment to proactively learning how to use new tools and complex systems effectively and safely. Finally, the long-term viability is ensured by Maintenance Behavior, regularly caring for and updating systems to prevent deterioration, technical debt, and sudden failure.

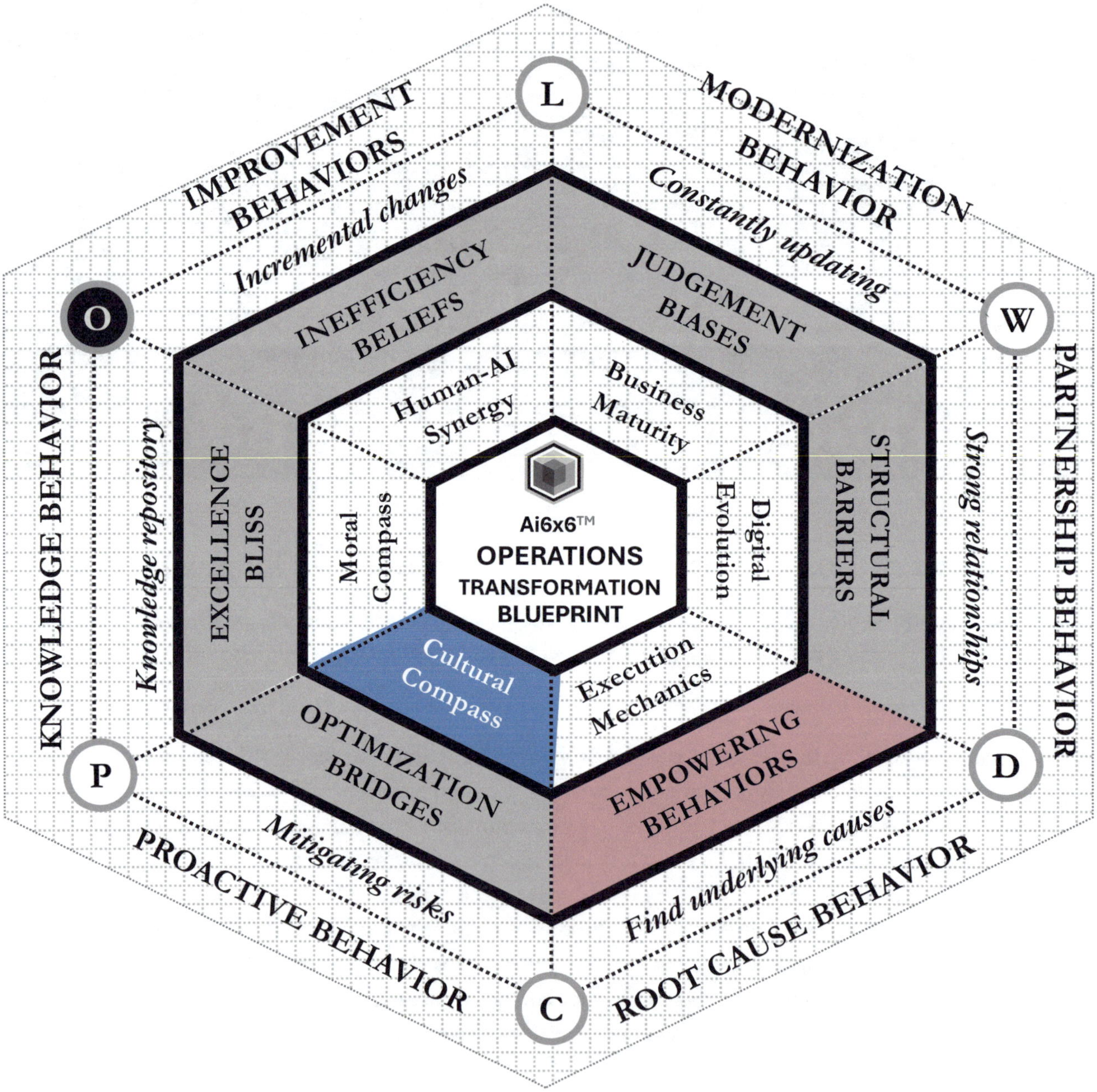

CULTURAL COMPASS FRAMEWORK FOR OPERATIONS LIFECYCLE

1. IMPROVEMENT BEHAVIOR | The Art of Increments

In a transformed operation, the practice of making "incremental changes" is a daily habit, not a once-a-year project. This is the "Kaizen" of the AI era. Daily, it looks like a team reviewing their "process logs" every morning to find one small friction to eliminate. By modeling improvement behavior, you ensure that your operations never become "stale" or "bloated." For an executive, this means rewarding "small wins" as much as major breakthroughs. This behavior builds a "compounding effect" of efficiency that is impossible to ignore. Over time, these daily improvements create a massive competitive gap that "static" organizations can never bridge, turning your operations into a self-evolving engine of excellence, speed, and market adaptability.

DO'S FOR THE LEADER	DON'TS FOR THE LEADER
Make incremental changes to processes.	Stick to the old way of doing things.
Use data to inform improvements.	Make changes based on gut feelings.
Encourage feedback for improvements.	Ignore feedback on slow processes.
Create a culture of continuous improvement.	Assume your processes are perfect.
Reward people for finding improvements.	Punish people for pointing out flaws.
Share success stories for improvement.	Keep all improvements a secret.

SITUATION	SUGGESTED ACTION
Teams identify recurring friction but postpone fixes because issues appear manageable in the moment	Prioritize small corrections, schedule rapid adjustments, and prevent minor inefficiencies from compounding into systemic problems
Operators notice workflow delays but assume they are unavoidable parts of daily operations	Document delays, test alternatives, and implement targeted improvements to restore operational flow and reduce waste
Leaders focus on major initiatives and overlook small operational wins that strengthen system stability	Track micro-improvements, celebrate progress, and reinforce continuous refinement across operational teams
Teams accept outdated procedures because updating them seems disruptive to current throughput expectations	Redesign steps, pilot changes, and integrate improvements without compromising operational continuity or delivery commitments
Organizations treat improvement as periodic rather than embedded in daily operational routines	Establish continuous review cycles, assign ownership, and maintain momentum through structured operational cadence
Executives assume improvement requires major investment rather than disciplined operational tuning	Identify friction points, implement targeted fixes, and scale effective refinements across workflows based on measurable impact

Consistently updating old systems to be more efficient is the behavior that prevents "Legacy Amnesia" and technical stagnation. It means having a "rolling update" strategy for all hardware and software. Daily, this looks like an IT and Operations team that is always "pruning" technical debt rather than letting it accumulate into a mountain. This behavior ensures that your AI models are always running on the "best possible foundation." For you, it means allocating budget for "continuous upgrades" rather than "one-off overhauls" every decade. Modernization behavior turns your infrastructure into a "competitive asset" rather than a "maintenance liability," ensuring you can adopt the latest AI innovations the moment they are ready to deliver value to the business.

DO'S FOR THE LEADER	DON'TS FOR THE LEADER
Consistently update old systems.	Stick with outdated, inefficient tech.
Invest in tech that solves problems.	Buy new tech just for the sake of it.
Train employees on new technology.	Throw new tech at people with no training.
Integrate new tech with old systems.	Create a mess of unconnected systems.
Use tech to automate repetitive tasks.	Have people do tasks a machine could do.
Get feedback on new systems.	Assume new systems are perfect.

SITUATION	SUGGESTED ACTION
Teams rely on aging systems because modernization feels risky during high operational load periods	Plan phased upgrades, test under controlled conditions, and modernize without disrupting throughput or stability
Leaders postpone modernization because legacy systems appear stable despite hidden operational fragility	Assess failure modes, quantify risk, and prioritize upgrades based on operational exposure and continuity requirements
Operators resist new tools because they fear losing efficiency built through legacy familiarity	Provide training, run parallel workflows, and transition gradually to maintain confidence and operational reliability
Teams underestimate technical debt accumulation because issues surface slowly over long operational cycles	Track debt indicators, schedule remediation, and integrate modernization into operational planning and budgeting
Organizations modernize reactively after failures rather than proactively addressing aging infrastructure	Build modernization roadmaps, allocate resources, and execute upgrades before failures occur under load
Executives assume modernization is purely technical rather than deeply tied to operational resilience	Align modernization with business continuity, risk reduction, and long-term operational capacity planning

Building strong, "collaborative relationships" with suppliers and partners is the daily behavior that secures your supply chain against global shocks. It means sharing your "demand forecasts" with suppliers so they can plan better, and actively listening to their "innovative ideas." Daily, this looks like "joint planning sessions" rather than just sending "purchase orders" through a portal. Partnership behavior creates a "buffer of goodwill" that is essential during crises. For you, it means treating your vendors as "strategic extensions of your own team." This behavior ensures that when the market gets tough, you have a loyal network that will go the extra mile to keep your operations running smoothly while others are left stranded and desperate.

DO'S FOR THE LEADER	DON'TS FOR THE LEADER
Build strong, collaborative relationships.	Treat suppliers as a single transaction.
Communicate openly with partners.	Keep your partners in the dark.
Work together to solve problems.	Blame suppliers for all issues.
Have fair, long-term contracts.	Use short-term, aggressive contracts.
Share information for better planning.	Hide your plans from your partners.
Celebrate shared successes.	Only point out partner failures.

SITUATION	SUGGESTED ACTION
Teams engage suppliers only during crises, creating reactive relationships lacking operational alignment	Establish steady communication, share forecasts, and build joint plans that stabilize performance under varying load
Leaders negotiate purely on cost without considering operational reliability or long-term capacity commitments	Evaluate resilience, review historical performance, and prioritize partners who strengthen operational continuity
Operators escalate issues late because supplier expectations were never clarified or reinforced consistently	Define service thresholds, document responsibilities, and enforce accountability through structured operational reviews
Teams treat suppliers as interchangeable despite significant differences in capability and operational maturity	Segment suppliers, tailor engagement, and allocate work based on proven performance under stress
Organizations overlook early supplier deterioration because reporting focuses only on contract compliance metrics	Track trend lines, monitor deviations, and intervene before reliability erosion becomes systemic
Executives assume partnership strength is static rather than requiring continuous operational investment	Maintain governance cadence, review joint performance, and refresh commitments based on evolving operational needs

Investigating failures to understand their "underlying causes"—rather than just treating the surface symptoms—is the behavior that breaks the "Failure Amnesia" loop. Daily, this looks like using the "5 Whys" method for every operational slip-up, no matter how small. It involves using AI to trace the data path of an error back to its original source. This behavior ensures that once a problem is fixed, it "stays fixed" forever. For the executive, this means creating a "safe-to-fail" culture where finding the truth is more important than avoiding blame. Root cause behavior builds an operation that is "resilient by design," where every setback is used as a valuable opportunity to harden the system against future errors.

DO'S FOR THE LEADER	DON'TS FOR THE LEADER
Investigate the root cause of issues.	Only treat the symptoms of a problem.
Use data to find and solve problems.	Solve problems based on gut feelings.
Encourage problem-solving at all levels.	Expect a single person to solve problems.
Create a blameless problem-solving culture.	Blame someone whenever a problem occurs.
Document solutions to prevent repeats.	Let the same problems happen again.
Learn from every failure.	Let failures just fade away.

SITUATION	SUGGESTED ACTION
Teams stop investigation once an obvious symptom is identified, missing deeper operational drivers	Push beyond first answers, test hypotheses, and validate findings through cross-functional review
Leaders demand rapid explanations that discourage thorough analysis during high-pressure incidents	Set investigation windows, protect analytical rigor, and ensure conclusions reflect verified evidence
Operators focus on individual errors rather than examining systemic workflow or design weaknesses	Map processes, inspect handoffs, and identify structural contributors to recurring failures
Teams rely on anecdotal observations instead of analyzing operational data patterns	Consolidate logs, review metrics, and base conclusions on measurable system behavior
Organizations repeat similar failures because corrective actions address symptoms rather than root causes	Track recurrence, reassess findings, and implement structural fixes that eliminate underlying drivers
Executives assume root-cause analysis is a technical exercise rather than an operational discipline	Integrate RCA into governance, assign ownership, and enforce follow-through on identified improvements

Identifying and "mitigating potential risks" before they occur is the hallmark of a truly transformed operation. Daily, this looks like running "what-if" simulations using AI to see how a port strike, a price spike, or a power outage would affect your bottom line. It's about setting "early warning alerts" on key operational metrics. This behavior shifts your team from "reactive firefighters" to "strategic navigators." For you, it means investing in "predictive maintenance" and "risk intelligence" tools. Proactive behavior ensures that your operations are always "ahead of the curve," allowing you to dodge disasters that sink your competitors and giving you the "peace of mind" to focus on growth rather than survival.

DO'S FOR THE LEADER	DON'TS FOR THE LEADER
Anticipate issues before they arise.	Wait for problems before addressing them.
Be proactive with your communication.	Only communicate after a problem occurs.
Use data to predict future needs.	React to every new request.
Plan for potential problems.	Have no plan for issues.
Create a proactive support system.	Have a reactive, fire-fighting system.
Get feedback on potential issues.	Ignore all potential problems.

SITUATION	SUGGESTED ACTION
Teams wait for visible failures before addressing known operational vulnerabilities	Monitor weak signals, escalate early, and implement safeguards before issues escalate
Leaders prioritize short-term throughput over addressing emerging operational risks	Balance production with resilience, allocate capacity, and reinforce risk-aware decision-making
Operators ignore subtle anomalies because they appear harmless during normal load conditions	Track deviations, investigate patterns, and validate whether anomalies indicate deeper system stress
Teams postpone maintenance until systems show clear signs of degradation	Schedule preventive work, test components, and maintain operational stability through disciplined upkeep
Organizations react to disruptions rather than anticipating predictable failure modes	Build risk models, simulate scenarios, and prepare mitigation plans aligned with operational realities
Executives assume resilience emerges naturally without structured risk management practices	Establish governance, assign ownership, and integrate proactive risk reviews into operational cadence

Fostering a culture of "openly sharing information" and best practices is the behavior that kills "Knowledge Amnesia" and empowers the entire team. Daily, this looks like a "wiki-first" culture where every process tweak or solution is documented in real-time. It involves using AI to "tag and surface" relevant knowledge to the people who actually need it at that moment. This behavior ensures that the "collective intelligence" of the organization is always accessible. For the executive, it means rewarding those who "teach and document" rather than those who "hoard information." Knowledge behavior turns your workforce into a "learning community," ensuring that your operational excellence is "durable" and can survive any amount of staff turnover.

DO'S FOR THE LEADER	DON'TS FOR THE LEADER
Document all institutional knowledge.	Let knowledge be lost due to turnover.
Create a central knowledge base.	Let each team have its own knowledge base.
Share information openly.	Hoard information in departments.
Use tech to store and share knowledge.	Rely on human memory for everything.
Make it easy to find information.	Have no documentation or a bad system.
Create a culture of open communication.	Create a culture of secrecy.

SITUATION	SUGGESTED ACTION
Teams hoard operational knowledge because sharing feels time-consuming or undervalued	Incentivize documentation, streamline tools, and reinforce knowledge sharing as core operational responsibility
Leaders underestimate onboarding complexity because institutional knowledge remains undocumented or fragmented	Build centralized repositories, standardize formats, and maintain accessible operational references
Operators repeat mistakes because insights from earlier incidents were never captured or distributed	Record learnings, update playbooks, and integrate improvements into training materials
Teams duplicate work because prior research and decisions were not preserved or easily discoverable	Consolidate archives, maintain searchable systems, and enforce documentation standards across functions
Organizations lose expertise during transitions because knowledge transfer is informal or inconsistent	Establish transition protocols, assign mentors, and ensure continuity through structured handovers
Executives assume knowledge naturally spreads across teams without deliberate operational mechanisms	Create forums, schedule reviews, and embed knowledge exchange into operational governance

Ai6x6™ Transformation Blueprint

Operations Optimization Bridges

"The secret of success is to do the common thing uncommonly well." - John D. Rockefeller Jr.

John D. Rockefeller Jr. (1874–1960) was an American financier and philanthropist. He is famous for managing his family's vast philanthropic efforts and playing a key role in the creation of Rockefeller Center in New York City.

Operations optimization bridges are practical frameworks and processes that connect a state of operational inefficiency to a future of excellence. They are the strategic tools that turn good intentions into a seamless and satisfying operational system. The Human-AI Bridge is a framework that ensures humans and AI work together effectively, leveraging the strengths of both, thereby overcoming algorithmic aversion. An Accountability Bridge is a clear framework for determining accountability when a system fails, ensuring that the root cause is investigated rather than a human operator being unfairly blamed. The Strategic Allocation Bridge is a system for delegating tasks based on objective criteria, rather than personal preference. As John D. Rockefeller Jr.'s quote suggests, the secret of success is to do "the common thing uncommonly well." These bridges enable an organization to do just that, ensuring that every operational task is executed with excellence.

Operations Optimization Bridges, investigates the mechanisms that connect and optimize processes, driving Optimization and accelerating Digital Evolution. System Integration Bridges are foundational, connecting core software systems (e.g., ERP, CRM, SCM) to eliminate manual data transfer and ensure a single source of truth across functional areas. Automation Bridges involve the implementation of RPA and other intelligent automation tools that seamlessly link previously manual tasks, eliminating bottlenecks and accelerating workflows. Supply Chain Bridges provide real-time visibility and connectivity across suppliers, logistics providers, and internal production, enabling adaptive inventory and fulfillment. Monitoring Bridges deploy tools for continuous performance tracking, providing instant alerts on system errors, bottlenecks, or degradation to maintain uptime and quality control. Process Mapping Bridges are visualization tools and techniques used to document and communicate end-to-end workflows, ensuring every stakeholder understands the system. Finally, Human-in-the-Loop Bridges establish clear, defined mechanisms for human oversight and intervention within automated processes, ensuring complex decision-making remains accountable and guided by human judgment.

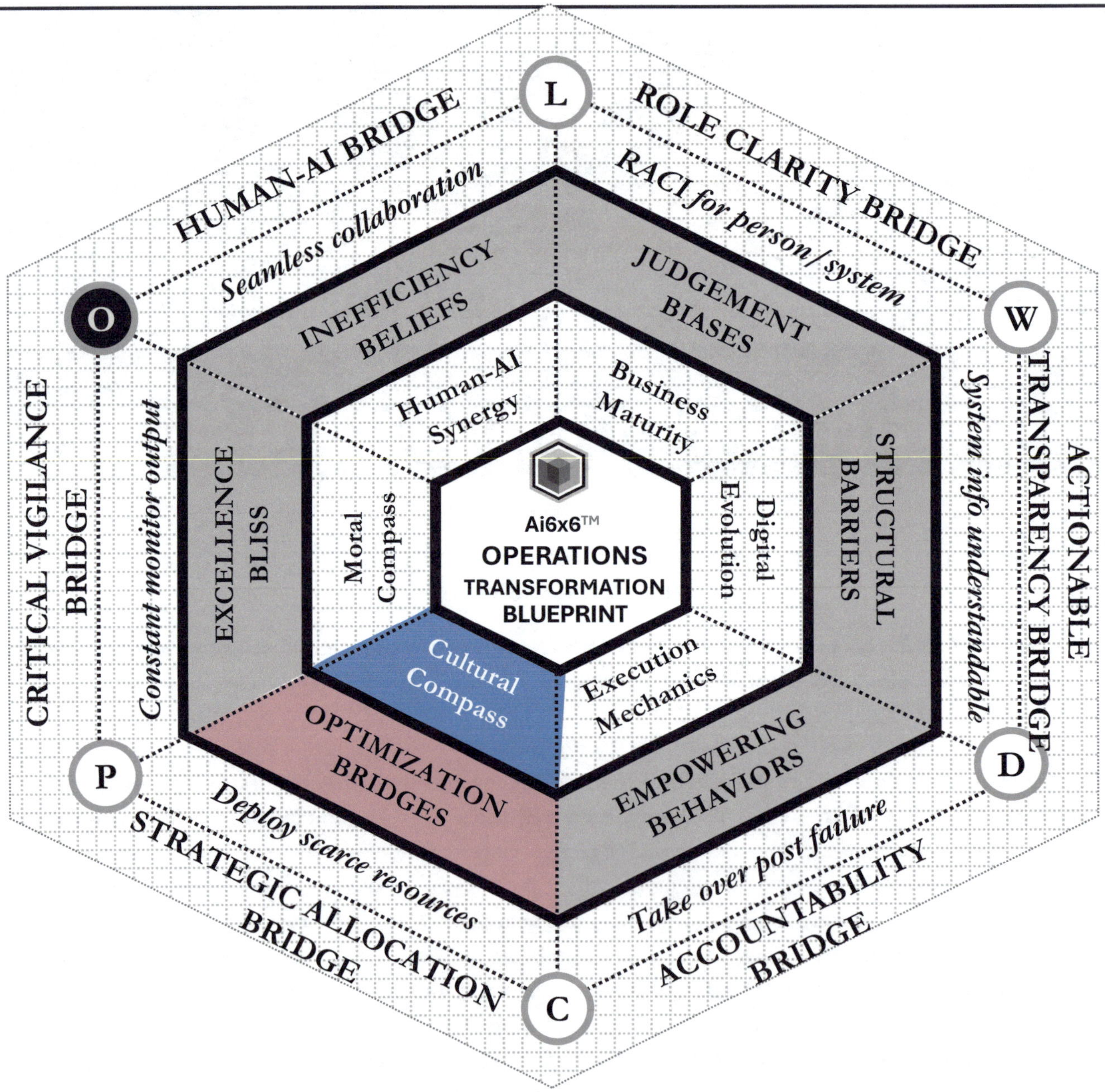

Ai6x6 CULTURAL COMPASS FRAMEWORK FOR OPERATIONS LIFECYCLE

1. HUMAN-AI BRIDGE | The Collaborative Loop

This bridge is a formal framework—like "Human-in-the-Loop" (HITL)—that ensures humans and AI work together in a synergistic way. It defines exactly when the AI should "take the lead" and when a human must "provide the final sign-off." Daily, this bridge looks like a user interface where the AI "explains its confidence level" and reasoning to the human operator. By building this bridge, you cross from "automated silos" to "collaborative intelligence." It ensures your AI is a "force multiplier" for human talent. For the executive, this bridge is the "safety net"; it prevents "runaway algorithms" from causing damage. This bridge provides the "operational harmony" required to run high-speed, high-stakes processes with total confidence.

DO'S FOR THE LEADER	DON'TS FOR THE LEADER
Ensure humans and AI work together.	Let AI work in a silo without humans.
Use AI as a tool to enhance humans.	Mistrust AI-driven processes.
Explain how the AI works.	Use AI as a black box system.
Train employees to work with AI.	Assume employees don't need training.
Validate AI output with human oversight.	Over-rely on AI output without checking it.
Use AI to increase efficiency.	Let fear of AI prevent efficiency.

SITUATION	SUGGESTED ACTION
Teams struggle when AI outputs conflict with frontline observations during high operational load	Define override rules, compare outcomes, and calibrate systems using verified operational evidence from human operators
Operators hesitate to trust AI recommendations without understanding model boundaries or failure modes	Expose limitations, train teams on edge cases, and reinforce shared responsibility for decision quality
Leaders deploy AI without clarifying when humans must intervene during critical workflows	Set intervention points, document escalation paths, and maintain human oversight for high-impact operational decisions
Teams rely on AI for tasks requiring contextual judgment the system cannot interpret reliably	Map task suitability, reassign responsibilities, and ensure AI supports rather than replaces contextual reasoning
Operators override AI inconsistently because guidance on acceptable deviation is unclear	Define override criteria, track deviations, and refine models based on validated operator feedback
Executives assume AI performance improves automatically without structured retraining or governance	Establish retraining cadence, monitor drift indicators, and enforce ownership for ongoing model upkeep

This bridge is a system that clearly defines the "roles and responsibilities" of each person and each automated system in a complex process. It eliminates the "who's doing what?" confusion that slows down operations. Daily, this looks like a "RACI matrix" that includes AI agents as key stakeholders. By building this bridge, you ensure that there are no "gaps" in the process where errors can hide or tasks can be forgotten. It prevents the "Moral Crumple Zone" by making it clear who is responsible for what. For the executive, this bridge is the "accountability engine." It ensures that every part of the operation has a "clear owner," creating a sense of "order and focus" that is essential for scaling.

DO'S FOR THE LEADER	DON'TS FOR THE LEADER
Clearly define roles and responsibilities.	Have vague or overlapping roles.
Create a clear accountability framework.	Blame a human operator for a system's failure.
Document who is responsible for what.	Let people guess who is responsible.
Use a system to track responsibilities.	Use no system for tracking responsibilities.
Be fair and transparent in delegation.	Delegate based on personal preference.
Get feedback on role clarity.	Assume roles are clear.

SITUATION	SUGGESTED ACTION
Teams duplicate work because responsibilities between humans and systems are not explicitly defined	Map ownership, assign boundaries, and eliminate overlap through structured operational design
Operators hesitate during incidents because decision authority is unclear across workflows	Define authority levels, document escalation triggers, and rehearse decision paths under simulated load
Leaders assume teams understand system responsibilities without formal documentation or training	Publish role definitions, validate comprehension, and reinforce clarity through operational reviews
Teams misinterpret system capabilities and assign tasks beyond technical limits	Clarify constraints, update training, and align task allocation with validated system behavior
Operators intervene unnecessarily because they misunderstand when automation should run independently	Define autonomy thresholds, test scenarios, and reinforce correct intervention timing
Executives assume role clarity emerges naturally as systems evolve	Review responsibilities regularly, adjust boundaries, and maintain alignment as operational conditions change

This bridge is a process that makes complex system information "understandable and useful" for the everyday operator. It's the bridge between "raw data" and "smart decisions." Daily, this looks like "contextual dashboards" that show not just "what" is happening, but "why" it's happening and "what to do next." By building this bridge, you eliminate the "Transparency Illusion." It ensures that your team isn't just "watching" the system, but actively "managing" it. For the executive, this bridge is the "clarity of vision" tool. It allows you to see the true "health" of your operations at a glance. This bridge provides the "operational awareness" that allows for fast, accurate pivots in a rapidly changing market environment.

DO'S FOR THE LEADER	DON'TS FOR THE LEADER
Make system info truly understandable.	Believe a system is transparent just because it has some data.
Provide clear context for data.	Provide raw data with no explanation.
Create a single source of truth.	Let each team have a different view of data.
Explain how the system works.	Let users assume they know how it works.
Use visualization to explain data.	Use only text and numbers.
Get feedback on transparency.	Assume your system is transparent.

SITUATION	SUGGESTED ACTION
Teams receive raw data without context, making operational decisions slow and error-prone	Add interpretation layers, highlight anomalies, and present information aligned with operational decision points
Operators misjudge system health because dashboards hide upstream dependencies or blind spots	Expose lineage, reveal dependencies, and validate whether displayed metrics reflect actual system conditions
Leaders rely on summaries that obscure critical operational signals during volatile periods	Provide drill-downs, surface leading indicators, and ensure visibility into early warning patterns
Teams misinterpret alerts because thresholds lack operational justification or historical grounding	Recalibrate thresholds, test scenarios, and align alerts with real operational risk
Operators trust green indicators without understanding what conditions they actually represent	Clarify definitions, test failure modes, and ensure indicators reflect meaningful operational states
Executives assume transparency exists because data is technically accessible	Translate complexity, simplify views, and ensure information supports timely operational decisions

This bridge is a clear framework for determining "who is responsible" when a system—especially an autonomous one—fails. It's the "legal and ethical" bridge for the AI era. Daily, this looks like "governance protocols" that are reviewed by both legal and operations teams. By building this bridge, you ensure that failures lead to "learning," not just "litigation." It prevents the unfair blaming of frontline workers for system-level flaws. For the executive, this bridge is the "trust-builder" with your workforce and your customers. It ensures that your organization "takes ownership" of its technology. This bridge provides the "ethical stability" required to deploy advanced AI in high-consequence areas like finance, healthcare, or heavy manufacturing.

DO'S FOR THE LEADER	DON'TS FOR THE LEADER
Create a clear accountability framework.	Blame a human operator for a system's failure.
Blame the system, not the person.	Hunt for a scapegoat after a failure.
Investigate the root cause of failures.	Only punish the person closest to the error.
Improve the system to prevent repeats.	Punish a person and do nothing else.
Create a blameless post-mortem.	Punish people for making mistakes.
Focus on the process, not the person.	Focus on the person, not the process.

SITUATION	SUGGESTED ACTION
Operators are blamed for failures caused by flawed automation or incomplete system safeguards	Trace failure origins, review design decisions, and assign accountability to responsible system owners
Teams avoid reporting issues because accountability defaults to the nearest human operator	Establish blameless reporting, investigate system factors, and reinforce psychological safety
Leaders assign blame before reviewing logs, configurations, or system behavior under load	Analyze evidence, validate conditions, and ensure conclusions reflect actual failure sources
Operators hesitate to intervene because unclear accountability creates operational risk	Define intervention authority, document responsibilities, and rehearse scenarios requiring decisive action
Teams escalate incidents inconsistently because accountability rules vary across workflows	Standardize escalation paths, align expectations, and enforce consistent accountability practices
Executives overlook systemic design flaws because accountability defaults to individual performance	Conduct systemic reviews, document design gaps, and align accountability with structural contributors

This bridge is a system for delegating tasks based on "objective criteria" like speed, accuracy, cost, and the need for human empathy. It's the bridge that decides: "Should a human do this, or should the AI?" Daily, this looks like a "decision engine" that routes tasks to the most efficient resource based on the current context. By building this bridge, you eliminate "Delegation Bias." It ensures your high-value humans are focused on "complex problem solving" while the AI handles the "routine data crunching." For the executive, this bridge is the "margin optimizer." It ensures you are getting the "highest ROI" from every resource, allowing you to do more with less, without burning out.

DO'S FOR THE LEADER	DON'TS FOR THE LEADER
Delegate tasks based on objective criteria.	Delegate tasks based on personal preference.
Use data to inform who does what.	Give tasks to a favorite or a friend.
Be fair and transparent in delegation.	Be unfair or secretive in delegation.
Delegate to empower people.	Delegate to avoid doing a task yourself.
Use clear rules for delegation.	Have no clear rules for who does what.
Match tasks with skills.	Give tasks to people who are not skilled.

SITUATION	SUGGESTED ACTION
Teams assign work based on convenience, not capability, ignoring task complexity and risk	Define allocation rules, assess skills objectively, and match responsibilities to proven capacity under realistic operational load and risk conditions across scenarios consistently
Operators receive tasks misaligned with strengths because assignment logic is inconsistent across workflows	Map competencies, rebalance distribution, and ensure assignments reflect actual operational requirements, constraints, and current system behavior under varying conditions
Leaders delegate critical tasks without reviewing workload, readiness, or situational risk profile properly	Validate capacity, review expertise, and assign responsibilities according to situational risk, urgency, and required operational judgment under current system conditions and constraints
Teams rely on outdated delegation patterns that ignore evolving demand signals and system behavior	Monitor performance trends, adjust assignments regularly, and align delegation decisions with live operational data and constraints across workflows and time horizons consistently
Operators hesitate to challenge misaligned assignments because escalation norms are unclear or inconsistent	Clarify escalation paths, communicate expectations, and correct misalignment quickly through structured reviews and transparent operational governance across teams and time periods consistently
Executives assume delegation quality improves naturally without explicit oversight, metrics, or feedback loops	Track delegation outcomes, review recurring patterns, and refine allocation rules using measurable performance data and post-incident analysis across operations over time horizons

This bridge is a framework that encourages operators to "constantly monitor" and "critically question" the output of every automated system. It's the "skeptic's bridge" that prevents Over-reliance Bias. Daily, this looks like "red-teaming" sessions where operators try to find flaws in the AI's logic or data sources. By building this bridge, you ensure your team remains "mentally sharp" and "operationally ready." It creates a culture of "professional skepticism." For the executive, this bridge is the "fail-safe." It ensures that your most important processes are always being "watched by a human eye." This bridge provides the "resilient oversight" that is the ultimate protection against the "unseen errors" of a purely algorithmic world.

DO'S FOR THE LEADER	DON'TS FOR THE LEADER
Encourage operators to monitor output.	Let the system run with no human oversight.
Train employees to question the system.	Assume the system is always correct.
Create a human-in-the-loop process.	Let a system run without human input.
Set clear limits on a system's use.	Let a system be used for everything.
Get feedback on system usage.	Assume a system is being used correctly.
Have a clear process for overrides.	Have no clear process for overrides.

SITUATION	SUGGESTED ACTION
Teams overlook subtle anomalies because monitoring tools hide early indicators of stress developing	Expose weak signals, refine thresholds, and train operators to recognize emerging instability before it degrades service under load and across systems consistently
Operators trust green indicators without verifying whether metrics reflect actual system conditions accurately	Validate indicator definitions, test failure modes, and ensure displayed states correspond to real operational health under different loads and scenarios reliably over time
Leaders assume monitoring coverage is complete without reviewing blind spots or dependency gaps	Audit observability, map dependencies, and extend monitoring to upstream, downstream, and cross-system failure points under normal and peak operating conditions consistently
Teams ignore intermittent anomalies because they appear harmless during normal throughput periods initially	Track anomaly frequency, analyze clustering, and escalate when patterns suggest emerging systemic degradation under varying loads and time windows across environments
Operators rely solely on alerts without questioning whether thresholds remain operationally relevant today	Recalibrate alerts regularly, test scenarios, and align thresholds with current risk tolerance and system behavior under different operating modes and conditions consistently
Executives assume vigilance emerges naturally without structured reinforcement or operational discipline mechanisms behind	Establish monitoring cadence, assign clear ownership, and embed vigilance expectations into daily operational routines and performance reviews across teams and time horizons

Ai6x6™ Transformation Blueprint

Operations Excellence Bliss

"Quality means doing it right when no one is looking." - Henry Ford

Henry Ford (1863–1947) was an American industrialist and founder of the Ford Motor Company. He is best known for pioneering the moving assembly line, which revolutionized manufacturing and made cars affordable for the public.

Operations excellence bliss is the ultimate harmonious state where operations are a source of profound satisfaction and success. This is the culmination of a successful operations transformation. Blissful Partnerships is the joy of having strong, trusted relationships with suppliers and partners. Blissful Advancement is the pride that comes from a system that is constantly being improved and modernized. Blissful Solutions is the satisfaction of finding clever and effective solutions to complex problems, thanks to a culture of root cause analysis and creative problem-solving. As Henry Ford's quote suggests, quality means doing it right when no one is looking. This state of bliss is not about a single successful project but about a culture where excellence is the norm. It is the reward for the hard work of introspection and systemic change, creating a company that is not just successful but also deeply respected for its efficiency, reliability, and innovation.

This framework secures the Execution domain by meticulously analyzing Operations Excellence Bliss, investigating the optimal outcomes that result from systems and automation, driving Optimization and accelerating Digital Evolution. The primary goal is Operational Harmony, ensuring a smooth, predictable, and highly efficient workflow across the entire value chain, reducing complexity and internal stress. Success minimizes waste, resulting in Minimal Waste throughout processes, inventory, and resources, maximizing cost-effectiveness and resource stewardship. Bliss seeks Zero Defects, achieving a system that consistently produces flawless products or services with minimal errors or rework required, dramatically improving quality and customer trust. The pinnacle of system management is Predictive Maintenance, where systems possess the intelligence to detect and correct potential failures before they disrupt operations, drastically improving uptime and reliability. The outcome must include High Scalability, ensuring systems and processes can grow and adapt rapidly and cost-effectively alongside the expansion of the business, which is the precondition for capitalizing on growth opportunities.

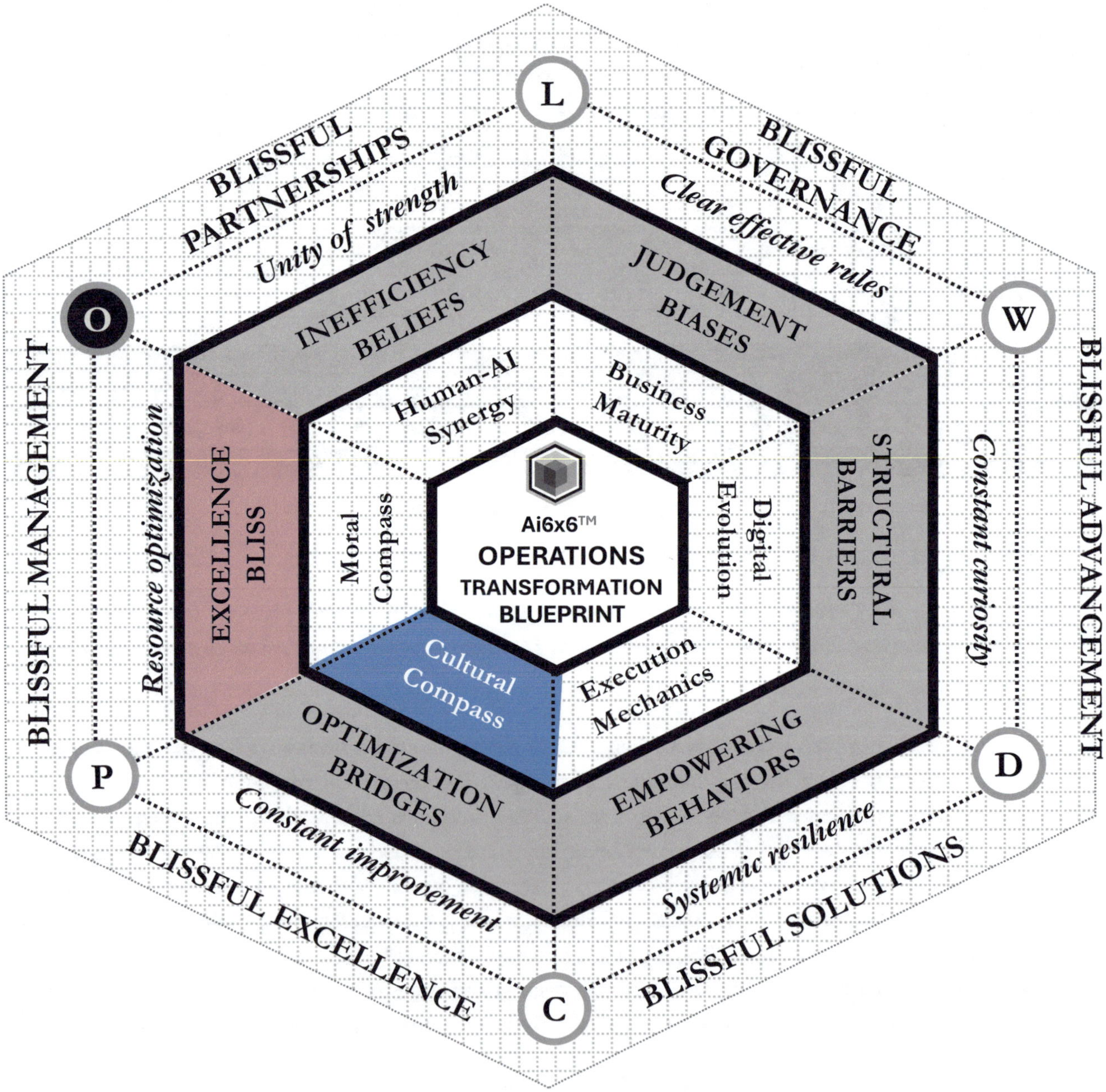

Ai6x6 CULTURAL COMPASS FRAMEWORK FOR OPERATIONS LIFECYCLE

There is a profound sense of "security and strength" that comes from having strong, trusted relationships with your key suppliers. This bliss is the "death of supply chain anxiety." In your daily life as an executive, it's the feeling of knowing that when a global crisis hits, your partners will "prioritize your needs" because of the bridge you've built. This bliss turns procurement from a "battle" into a "bond." It provides a sense of "belonging to a larger, successful ecosystem." This feeling of "not being alone" in a volatile market is a powerful motivator. This bliss is the reward for "Partnership Behavior"—it transforms your supply chain into a "community of success" that is resilient, innovative, and loyal to your vision.

DO'S FOR THE LEADER	DON'TS FOR THE LEADER
Build strong, trusted relationships.	Treat suppliers as a single transaction.
Communicate openly with partners.	Keep your partners in the dark.
Work together to solve problems.	Blame suppliers for all issues.
Have fair, long-term contracts.	Use short-term, aggressive contracts.
Share information for better planning.	Hide your plans from your partners.
Celebrate shared successes.	Only point out partner failures.

SITUATION	SUGGESTED ACTION
Teams feel confident when suppliers consistently deliver under pressure without constant supervision requirements	Share demand forecasts, align expectations, and reinforce reliability through structured reviews and joint operational planning across cycles and regions where relevant
Operators appreciate suppliers who respond quickly during disruptions and communicate transparently throughout events	Establish direct channels, define response expectations, and review incident handling after each disruption to improve coordination and future performance
Leaders value suppliers who flag risks early before they impact operations materially downstream	Encourage early escalation, reward transparency, and integrate supplier risk signals into internal planning and mitigation across functions and time horizons
Teams celebrate when suppliers flex capacity to match volatile demand without degrading quality	Share variability patterns, coordinate scaling plans, and formalize mechanisms for rapid capacity adjustments during peaks and troughs in demand
Organizations thrive when suppliers propose improvements that remove friction from shared workflows continuously	Document joint wins, expand successful patterns, and embed supplier innovation into ongoing operational design and governance discussions
Executives trust partnerships that consistently strengthen resilience across volatile operating conditions and markets	Review performance regularly, renew commitments selectively, and prioritize partners who demonstrably reduce operational risk over time and events

The satisfaction of having "clear and effective rules" that are easy for everyone to follow is the bliss of "ordered freedom." It's the feeling of an operation that "runs like clockwork" because everyone knows the boundaries. This bliss is the "death of confusion." Daily, it means your team spends their energy on "meaningful work," not on "navigating bureaucracy." It provides a sense of "fairness and clarity" that lowers stress. This bliss is the realization that "great rules" don't slow you down; they "speed you up" by removing doubt. When your AI governance is so clear that compliance is automatic, you achieve a level of "blissful governance" that allows for rapid, low-risk innovation and long-term growth.

DO'S FOR THE LEADER	DON'TS FOR THE LEADER
Have clear and effective rules.	Have vague or non-existent rules.
Make compliance a seamless process.	Make compliance a chore or a burden.
Use technology to automate compliance.	Rely on manual compliance checks.
Train staff on all regulations.	Assume everyone knows the rules.
Update processes for new laws.	Fail to adapt to changing regulations.
Be transparent with regulators.	Hide or obscure compliance issues.

SITUATION	SUGGESTED ACTION
Teams operate confidently when rules are unambiguous and aligned with real operating conditions	Clarify expectations, remove contradictions, and ensure governance reflects how systems actually behave under load and failure scenarios observed historically
Operators appreciate governance that reduces ambiguity during high-pressure decision windows significantly operationally speaking	Define decision paths, document triggers, and rehearse scenarios where rapid judgment is required under stress and limited information availability
Leaders value governance that prevents escalation bottlenecks during operational incidents and crises alike	Streamline approvals, assign clear authority, and maintain flow even when multiple teams are involved across sites and time zones
Teams perform better when governance removes unnecessary steps that slow throughput significantly downstream	Identify redundant controls, simplify workflows, and preserve only checks that materially reduce operational risk under normal and peak conditions
Organizations benefit when governance adapts quickly to changing operational realities and constraints dynamically	Review rules regularly, retire obsolete requirements, and update constraints as systems and markets evolve over time and regions
Executives trust governance that strengthens resilience without adding unnecessary bureaucracy or delay everywhere	Balance control with agility, measure governance impact, and refine structures using operational feedback and incident reviews from multiple functions over time

The pride that comes from a system that is "constantly being improved" and updated is the bliss of "evolutionary momentum." It's the feeling of knowing that your operations are "better today than they were yesterday." This bliss is the "death of stagnation." Daily, it manifests as a team that is "excited by change" rather than "fearful of it." It provides a sense of "forward motion and mastery." This bliss is the reward for "Modernization Behavior." When you know your tech stack and your processes are at the "cutting edge," you lead with a sense of "competitive invincibility." This bliss turns the daily grind into a "journey of excellence," where every update is a victory and every tool is a breakthrough.

DO'S FOR THE LEADER	DON'TS FOR THE LEADER
A system that is constantly improved.	A system that never gets updated.
Use data to inform improvements.	Make changes based on gut feelings.
Encourage feedback for improvements.	Ignore feedback on slow processes.
Reward people for finding improvements.	Punish people for pointing out flaws.
Modernize old systems.	Stick with outdated, inefficient tech.
Use tech to automate repetitive tasks.	Have people do tasks a machine could do.

SITUATION	SUGGESTED ACTION
Teams gain confidence when small improvements accumulate into meaningful operational performance gains	Track incremental wins, reinforce momentum, and integrate improvements into daily routines to maintain continuous operational progress across workflows
Operators appreciate when outdated steps are removed and workflows become cleaner and faster overall	Identify friction points, eliminate unnecessary actions, and validate improvements through measurable throughput gains across different operating conditions
Leaders value seeing improvement cycles shorten as teams build stronger operational discipline	Establish cadence, review progress, and reinforce behaviors that accelerate improvement without compromising system stability or reliability
Teams feel momentum when improvements reduce recurring incidents and stabilize system behavior consistently	Analyze incident patterns, implement targeted fixes, and verify impact through structured post-change monitoring across environments
Organizations thrive when improvement becomes habitual rather than a periodic initiative requiring resets	Embed improvement expectations, assign ownership, and maintain visibility through structured operational reviews and performance cycles
Executives appreciate when improvement efforts directly strengthen resilience under volatile operating conditions	Prioritize changes that reduce risk, validate outcomes, and scale successful patterns across teams and workflows

The satisfaction of finding "clever and effective solutions" to complex operational problems is the bliss of "creative mastery." It's the "Aha!" moment when an AI insight and human intuition perfectly merge to solve a difficult bottleneck. This bliss is the "death of frustration." Daily, it manifests as "high-fives" on the warehouse floor or in the virtual meeting room. It provides a sense of "intellectual triumph." This bliss is the reward for "Root Cause Behavior." When you solve a problem at its source, you feel the "bliss of permanent resolution." This feeling of "being smarter than the problem" is the ultimate reward for an operations team, turning difficult challenges into the most satisfying part of the workday.

DO'S FOR THE LEADER	DON'TS FOR THE LEADER
Find clever and effective solutions.	Only treat the symptoms of a problem.
Investigate the root cause of issues.	Solve problems based on gut feelings.
Use data to find and solve problems.	Solve problems based on intuition.
Encourage problem-solving at all levels.	Expect a single person to solve problems.
Create a blameless culture.	Blame someone whenever a problem occurs.
Document solutions to prevent repeats.	Let the same problems happen again.

SITUATION	SUGGESTED ACTION
Teams feel rewarded when persistent operational issues are resolved with simple, elegant fixes	Document solution logic, share insights, and replicate effective patterns across similar workflows and operating conditions
Operators appreciate solutions that remove unnecessary steps and reduce cognitive load during execution	Redesign workflows, validate usability, and ensure changes reduce operational burden without introducing new risks or dependencies
Leaders value solutions that eliminate recurring bottlenecks and restore predictable throughput consistently	Analyze constraints, test alternatives, and implement changes that stabilize flow under varying load conditions and demand patterns
Teams celebrate when solutions reduce incident frequency and improve system reliability measurably	Track post-solution performance, confirm durability, and refine design based on real-world system behavior
Organizations benefit when solutions address root causes rather than surface symptoms alone	Validate findings, implement structural fixes, and monitor for recurrence across operational cycles and environments
Executives appreciate solutions that strengthen resilience without requiring major investment or disruption	Identify high-leverage adjustments, measure impact, and scale improvements across operations where appropriate

The deep satisfaction of a system that is "running smoothly and efficiently" with zero waste and high morale is the bliss of "operational flow." It's the feeling of "quiet, sustained power." This bliss happens when the "Human-AI Bridge" and the "Role Clarity Bridge" are working perfectly together. Daily, it looks like a dashboard that is "all green" and a team that is "at peace and productive." This bliss is the "death of the emergency." It provides a sense of "unshakeable competence." This bliss is the ultimate goal of "Operations Transformation." When your operations are so excellent they become "invisible," you have achieved a level of mastery that allows you to focus entirely on "strategic vision" and "delighting the customer."

DO'S FOR THE LEADER	DON'TS FOR THE LEADER
A system that runs smoothly.	A system that is constantly breaking.
Eliminate waste in all processes.	Waste time on low-value tasks.
Use lean methodologies.	Let processes be inefficient.
Optimize resource allocation.	Waste resources on redundant work.
Use tech to find and fix waste.	Fail to measure or track waste.
Build quality into every step.	Only check for quality at the end.

SITUATION	SUGGESTED ACTION
Teams feel confident when workflows run predictably without constant intervention or firefighting	Maintain discipline, monitor signals, and reinforce behaviors that preserve operational stability across conditions
Operators appreciate when systems respond consistently under varying load conditions and constraints	Test performance, validate thresholds, and ensure systems remain reliable across scenarios and demand patterns
Leaders value environments where issues are rare and quickly contained when they occur	Strengthen detection, refine escalation, and maintain readiness for unexpected events across workflows
Teams perform better when operational noise is low and focus remains on execution	Reduce distractions, streamline processes, and maintain clarity around priorities and responsibilities
Organizations thrive when excellence becomes the default rather than an occasional achievement	Embed standards, enforce accountability, and reinforce excellence through consistent governance and review
Executives appreciate environments where operational excellence directly supports strategic growth objectives	Align performance goals, measure impact, and scale excellence across business units and regions

The joy of having a system that can "easily adapt" to new challenges and demands without breaking is the bliss of "adaptive resilience." It's the feeling of "being ready for anything the world throws at you." This bliss happens when your "Proactive Behavior" and "Strategic Allocation Bridge" have created a truly elastic operation. Daily, it means you don't "panic" when a new competitor enters or a market shifts. It provides a sense of "limitless potential." This bliss is the realization that your operations are no longer a "constraint" on your growth, but the "engine" of it. When you can "manage the future" as easily as you manage the present, you have reached the pinnacle of operational excellence in the AI era.

DO'S FOR THE LEADER	DON'TS FOR THE LEADER
A system that can easily adapt.	A system that is rigid and inflexible.
Be flexible with resource allocation.	Have rigid, inflexible resource plans.
Cross-train employees for flexibility.	Let people get stuck in a single role.
Plan for sudden resource changes.	Be unprepared for shifts in demand.
Adapt processes to new information.	Refuse to change old processes.
Use tech to optimize resource use.	Manage resources manually.

SITUATION	SUGGESTED ACTION
Teams feel empowered when systems adjust smoothly to shifting demand without major reconfiguration	Build flexibility, test scaling behavior, and maintain readiness for rapid load changes across environments
Operators appreciate when tools adapt to new workflows without requiring extensive retraining	Design modular interfaces, validate usability, and update training materials incrementally as workflows evolve
Leaders value systems that absorb change without destabilizing core operations or throughput	Assess impact, reinforce safeguards, and ensure changes align with operational resilience goals and constraints
Teams perform confidently when management processes remain stable during transitions or reorganizations	Clarify expectations, maintain communication, and support teams through structured change cycles and reviews
Organizations benefit when adaptability reduces downtime during market or operational shifts	Monitor conditions, adjust plans, and maintain agility through proactive scenario planning and forecasting
Executives appreciate systems that evolve without requiring disruptive overhauls or major redesigns	Invest in modular architecture, track performance, and scale improvements gradually across operational domains

BONUS CHAPTER

Ai6x6™ Transformation Blueprint

Operations Diversity

"The value of a diverse team is its capacity to challenge the norm of groupthink and thus boost organizational performance and improve *decision-making."*

Yrthya Dinzey-Flores, Head of Diversity, Equity, and Inclusion at General Mills. Diverse teams introduce conflicting but constructive viewpoints that are essential for successful transformation and overcoming groupthink, a major impediment to innovation.

In **Operations Transformation**, diversity is critical for designing resilient, globally compliant, and maximally efficient digital operating models. Teams with diverse operational experiences—drawn from various global regions, different industries, multiple regulatory environments, and a range of functional roles—are uniquely qualified to challenge legacy procedures and pinpoint **systemic inefficiencies** that may be deeply embedded and thus invisible to a single-domain team. This broad perspective is essential for conducting a comprehensive risk analysis, ensuring that the new, digitally-enabled operations (built on cloud, automation, and AI) are scalable and secure across all jurisdictions. Furthermore, diversity ensures that the new workflows are designed with the end-user in mind, maximizing adoption rates among employees with varied digital literacy levels and cultural expectations. By integrating this vast range of experience and knowledge, the company ensures that its transformed operating model is not only cost-effective and faster but also **highly adaptable, compliant**, and capable of maintaining exceptional business continuity in a complex global environment.

In today's interconnected world, diversity is a fundamental pillar of strength and innovation. Often simplified to visible differences, true diversity is a rich tapestry of attributes shaping how individuals perceive and contribute. Understanding this multifaceted nature is crucial for fostering environments where every voice thrives. This essay explores diversity through six critical dimensions: **demographic, cognitive, experiential, geographic, functional, and perspective diversity**. By examining these distinct facets, we can better appreciate how a holistic approach to diversity enriches teams and communities, driving robust solutions.

First, **Demographic Diversity** encompasses varied inherent and acquired characteristics. This includes age, bringing diverse generational insights, and gender, race, ethnicity, sexual orientation, ensuring a wide range of lived experiences. Socioeconomic status, religion, disability, and nationality further contribute to a richer collective understanding. Valuing demographic diversity ensures equitable representation and leverages unique insights from varied personal journeys.

Secondly, **Cognitive Diversity** highlights how people think, process information, and approach problems. This includes varied thinking styles like analytical vs. intuitive approaches, and different problem-solving or learning methods. When a group embraces cognitive diversity, it tackles problems from multiple angles, leading to more innovative solutions and a thorough understanding of complex issues, free from a single dominant thought process.

Next, **Experiential Diversity** focuses on the breadth of life and work experiences individuals bring. This covers varied work histories across industries and roles, providing a rich pool of professional knowledge. Diverse educational backgrounds, from different disciplines or institutions, also offer distinct frameworks. Beyond formal credentials, personal life experiences like travel and cultural immersion contribute significantly, equipping teams with a wide array of skills and adaptable approaches.

Geographic Diversity emphasizes the impact of place on an individual's perspective. This includes differences from regional or local backgrounds, shaping values and priorities. National backgrounds introduce distinct cultural norms, while urban versus rural upbringings offer varying insights into community dynamics and societal needs. These varied geographical lenses provide a broader understanding of global and local markets and behaviors.

Furthermore, **Functional Diversity** pertains to the variety of professional roles, specialized areas, and departmental expertise within a team. This dimension recognizes contributions from different functions like engineering, marketing, or finance, each bringing specific domain knowledge. Variations in hierarchical roles and industry experience also inform cross-pollination of best practices. Leveraging functional diversity ensures all necessary technical skills are present, fostering comprehensive problem-solving.

Finally, **Perspective Diversity** encompasses the fundamental differences in viewpoints and mental models individuals hold, often influenced by other diversity forms. This includes varied ideological stances, cultural interpretations, and distinct value systems. Understanding diverse stakeholder perspectives—from customers to employees—is crucial for holistic decision-making. Ultimately, perspective diversity allows for problems to be framed in multiple ways, challenging assumptions and encouraging empathetic, effective outcomes.

By intentionally cultivating and integrating these six dimensions—demographic, cognitive, experiential, geographic, functional, and perspective—organizations and societies can harness the profound power of true diversity. This comprehensive approach not only fosters equity but also enhances creativity, resilience, and innovation, demonstrating that a multitude of viewpoints is the clearest path to collective success.

Embracing **demographic diversity** within operational teams, from factory floors to logistics hubs and customer service centers, means intentionally cultivating a talent pool that brings varied life experiences to process optimization, problem-solving, and service delivery. This goes beyond mere representation; it's about leveraging the inherent value that comes from operations professionals with varied ages, genders, racial and ethnic backgrounds, physical abilities, socio-economic upbringings, and sexual orientations. These visible and non-visible differences are crucial for identifying diverse operational challenges, preventing unintended biases in process design, ensuring inclusive service delivery, and developing operational solutions that work effectively across varied customer bases and employee groups. A demographically diverse operations team brings distinct insights into workflow efficiencies, cultural nuances in process adoption, and potential blind spots in supply chain management, which are crucial for effective operations transformation. When operational teams lack this breadth, they risk developing processes that are narrow in scope, fail to address the needs of significant employee or customer populations, or perpetuate existing inefficiencies. True demographic diversity fosters more efficient, resilient, and ultimately, more adaptable operations, better equipped to address complex logistical challenges and serve diverse markets effectively.

Topic	Tangible steps	Challenges you may face
AGE DIVERSITY IN PROCESS OPTIMIZATION		
Varied age groups bring different perspectives on legacy processes vs. new technologies. Enhances innovation and broadens solutions for operational efficiency.	Form cross-generational process improvement teams. Implement intergenerational mentorship for operational roles. Provide continuous learning on new tech accessible to all ages.	Resistance to new operational methods from established teams. Overcoming biases favoring specific age groups in operational roles. Managing different work paces and communication styles.
GENDER DIVERSITY IN MANUFACTURING/LOGISTICS		
Balanced gender representation influences team dynamics, safety protocols, and problem-solving on operational floors. Diverse gender perspectives lead to comprehensive solutions.	Set targets for gender balance in manufacturing/logistics teams. Promote equal pay and promotion opportunities in operational roles. Implement inclusive language and communication training for operational staff.	Unconscious biases in hiring/promotion for physically demanding roles. Lack of support for work-life balance hindering participation. Resistance to changing established operational norms.
RACIAL & ETHNIC DIVERSITY IN SUPPLY CHAIN MANAGEMENT		
Diverse racial & ethnic backgrounds foster critical insights into cultural relevance in global supply chains, stakeholder management, and ethical sourcing. Improves global operational resilience.	Develop cultural awareness training for supply chain professionals. Establish ERGs focused on operational inclusion. Prioritize diverse representation in supply chain governance committees.	Systemic bias in vendor selection. Difficulty fostering psychological safety in diverse operational environments. Insufficient understanding of cultural implications for logistics.

2. COGNITIVE DIVERSITY

Cognitive diversity within operational teams is paramount for navigating the inherent complexities of operations transformation. It acknowledges that true innovation and robust problem-solving in operations stem not just from varied experiences or demographics, but fundamentally from differences in how people think, process information, and approach efficiency and logistical challenges. An operational team composed solely of analytical thinkers might excel at optimizing existing processes but could struggle with identifying novel process improvements or designing truly disruptive operational models. Conversely, a team too reliant on intuitive thinking might miss critical data points or overlook practical constraints in operational implementation. By fostering a mix of analytical, creative, systemic, and neurodiverse cognitive styles across all operational roles, organizations are better equipped to challenge assumptions in process design, identify unseen bottlenecks, predict unforeseen supply chain disruptions, and design more comprehensive and resilient operational strategies. It's about building a collective operational intelligence where different cognitive strengths complement each other, leading to more thorough process evaluations, more innovative operational solutions, and a greater capacity for adaptive learning in the face of evolving market demands. This richness in thought ensures that operational decisions are thoroughly vetted from multiple angles, preventing blind spots and fostering intellectual agility crucial for sustained operations transformation success.

Topic	Tangible steps	Challenges you may face
ANALYTICAL VS. INTUITIVE THINKING IN PROCESS DESIGN		
Balancing data-driven & gut-feeling approaches in process design shapes workflow and outcomes. Leads to nuanced decision-making, combining factual rigor with insightful foresight.	Design process improvement workflows requiring diverse cognitive input. Encourage Ops. managers to articulate both data-backed and intuitive reasoning. Provide training on thinking styles.	Friction between data rigor & rapid intuition in operations. Difficulty quantifying intuitive operational insights. Culture valuing one operational style over another hinders collaboration.
SYSTEMIC THINKERS IN SUPPLY CHAIN OPTIMIZATION		
Operations professionals seeing interconnectedness of systems & long-term implications drive comprehensive supply chain improvement. Ensures holistic, sustainable operational changes.	Provide systems thinking training for supply chain strategists. Assign systemic thinkers to cross-functional operational integration initiatives. Use "value stream mapping" to identify impacts in operations.	Difficulty identifying/developing systemic thinking in operational roles. Resistance to long-term operational views. Overwhelm from too much interconnected operational information.
CREATIVE PROBLEM-SOLVERS IN LOGISTICS SOLUTIONS		
Operations professionals with imaginative thinking inspire novel solutions to complex logistical challenges. Crucial for breakthrough operational innovations.	Establish operations innovation labs or hackathons. Implement brainstorming sessions for unconventional logistics ideas. Create dedicated time for "blue-sky" operational thinking.	Risk aversion stifling new Ops. ideas. Difficulty translating creative operational solutions into actionable implementation plans. Perceived lack of "seriousness" for certain functions.

Experiential diversity among operations professionals is a powerful catalyst for successful operations transformation, providing a rich tapestry of practical insights drawn from varied career paths, industries, educational backgrounds, and personal life stages. Operations professionals who have navigated different organizational cultures, worked across diverse sectors (e.g., manufacturing to service, public to private), or gained international exposure bring a broader repertoire of operational solutions and a more nuanced understanding of complex logistical challenges. This breadth of experience prevents insular thinking in operations strategy, encourages the adoption of "best practices" from seemingly unrelated operational fields, and fosters resilience by drawing on lessons learned from past successes and failures in varied contexts. For instance, an operations professional with experience in disaster relief can bring invaluable composure and clear-headedness during periods of intense supply chain disruptions. By intentionally cultivating an operations team with diverse professional and personal journeys, an organization gains a formidable collective wisdom, enhancing its ability to foresee operational opportunities, mitigate operational risks, and execute operations transformation initiatives with greater adaptability and effectiveness. It promotes a learning culture where varied practical operational wisdom is valued and applied to new strategic operational challenges.

Topic	Tangible steps	Challenges you may face
CROSS-INDUSTRY INSIGHTS IN PROCESS DESIGN		
Operations professionals from different industries bring fresh perspectives & transferable solutions to operational challenges. Broadens organizational problem-solving in operations.	Recruit operations talent with diverse industry backgrounds. Organize peer-learning forums for cross-industry operational insights. Encourage participation in external industry operations associations.	Difficulty translating operational insights to new industry contexts. Resistance to "outsider" operational approaches. Lack of understanding of new industry operational nuances for integration.
STARTUP VS. CORPORATE OPERATIONS SCALABILITY		
Impact of operations professionals from agile startup vs. structured corporate environments on operational scalability & innovation. Blending these fosters agility & stability.	Integrate startup operations talent into corporate innovation units. Create joint ventures pairing corporate/startup operations teams. Foster internal "intrapreneurship" programs for operations.	Culture clash between agile & traditional operational structures. Difficulty integrating fast operational decisions into slower processes. High expectations for immediate operational ROI.
GLOBAL VS. LOCAL SUPPLY CHAIN MANAGEMENT		
Operations professionals with diverse international & local market exposure balance global operational standardization & regional adaptation. Optimizes operational efficiency & market relevance.	Implement global operational talent rotation programs. Establish regional operations knowledge hubs with local market insights. Encourage dual reporting lines for international operational projects.	Complex balance of global vs. local operational needs. Resistance from local operational teams to global mandates. Misinterpretation of cultural nuances in operational logistics.

Geographic diversity within operational teams is indispensable for organizations operating in a globalized world, especially during periods of significant operations transformation. Operations professionals deeply rooted in varied regions, countries, or even urban versus rural settings bring critical local market understanding, cultural nuances, and insights into unique regulatory environments and logistical challenges of operations. Without this localized operational perspective, global operational strategies risk being homogenous, culturally insensitive, or simply impractical in execution. For example, an operations professional experienced in emerging markets might bring crucial insights into scalable, low-cost operational solutions, while someone from a highly regulated economy understands the complexities of operational compliance. This diversity ensures that operations transformation initiatives are not just conceptualized centrally but are designed with built-in adaptability for regional operational variations. It empowers the organization to navigate geopolitical risks affecting operations, understand localized customer behaviors for service delivery, and foster stronger relationships with diverse regional operational stakeholders. Operations professionals with varied geographic exposure contribute to operational strategies that are globally coherent yet locally relevant, maximizing impact and minimizing missteps in a fragmented operational world.

Topic	Tangible steps	Challenges you may face
REGIONAL OPERATIONAL MARKET UNDERSTANDING		
Operations professionals from different regions offer unique insights into local market dynamics, consumer behaviors, & competition affecting operations. Vital for global operational strategies.	Recruit operations talent with local market knowledge. Mandate global operational strategies be reviewed/ adapted by local operational teams. Support employee travel/immersion in diverse regions for Ops. insights.	Overlooking operational nuances for standardization. Difficulty adapting to rapid changes in local markets. Resistance from central operations teams to delegate.
NAVIGATING GLOBAL REGULATORY LANDSCAPES FOR OPERATIONS		
Operations professionals familiar with diverse international legal/compliance frameworks are crucial for multi-national operations. Ensures adherence & mitigates operational risks.	Create global operational compliance guidelines for teams. Invest in continuous education on international operational regulations. Consult local legal experts for every operational project.	Keeping up with rapidly changing operational laws. Reconciling conflicting requirements. Risk of non-compliance due to inadequate local operational understanding.
LOCAL INFRASTRUCTURE AND LOGISTICS KNOWLEDGE		
Operations professionals understanding local infrastructure (logistics, digital access) & resources impact operational efficiency. Optimizes local operational implementation & delivery.	Empower local operational teams for infrastructure investments. Develop centralized knowledge of regional operational capabilities. Foster local service provider partnerships for operations.	Underestimating varied local infrastructure impact on operations. Inaccurate resource assumptions for operational distribution. Difficulty scaling global operational solutions.

Functional diversity within operational teams is critical for executing comprehensive and interconnected operations transformations. It ensures that operational strategic and tactical decisions are informed by the deep expertise and unique perspectives of various professional disciplines, rather than being confined to a single departmental silo. An operational team comprising experts from finance, technology, human resources, product, marketing, and legal functions can collectively identify operational opportunities, anticipate operational challenges, and design operational solutions that are holistic, practical, and sustainable. For instance, a finance analyst ensures fiscal viability of operational investments, while a technology specialist champions automation adoption, and an HR generalist focuses on people-centric change management for operational teams. Without this blend, operations transformation efforts might optimize one area at the expense of another, leading to fragmented implementation or unintended negative consequences in process execution. By fostering cross-functional collaboration and valuing diverse functional expertise, an organization can break down traditional barriers, enhance teamwork, and create truly integrated operational strategies that leverage the strengths of every part of the business, ensuring that the operations transformation is not just conceptually sound but also executable across all organizational facets.

Topic	Tangible steps	Challenges you may face
FINANCE-LED OPERATIONAL EFFICIENCY MEASUREMENT		
Operations professionals with strong financial backgrounds prioritize cost optimization & ROI as primary operational transformation drivers. Ensures fiscal responsibility in operational projects.	Appoint finance professionals to operational project teams to define ROI. Establish clear financial metrics & ROI targets for operational initiatives. Implement rigorous cost-benefit analyses for all operational investments.	Over-emphasis on short-term operational cuts. Resistance if financial goals overshadow operational innovation. Difficulty projecting ROI for intangible operational benefits.
TECHNOLOGY-DRIVEN OPERATIONAL AUTOMATION		
Operations professionals with deep tech expertise champion automation & digital transformation of processes. Essential for driving tech-enabled operational change.	Elevate technologists to strategic roles in operational projects. Foster "tech literacy" for operational teams. Create cross-functional innovation teams combining operational tech/business experts.	Resistance to new operational technologies. Rapid evolution challenges operational skills. Difficulty integrating new operational tech with legacy systems and traditional workflows.
HR'S ROLE IN OPERATIONAL SKILLS DEVELOPMENT		
HR professionals focus on talent management, reskilling, & employee well-being during operations transformation. Ensures human capital is central to operational success.	Position HR teams as strategic partners in all operational change initiatives. Develop robust change management/communication plans for operational adoption. Implement reskilling programs.	Business units resist prioritizing people over operational technology. Difficulty measuring HR ROI in operational initiatives. Challenges addressing employee anxiety/ resistance to new ops. processes.

Perspective diversity within operational teams is arguably one of the most subtle yet profound drivers of successful operations transformation. It goes beyond demographics, experience, or function, delving into the varied worldviews, core values, ethical frameworks, and risk tolerances that shape how operations professionals interpret information and prioritize operational objectives. An operational team with diverse perspectives challenges inherent biases in process design, avoids groupthink in problem-solving, and considers a wider spectrum of operational possibilities and potential consequences. For example, balancing operations professionals who prioritize short-term efficiency gains with those focused on long-term sustainability, or those who are risk-seeking with process experimentation with those who are risk-averse with safety protocols, leads to more robust and balanced operational strategic decisions. This diversity ensures that operations transformation is not just about efficiency or cost reduction, but also about ethical responsibility, employee safety, environmental impact, and genuine stakeholder value from operations. It encourages operations professionals to approach operational problems from fundamentally different angles, leading to more resilient operational strategies, better employee engagement, and a more adaptive and morally grounded organization capable of navigating complex and ambiguous future operational landscapes. It cultivates a culture of constructive challenge and holistic operational decision-making.

Topic	Tangible steps	Challenges you may face
COST-CENTRIC VS. VALUE-CENTRIC VIEWS IN OPERATIONS		
Operations professionals balance cost efficiency with long-term value creation. Ensures processes optimize both financial health and strategic outcomes.	Implement value stream mapping to identify non-value-added activities. Conduct workshops to shift mindset from cost-cutting to value-adding processes. Establish KPIs balancing cost quality satisfaction.	Internal resistance to shifting from purely cost-centric. Difficulty quantifying long-term value from operational changes. Operations teams struggle balancing immediate savings vs. future investment.
SHORT-TERM VS. LONG-TERM OPERATIONAL EFFICIENCY		
Operations professionals' differing time horizons influence project pace, scope, & success of operational initiatives. Balancing these ensures both immediate gains & future viability.	Develop multi-year operational roadmaps with short-term milestones. Assign operational teams with complementary focuses. Use scenario planning to visualize impacts of operational strategies.	Pressure for immediate operational results. Operations professionals may prioritize short-term gains. Difficulty articulating long-term operational value to all stakeholders.
RISK-SEEKING VS. RISK-AVERSE APPROACHES IN PROCESS INNOVATION		
Varied risk tolerance impacts pursuit of disruptive vs. incremental operational improvements. A healthy balance drives calculated operational growth and controlled experimentation.	Foster culture where calculated operational risks are encouraged. Establish "fail fast, learn faster" practices for operational projects. Create project portfolios balancing high-risk/reward with safer bets.	Decision paralysis when operational risk tolerances conflict. Dominance of extreme operational risk views. Difficulty setting clear, unified operational risk parameters across teams.

BLUEPRINT SERIES

SECTION G

Moral Compass

Ai6x6™ Transformation Blueprint

Story 1 — The Nuremberg Trials and Ethical Accountability (1945–1946): After World War II, the world confronted atrocities committed under the guise of legal authority. The Nuremberg Trials established that individuals and institutions are accountable for actions that violate fundamental human rights. Legal frameworks were strengthened by moral reasoning, not replaced by it. Judges balanced evidence with ethical principles to define new global norms. The trials demonstrated that morality must guide power. They became a foundation for modern ethical governance.

Story 2 — The Tuskegee Syphilis Study Exposure (1932–1972): For decades, researchers withheld treatment from Black men to observe disease progression, violating consent and dignity. Public exposure triggered national outrage and forced reforms in medical ethics. Oversight boards, informed consent, and patient rights became mandatory. Machines and data had played roles in tracking outcomes, but human ethics had been absent. The scandal showed how systems fail when morality is excluded. It reshaped global standards for ethical research.

Story 3 — The Environmental Movement and Earth Day (1970): Industrial growth created pollution that threatened ecosystems and public health. Scientists provided data, but moral arguments mobilized action. Citizens demanded accountability, and governments introduced regulations to protect air, water, and wildlife. The movement reframed environmental protection as a shared ethical duty. It demonstrated how moral clarity can transform policy, industry, and public behavior. The cultural shift established sustainability as a long-term societal obligation.

Story 4 — The Anti-Apartheid Global Solidarity Movement (1950s–1994: Apartheid enforced racial segregation and oppression in South Africa. Activists, governments, and institutions worldwide united to challenge the regime through sanctions, protests, and moral pressure. Human stories amplified the injustice, while global networks coordinated action. The movement showed how collective moral conviction can influence political systems. It proved that ethical solidarity can reshape international norms. The end of apartheid became a landmark in moral mobilization.

Story 5 — The Rise of Corporate Social Responsibility (1990s–2000s): Globalization exposed labor abuses, environmental harm, and opaque supply chains. Consumers demanded transparency, and companies adopted codes of conduct, sustainability reporting, and ethical sourcing. Machines tracked compliance, but humans defined the standards. CSR shifted from public relations to strategic responsibility. Organizations learned that long-term value depends on ethical behavior. It marked a turning point where morality became integral to corporate identity.

Story 6 — The AI Ethics Movement (2015–2025): As AI systems influenced decisions in hiring, policing, healthcare, and finance, concerns grew about bias, transparency, and accountability. Researchers, policymakers, and civil society pushed for ethical frameworks to guide development. Machines provided capability, but humans defined boundaries. The movement emphasized fairness, explainability, and human oversight. It demonstrated that emerging technologies require moral governance as much as technical innovation. AI ethics became a global priority shaping the future of intelligent systems.

LESSONS FROM AN OPERATIONS TRANSFORMATION PERSPECTIVE

Operational maturity in moral domains develops when organizations embed ethics into daily workflows, not just policies. Over time, institutions learned that harm often arises from routine decisions made without ethical reflection. Mature operations integrate transparency, accountability, and fairness into processes—from hiring and procurement to risk management and customer service. Machines enforce consistency, but humans ensure moral judgment. Maturity is achieved when operations prevent misconduct proactively, respond to issues responsibly, and reinforce ethical norms through practice. Ethical operations become a stabilizing force that protects people, strengthens trust, and sustains long-term institutional integrity.

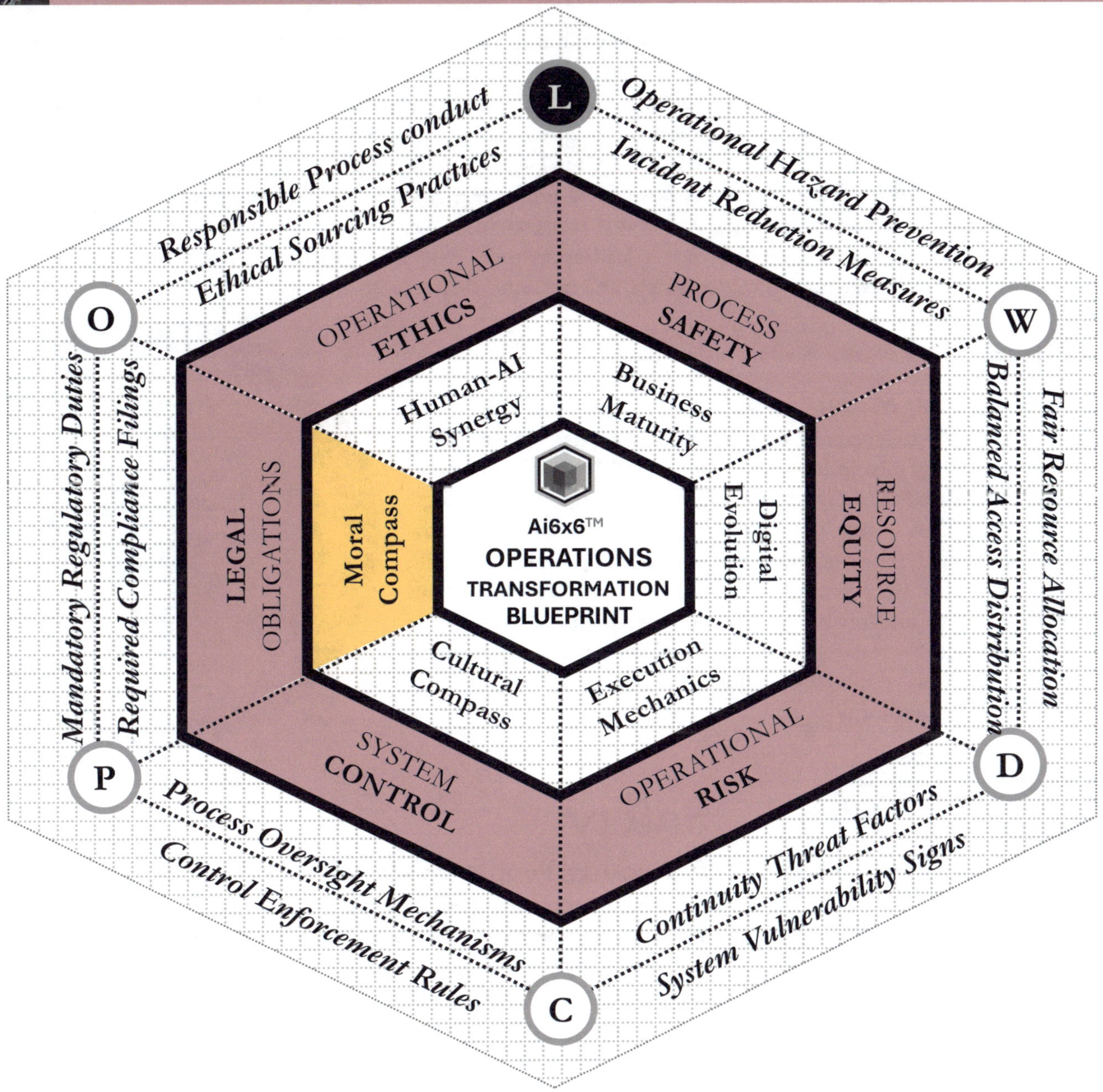

The **Ai6x6 Research Lab** identified the Moral Compass as the most vital lens because it serves as the final guardian of the organization's honor and the safety of its world. This is the ultimate filter for every decision, ensuring the drive for efficiency never outpaces the bedrock of **Sourcing Ethics** and **Process Safety**. We chose this because an organization's legacy is built on its **Continuity** and **Integrity**, not just its throughput. It is vital because it provides the moral authority required to lead through the ethical "fog" of autonomous agents and global sourcing complexities.

This lens is the heart's final check—a rigorous audit of **Operational Risk**, **Equity**, and **System Control**. It ensures the transformed operation is not only a titan of efficiency but a beacon of public trust. The Lab believes that in the 2026 landscape, the greatest risk to an operation is not a system failure, but a moral breach in the supply chain or a hazard ignored. By focusing on **Legal Obligations** and **Guardrail Control**, we provide the tools to navigate the heavy responsibility of movement. This lens is about the courage to be honorable when the temptation for speed is highest, ensuring the legacy of the mission is one of integrity and human value.

Morality has always shaped how organizations behave, but in the age of AI, it becomes a structural requirement rather than an aspirational ideal. The **Moral Compass** lens forces organizations to confront a truth that is often ignored: intelligence—whether human or machine—does not create ethical behavior. It amplifies whatever moral foundation already exists. When an organization is grounded in integrity, clarity, and principled action, AI strengthens it. When the foundation is inconsistent, opaque, or misaligned, AI exposes those weaknesses with unforgiving speed. Morality is no longer a philosophical preference; it is the architecture that determines whether accelerated decisions, automated systems, and scaled consequences lead to progress or harm. This lens reframes morality as an enterprise-wide discipline, not a leadership trait. It applies equally to workforce practices, data governance, customer interactions, product design, and operational systems.

Ethics

Establishing the Moral Center

Ethics forms the first dimension of the Moral Compass. Ethics is not simply about avoiding misconduct; it is about establishing a moral center that guides every decision, every interaction, and every use of technology. Across all six pillars, ethics determines how data is collected, how employees are treated, how customers are served, how products are built, and how operations are executed. Organizations that prioritize ethics create cultures where integrity is expected, not encouraged. Ethical clarity becomes the anchor that ensures AI is deployed with intention, restraint, and respect. In this dimension, principled behavior becomes the foundation of moral authority.

Safety

Protecting People, Systems, and Society

Safety represents the second dimension. As AI accelerates decision-making and expands operational reach, the potential for harm increases. Safety is not limited to physical environments; it includes psychological safety for the workforce, data safety for customers, model safety for AI systems, and operational safety for mission-critical processes. Safety ensures that no individual, team, or automated system can cause harm through unchecked influence or unintended consequences. It is not about limiting innovation; it is about ensuring that innovation does not outpace responsibility. In this dimension, responsible protection becomes essential to moral operation.

Equity

Ensuring Fairness Across Stakeholders

Equity forms the third dimension. Influence in modern organizations is no longer tied solely to hierarchy; it flows through data pipelines, algorithmic recommendations, and automated workflows. Equity ensures that these systems do not privilege certain groups, perspectives, or outcomes at the expense of others. It applies to workforce advancement, customer treatment, product accessibility, and data representation. Equity is not about equalizing results; it is about ensuring that the system itself is not biased. In this dimension, fairness becomes a structural requirement rather than a social aspiration.

Risk

Navigating Uncertainty With Discipline

Risk represents the fourth dimension. AI-enabled decisions carry greater speed, greater reach, and greater uncertainty. Risk is not about avoiding uncertainty; it is about navigating it with discipline. Organizations must anticipate consequences, stress-test assumptions, and understand the risks embedded in both human and machine judgment. This applies to operational risks, product risks, customer risks, workforce risks, and data risks. Organizations that embrace this dimension act decisively without acting blindly. In this dimension, risk-aware decision-making becomes a moral responsibility.

Control

Defining Governance and Accountability

Control forms the fifth dimension. As organizations adopt AI, authority becomes more complex: who decides, who approves, who monitors, and who intervenes when systems drift? Control ensures that governance structures are clear, decision rights are defined, and accountability is transparent. It applies to data governance, product governance, operational governance, and workforce governance. Control is not about bureaucracy; it is about ensuring that power—human or machine—is exercised with clarity and accountability. In this dimension, structured oversight becomes the backbone of moral operation.

Legal

Upholding Lawful and Responsible Conduct

Legal completes the Moral Compass. In a world where data, automation, and intelligence shape every action, legal obligations become inseparable from organizational obligations. Compliance is not a constraint; it is a safeguard. Legal integrity ensures that decisions align with regulatory expectations, societal norms, and ethical standards across all six pillars. It protects the organization from missteps and reinforces public trust. In this dimension, lawful stewardship becomes the final test of moral credibility.

The Moral Compass lens reminds organizations that moral authority is not a byproduct of good intentions; it is the result of deliberate, disciplined choices. Morality must be engineered into systems, not assumed in people. When organizations prioritize ethics, enforce safety, ensure equity, manage risk, establish control, and uphold legal accountability, they create environments capable of navigating complexity with confidence and integrity. They build systems where people feel protected, customers feel respected, and society feels considered. They create the moral conditions for transformation to take root and endure. In this maturity, morality becomes an enterprise capability—one that strengthens trust, accelerates adoption, and stabilizes transformation. It becomes the differentiator that determines whether AI amplifies value or amplifies harm. Organizations that master the Moral Compass do not merely deploy intelligent systems; they deploy them responsibly. They create futures where intelligence and integrity advance together, ensuring that progress is not only possible but principled.

Family Values — ETHICS

The foundational moral code—religious, cultural, and social—that dictates the ground rules for "Right vs. Wrong." Establishing the core integrity of the home so that all who belong understand the non-negotiable standards of conduct.

Home Security — SAFETY

Integrated protection—cameras, sensors, and smoke detectors—that provide total awareness of the premises. Deploying proactive warning systems and delegation guardrails to protect the inhabitants from "hidden fires."

Balanced Contribution — EQUITY

Ensuring everyone helps with the cooking and upkeep, with expenses shared equitably by each earning member. Maintaining participation and representation so that no member feels excluded from the collective success.

The Insurance Shield — RISK

Comprehensive fire, flood, and auto coverage to provide long-term peace of mind for the entire household. Using stress testing and "what-if" scenarios to protect the family assets from unforeseen market storms.

Access & Keys — CONTROL

Ensuring all adults have keys and codes, with clear emergency rules and a defined line of command for disasters. Establishing rigid RACI controls to ensure that a human always has the final word on the most critical locks.

The Estate Will — LEGAL

Recording Wills, Trusts, and Medical Mandates with a legal custodian to safeguard the valuables for the future. Securing the organization's legacy through regulatory accountability, ensuring the estate is legally unassailable for generations.

Ai6x6™ Transformation Blueprint

Operational Ethics

"Trustworthy AI is built on humility — the recognition that we do not know everything."

Stephen Hawking, theoretical physicist known for work on cosmology.

Operational Ethics requires principled execution, and Responsible Process Conduct defines how operational activities must be carried out with integrity, transparency, and accountability. It establishes expectations that prevent shortcuts, concealment, or unethical decision making. This matters because operational processes shape daily outcomes and long-term organizational credibility. Responsible Process Conduct prevents misconduct by ensuring actions align with stated values. It also reinforces fairness by ensuring all teams follow the same ethical rules. When ignored, operations become inconsistent and vulnerable to abuse. Teams experience confusion as ethical boundaries blur. Over time, Operational Ethics deteriorates as behavior becomes misaligned. Ethical Sourcing Practices ensure materials, partners, and suppliers meet responsible and sustainable standards.

The safety valves of the enterprise engine. Leaders use Ethical Sourcing Reports and Vendor Ethics Screens to ensure the organization's scale doesn't come at the cost of its soul. This topic involves using "Conflict-Free Procurement Records" to protect the supply chain from moral drift. By sitting with "Ethical Automation Logs," the leader ensures that AI-driven efficiency doesn't sacrifice human oversight. Operational ethics is the "Map" of the organization's "Kill Zones." It ensures that as you scale, the "Why" of your operations remains as clear and untarnished as your original leadership intent. In the print series, you can engross yourself in "Ethical Oversight Summaries," seeing how your operational choices influence your brand's moral legacy. This clinical focus on "Operational Ethics" ensures that your "Operations Transformation" is as scalable as it is ethical, preventing your engine from creating "Kill Zones" in your own supply chain and ensuring that every operational shift is a statement of brand-led integrity.

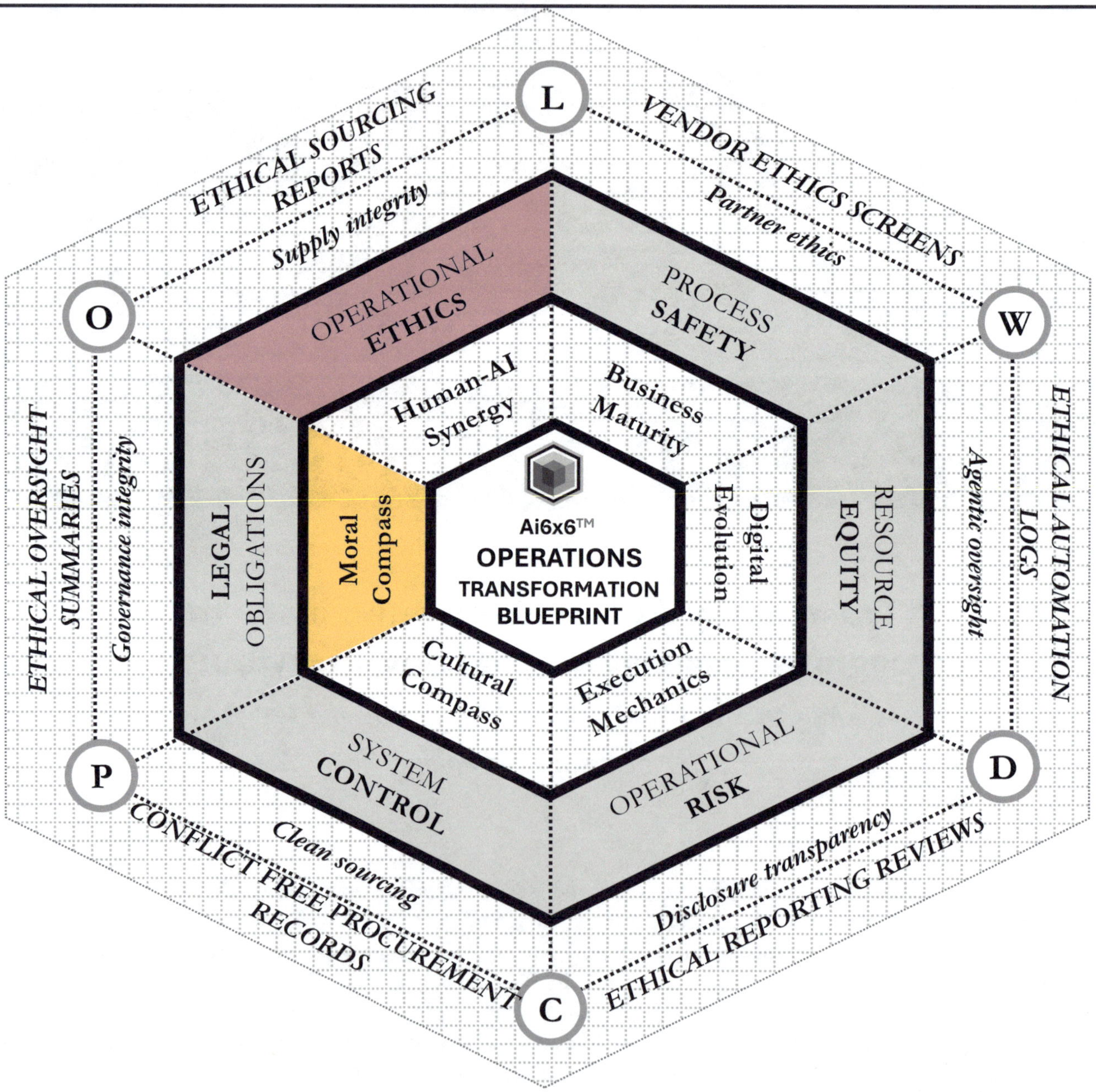

Ai6x6 OPERATIONAL ETHICS FOR MORAL COMPASS | OPERATIONS TRANSFORMATION

1. ETHICAL SOURCING REPORTS

Operational Ethics requires integrity in procurement, and Ethical Sourcing Reports document whether materials, labor, and upstream partners meet defined ethical standards. They evaluate environmental impact, labor conditions, human rights compliance, and transparency across the supply chain. This matters because unethical sourcing introduces hidden moral liabilities into operations. Ethical Sourcing Reports prevent exploitation by making upstream behavior visible and accountable. They also reinforce fairness by ensuring all suppliers meet the same ethical expectations. When ignored, unethical practices infiltrate operations unnoticed. Customers and regulators lose trust as supply chain risks surface. Over time, Operational Ethics collapses as sourcing decisions drift toward convenience over conscience.

When executed well, Ethical Sourcing Reports strengthen Operational Ethics by embedding moral scrutiny into procurement. Stakeholders trust that the organization sources responsibly. This improves brand credibility, reduces risk, and enhances long term resilience. Reports reinforce the principle that ethics must extend beyond internal operations. They remind leaders that supply chains reflect organizational values. Ultimately, they build an environment where sourcing decisions uphold integrity at every tier.

MORAL COMPASS SCENARIO	LEADERSHIP ACTION
Amira reviews a quarterly sourcing report and sees a sudden increase in purchases from a region flagged for labor abuses. She realizes the cost savings may be tied to unethical practices.	The leader freezes new contracts in that region and commissions an independent audit. They also require human-rights risk scoring in all sourcing reports.
Lennox notices that a supplier's environmental impact data is "self-certified" with no third-party verification. He senses greenwashing.	The leader demands verified environmental certifications and suspends preferred-vendor status until received. They also add verification fields to sourcing reports.
Zainab finds that sourcing reports aggregate all subcontractors under a single parent vendor. She recognizes that hidden tiers may conceal abuses.	The leader requires full tier-2 and tier-3 visibility in reports. They also make multi-tier transparency a condition of contract renewal.
Callum sees that a supplier with repeated ethics flags still receives the majority of high-margin orders. He realizes commercial incentives are overriding ethics.	The leader rebalances volume toward compliant suppliers and revises incentive structures. They also add ethics-weighting to allocation decisions.
Priyanka notices that sourcing reports omit worker-safety metrics entirely. She senses a blind spot in operational ethics.	The leader adds safety indicators—injury rates, inspections, violations—to every report. They also require ethics reviews alongside cost reviews.
Hugo reads a sourcing report that lists "alternative labor arrangements" without detail. He suspects disguised temporary or underage labor.	The leader demands full disclosure of labor categories and conducts surprise audits. They also ban opaque labor classifications in contracts.

Operational Ethics requires principled partnerships, and Vendor Ethics Screens evaluate whether external partners align with the organization's ethical standards. They assess labor practices, environmental stewardship, governance maturity, and compliance history. This matters because vendors often act as extensions of the organization. Vendor Ethics Screens prevent ethical breaches by filtering out partners who introduce unacceptable risk. They also reinforce fairness by ensuring all vendors are evaluated using the same criteria. When ignored, unethical partners compromise operational integrity. Customers and employees question the organization's judgment. Over time, Operational Ethics deteriorates as misaligned vendors shape the operating environment.

When executed well, Vendor Ethics Screens strengthen Operational Ethics by ensuring partnerships reflect shared values. Stakeholders trust that the organization collaborates responsibly. This improves operational stability, reduces reputational exposure, and enhances ethical alignment. Screens reinforce the principle that ethics must be enforced at the boundary of the enterprise. They remind leaders that vendor selection is a moral decision. Ultimately, they build an environment where external relationships uphold ethical standards.

MORAL COMPASS SCENARIO	LEADERSHIP ACTION
Daria discovers that vendor onboarding only checks creditworthiness and delivery history. She realizes ethics screening is absent.	The leader adds human-rights, environmental, and governance checks to onboarding. They also require ethics approval before vendor activation.
Malik sees that a vendor passed ethics screening five years ago but has since been fined for corruption. He senses stale due diligence.	The leader institutes recurring ethics re-screening and reviews all vendors with recent sanctions. They also add expiry dates to ethics approvals.
Sienna finds that ethics questionnaires allow vendors to skip "not applicable" on critical questions. She recognizes a loophole.	The leader redesigns the questionnaire with mandatory responses and evidence uploads. They also add random verification checks.
Tomas notices that local vendors in emerging markets are exempted from ethics screens "to speed up operations." He senses a double standard.	The leader removes exemptions and applies the same ethics bar globally. They also provide support to help smaller vendors comply.
Imani learns that a vendor uses a subcontractor previously banned by the company. She realizes the screen stops at the first tier.	The leader extends ethics screening to all subcontractors. They also require contractual disclosure of the full supply chain.
Viktor sees that ethics screening results are stored in email threads. He worries decisions can't be audited.	The leader centralizes ethics screens in a controlled system with audit trails. They also require documented rationale for high-risk approvals.

Operational Ethics requires responsible technology use, and Ethical Automation Logs document how automated systems make decisions, allocate resources, and impact people. They track algorithmic behavior, exceptions, overrides, and unintended consequences. This matters because automation can amplify bias or create opaque decision pathways. Ethical Automation Logs prevent harm by making automated processes transparent and auditable. They also reinforce fairness by ensuring automation treats all users and employees consistently. When ignored, automation becomes a black box that hides ethical failures. Teams lose visibility into how decisions are made. Over time, Operational Ethics collapses as automation operates without oversight.

When executed well, Ethical Automation Logs strengthen Operational Ethics by ensuring technology aligns with human values. Stakeholders trust that automation is monitored and governed. This improves accountability, reduces bias, and enhances operational integrity. Logs reinforce the principle that automation must be ethical by design and by operation. They remind leaders that technology requires continuous moral supervision. Ultimately, they build an environment where automation supports, rather than undermines, ethical conduct.

MORAL COMPASS SCENARIO	LEADERSHIP ACTION
Noor reviews automation logs and sees that an algorithm auto-rejects small suppliers from low-income regions. She realizes the rule encodes bias.	The leader disables the rule and runs a fairness audit on all automation logic. They also require ethics review for new automation rules.
Jasper notices that a scheduling bot consistently assigns night shifts to contractors instead of employees. He senses exploitation hidden in automation.	The leader reconfigures the bot to follow equitable scheduling rules. They also add human oversight for high-impact assignments.
Elin finds that an automated approval flow bypasses manual review for large orders from “trusted” vendors. She recognizes a corruption risk.	The leader reinstates human checks above defined thresholds. They also log all overrides for ethics review.
Raj sees that an inventory algorithm prioritizes regions with higher profit margins over regions with urgent humanitarian need. He realizes the logic ignores ethical priority.	The leader adds ethical weighting factors—need, vulnerability, impact—to the algorithm. They also require periodic ethics recalibration.
Camila discovers that a chatbot deflects worker safety complaints into generic HR tickets. She senses ethical issues being buried.	The leader routes safety-related automation events directly to a dedicated safety team. They also tag and track all safety complaints separately.
Owen notices that automation logs are purged after 30 days. He worries there’s no long-term visibility into ethical failures.	The leader extends retention for ethics-relevant logs and restricts deletion. They also require periodic ethics log reviews.

Operational Ethics requires truthful communication, and Ethical Reporting Reviews evaluate whether internal and external reports reflect accurate, complete, and unbiased information. They examine data integrity, narrative framing, and the ethical implications of what is included or omitted. This matters because reporting shapes decisions, perceptions, and accountability. Ethical Reporting Reviews prevent manipulation by ensuring information is presented honestly. They also reinforce fairness by ensuring all stakeholders receive truthful insights. When ignored, reports become tools for distortion. Leaders make decisions based on incomplete or misleading data. Over time, Operational Ethics deteriorates as truth becomes negotiable.

When executed well, Ethical Reporting Reviews strengthen Operational Ethics by embedding honesty into organizational communication. Stakeholders trust that reports reflect reality. This improves decision quality, reduces risk, and enhances ethical credibility. Reviews reinforce the principle that transparency is a moral obligation. They remind leaders that reporting integrity defines organizational integrity. Ultimately, they build an environment where truth governs operational narratives.

MORAL COMPASS SCENARIO	LEADERSHIP ACTION
Layla sees that operational dashboards highlight cost savings but hide metrics on worker injuries and overtime. She realizes reporting is morally lopsided.	The leader mandates balanced scorecards that include safety, fairness, and environmental impact. They also require ethics sign-off on executive reports.
Henrik notices that whistleblower reports are summarized into a single "issues" metric. He senses that severity and patterns are being obscured.	The leader separates whistleblower data into its own reporting stream with detailed categorization. They also ensure direct board visibility.
Noura finds that ESG metrics are manually "smoothed" before publication. She recognizes reputational polishing.	The leader bans manual smoothing and requires traceable data lineage. They also commission an external assurance review.
Elias sees that regional reports omit local regulatory violations to avoid "confusing headquarters." He realizes ethics are being filtered.	The leader requires full disclosure of all violations in global reports. They also protect regional leaders who report uncomfortable truths.
Zahra notices that supplier-audit failures are summarized as "areas for improvement." She senses euphemistic reporting.	The leader requires explicit labeling of failures, risks, and corrective actions. They also standardize severity language.
Conrad discovers that internal ethics survey results are shared only with HR. He recognizes that operational leaders never see systemic issues.	The leader integrates ethics survey findings into operational reviews. They also assign action owners for recurring themes.

5. CONFLICT-FREE PROCUREMENT RECORDS

Operational Ethics requires responsible sourcing of sensitive materials, and Conflict Free Procurement Records document whether products avoid materials linked to violence, exploitation, or geopolitical harm. They verify origins, certifications, and chain of custody documentation. This matters because conflict linked materials create severe ethical and reputational risks. Conflict Free Procurement Records prevent complicity by ensuring materials do not fund or support harmful activities. They also reinforce fairness by ensuring all suppliers meet the same conflict free standards. When ignored, unethical materials enter the supply chain unnoticed. Customers and regulators respond with scrutiny and distrust. Over time, Operational Ethics collapses as procurement loses moral discipline.

When executed well, Conflict Free Procurement Records strengthen Operational Ethics by ensuring materials reflect ethical responsibility. Stakeholders trust that the organization avoids harmful supply sources. This improves brand integrity, reduces regulatory exposure, and enhances global citizenship. Records reinforce the principle that procurement must protect human rights. They remind leaders that material origins carry moral weight. Ultimately, they build an environment where sourcing decisions uphold global ethical standards.

MORAL COMPASS SCENARIO	LEADERSHIP ACTION
Mira reviews procurement records and sees components sourced from regions associated with conflict minerals. She realizes the chain may not be conflict-free.	The leader demands conflict-mineral declarations and third-party audits. They also prioritize alternative suppliers with verified clean sourcing.
Dorianne notices that procurement codes lump conflict-sensitive materials into generic categories. She senses deliberate opacity.	The leader creates specific codes for high-risk materials. They also require enhanced scrutiny for those categories.
Kasim finds that a broker consolidates materials from multiple mines without disclosing origins. He recognizes a traceability gap.	The leader requires mine-level traceability as a condition of doing business. They also phase out opaque brokers.
Elodie sees that conflict-free certifications are expired but still accepted. She realizes no one is tracking validity.	The leader adds expiry tracking and automated alerts for renewals. They also suspend purchases when certifications lapse.
Bruno notices that procurement records show a sudden price drop for a high-risk mineral. He suspects unethical sourcing behind the discount.	The leader investigates the supply chain and pauses new orders. They also add anomaly-based ethics alerts to procurement analytics.
Hanae discovers that conflict-free commitments are stated in policy but not reflected in contracts. She senses a legal and ethical mismatch.	The leader updates all contracts with binding conflict-free clauses. They also require supplier acknowledgment and periodic re-attestation.

6. ETHICAL OVERSIGHT SUMMARIES

Operational Ethics requires governance, and Ethical Oversight Summaries consolidate findings from audits, reviews, and monitoring activities across the operational ecosystem. They highlight ethical risks, corrective actions, and systemic patterns. This matters because ethical oversight must be visible to drive accountability. Ethical Oversight Summaries prevent stagnation by ensuring ethical issues receive executive attention. They also reinforce fairness by ensuring all teams are held to the same ethical expectations. When ignored, oversight becomes fragmented and ineffective. Ethical risks accumulate without resolution. Over time, Operational Ethics deteriorates as governance loses coherence.

When executed well, Ethical Oversight Summaries strengthen Operational Ethics by creating a unified view of ethical performance. Stakeholders trust that the organization monitors ethics with discipline. This improves accountability, reduces risk, and enhances cultural integrity. Summaries reinforce the principle that ethics must be governed systematically. They remind leaders that oversight is essential to sustained ethical behavior. Ultimately, they build an environment where ethical performance is visible, managed, and continuously improved.

MORAL COMPASS SCENARIO	LEADERSHIP ACTION
Idril reads an oversight summary that lists "no major issues" despite multiple recent ethics incidents. She realizes the summary is sanitized.	The leader demands unfiltered incident counts, severity, and trends in every summary. They also require cross-checking against raw logs.
Maxton notices that oversight summaries focus on compliance checkboxes rather than lived impacts on workers and communities. He senses ethics reduced to paperwork.	The leader adds qualitative case studies and impact narratives to each summary. They also invite worker representatives to review findings.
Sorrel finds that oversight summaries are produced annually, long after issues emerge. She recognizes a dangerous lag.	The leader moves to quarterly oversight cycles with interim updates for severe issues. They also set response-time targets.
Yara sees that oversight summaries exclude high-risk joint ventures "outside operational scope." She realizes ethics are being compartmentalized.	The leader expands oversight to all entities tied to the brand. They also require joint-venture ethics clauses and reporting.
Pavel notices that oversight summaries are shared only with a small executive circle. He senses a culture of secrecy.	The leader broadens distribution to operational leaders and relevant teams. They also create anonymized versions for wider internal transparency.
Ines discovers that recommendations in oversight summaries are rarely tracked to completion. She recognizes ethics theater.	The leader assigns owners, deadlines, and status tracking for every recommendation. They also report completion rates alongside new findings.

Ai6x6™ Transformation Blueprint

Process Safety

"Ethics is the light that keeps technology from becoming a shadow."
Maya Angelou, poet and civil rights icon.

Process Safety requires proactive protection, and Operational Hazard Prevention defines the safeguards that keep processes stable, predictable, and free from preventable harm. It establishes the standards that identify and mitigate risks before they escalate. This matters because unsafe processes create immediate operational and human risk. Operational Hazard Prevention prevents incidents by ensuring systems are designed and maintained responsibly. It also reinforces fairness by ensuring all employees benefit from safe working conditions. When ignored, hazards accumulate and become more difficult to control. Teams experience uncertainty as processes become unpredictable. Over time, Process Safety deteriorates as vulnerabilities multiply. Incident Reduction Measures ensure risks are systematically addressed to prevent future failures.

This is the ultimate insurance policy for an antifragile organization. Emergency Protocol Files and Failure Containment Plans ensure the system can detect and fix its own operational gaps during market shocks. This topic involves mapping "Process Safety Checklists" and "Safety Monitoring Dashboards" to create a "Self-Healing" ecosystem. By engrossing yourself in "Hazard Control Sheets" on a printed page, you can visualize the organization's resilience. Process safety is the "Map" that allows you to navigate the "Kill Zone" of operational failure, turning stagnation into a state of permanent, self-disrupting operational mastery. Using the print blueprint, you can physically flip between "Safety Protocol Files" and "Incident Logs," seeing the whole map of operational security. This clinical focus transforms your operations from a source of fragility into a bedrock of "Bliss," ensuring that your "Operations Transformation" moves as fast as your thoughts, protected by the self-healing engine of the Ai6x6™ framework.

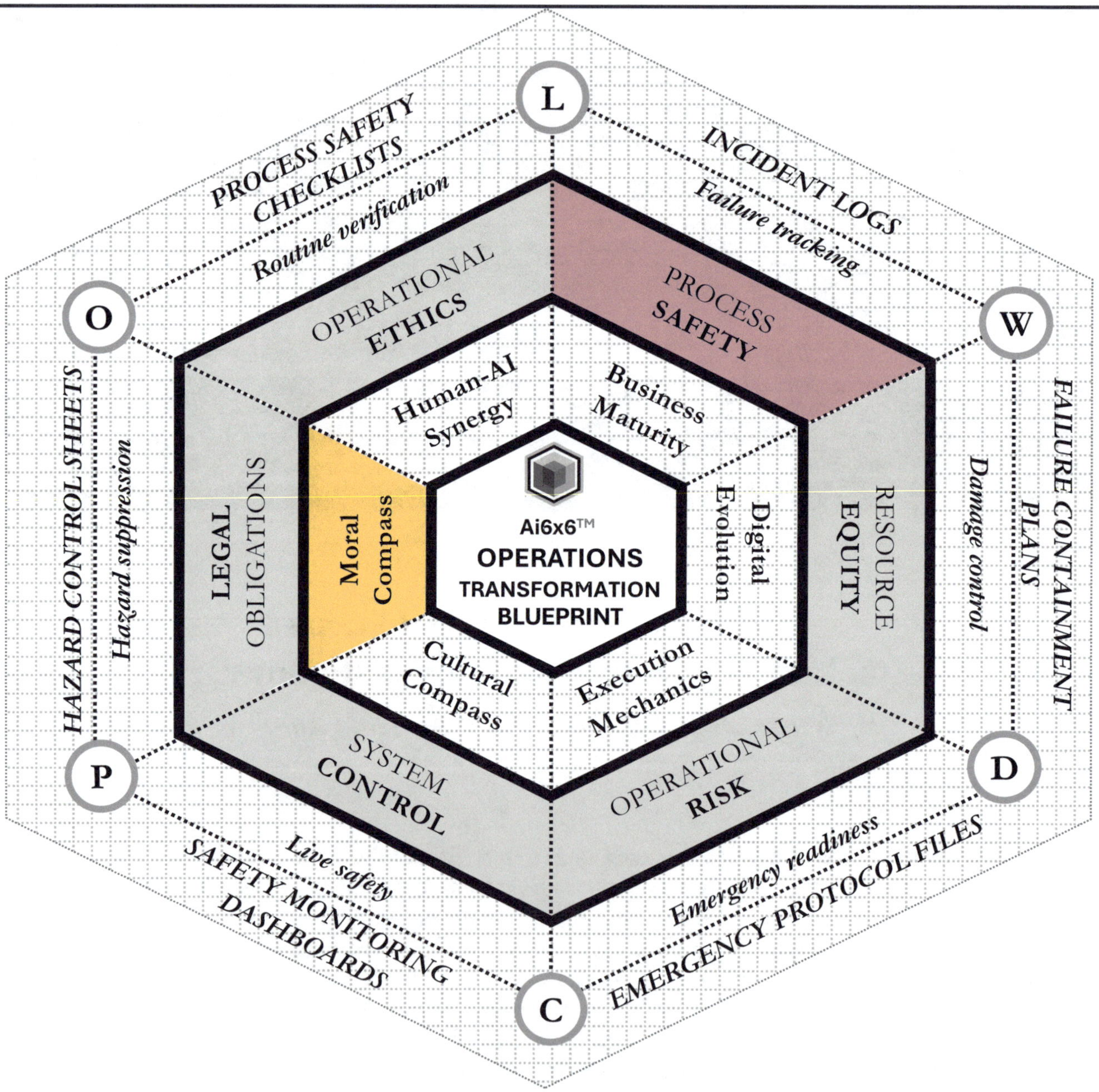

Ai6x6 PROCESS SAFETY FOR MORAL COMPASS | OPERATIONS TRANSFORMATION

1. PROCESS SAFETY CHECKLISTS

Process Safety requires structured discipline, and Process Safety Checklists ensure that every operational step is executed in a safe, controlled, and predictable manner. They translate complex safety requirements into actionable, repeatable tasks that eliminate ambiguity. This matters because safety failures often originate in overlooked details. Process Safety Checklists prevent incidents by forcing teams to verify critical conditions before proceeding. They also reinforce fairness by ensuring all employees follow the same safety expectations. When ignored, safety becomes inconsistent and dependent on individual memory. Teams introduce avoidable risk into routine operations. Over time, Process Safety deteriorates as discipline erodes.

When executed well, Process Safety Checklists strengthen Process Safety by embedding rigor into daily workflows. Employees trust that the organization prioritizes their protection. This improves reliability, reduces accidents, and enhances operational stability. Checklists reinforce the principle that safety must be procedural, not optional. They remind leaders that consistency is the foundation of safe operations. Ultimately, they build an environment where safety is executed with precision every time.

MORAL COMPASS SCENARIO	LEADERSHIP ACTION
Kalen discovers that operators are skipping the final verification step during equipment startup. He realizes the checklist isn't embedded in the workflow.	The leader integrates mandatory digital sign-offs into the control panel. They also require supervisors to review skipped steps daily.
Mireya notices that the checklist still references a valve model retired two years ago. She senses outdated safety assumptions.	The leader updates all equipment references and enforces version-controlled checklists. They also require engineering review before any checklist change.
Torin finds that teams complete checklists after the task instead of before it. He recognizes backward compliance.	The leader locks checklist submission until all pre-task steps are completed. They also retrain teams on pre-task safety discipline.
Asha sees that the checklist doesn't include PPE verification for new chemical handling tasks. She realizes hazard coverage is incomplete.	The leader adds PPE-specific steps and validates them with safety officers. They also require hazard-aligned checklist updates.
Joren notices that contractors use a simplified checklist missing critical steps. He senses a two-tier safety culture.	The leader standardizes checklists across employees and contractors. They also require contractor onboarding on full safety protocols.
Selka discovers that checklist items are written ambiguously, leading to inconsistent interpretation. She recognizes clarity risk.	The leader rewrites items with precise, measurable language. They also test comprehension with frontline operators.

Process Safety requires visibility into failures, and Incident Logs document every safety related event, near miss, and operational anomaly. They capture context, contributing factors, and potential consequences. This matters because unrecorded incidents repeat themselves. Incident Logs prevent recurrence by creating a structured record of what went wrong and why. They also reinforce fairness by ensuring all incidents—large or small—receive equal attention. When ignored, patterns remain hidden, and risks escalate. Teams normalize unsafe conditions without realizing it. Over time, Process Safety collapses as lessons go unlearned.

When executed well, Incident Logs strengthen Process Safety by transforming failures into actionable insight. Employees trust that issues will be addressed, not ignored. This improves accountability, reduces harm, and enhances organizational learning. Logs reinforce the principle that transparency is essential to safety. They remind leaders that every incident is a signal. Ultimately, they build an environment where safety improves through continuous documentation and analysis.

MORAL COMPASS SCENARIO	LEADERSHIP ACTION
Darin sees multiple minor chemical spills logged with identical descriptions. He realizes operators are copy-pasting entries.	The leader enforces structured fields requiring specific details. They also train teams on accurate incident documentation.
Liora notices that incidents spike during equipment calibration days. She senses a pattern tied to maintenance cycles.	The leader adds calibration-day safety protocols and supervises high-risk windows. They also track calibration-related incidents separately.
Fenya finds that near-misses are rarely logged. She recognizes that teams only report actual harm.	The leader mandates near-miss reporting and rewards early detection. They also add near-miss metrics to safety dashboards.
Corvin sees that incident logs from night shifts are sparse despite known issues. He suspects underreporting.	The leader conducts anonymous surveys and increases night-shift supervision. They also audit night-shift logs weekly.
Amara notices that incident logs lack environmental conditions like humidity or temperature. She senses missing causal data.	The leader adds environmental fields to every log. They also analyze correlations with incident patterns.
Jax discovers that logs are stored locally on each site, preventing cross-site learning. He recognizes siloed safety knowledge.	The leader centralizes incident logs and enables cross-site analytics. They also require monthly cross-site safety reviews.

3. FAILURE CONTAINMENT PLANS

Process Safety requires preparation for breakdowns, and Failure Containment Plans define how the organization limits harm when processes deviate from expected behavior. They outline steps for isolating failures, protecting personnel, and preventing escalation. This matters because even well-designed processes can fail under stress. Failure Containment Plans prevent widespread damage by ensuring failures remain controlled and predictable. They also reinforce fairness by ensuring all employees receive the same level of protection during incidents. When ignored, failures spread unchecked and create cascading hazards. Teams face unpredictable danger. Over time, Process Safety deteriorates as containment discipline disappears.

When executed well, Failure Containment Plans strengthen Process Safety by ensuring the organization is ready for the unexpected. Employees trust that failures will be managed responsibly. This improves resilience, reduces harm, and enhances ethical alignment. Plans reinforce the principle that safety requires both prevention and response. They remind leaders that containment is a critical component of risk management. Ultimately, they build an environment where failures are controlled with precision and care.

MORAL COMPASS SCENARIO	LEADERSHIP ACTION
Nerin discovers that a containment plan assumes manual shutdown, but the control room is unstaffed during certain hours. He realizes the plan is impossible to execute.	The leader adds automated shutdown triggers and adjusts staffing. They also test containment feasibility under all shift patterns.
Yara notices that containment plans don't include communication protocols for multi-site failures. She senses coordination breakdown risk.	The leader adds cross-site escalation channels and drills multi-site scenarios. They also assign regional containment leads.
Brann finds that containment plans rely on outdated floor layouts. He recognizes that evacuation routes no longer match reality.	The leader updates all layouts and re-marks physical routes. They also require layout reviews after any facility modification.
Selin sees that containment plans assume unlimited backup power. She realizes generator capacity is overstated.	The leader recalculates power budgets and updates contingency steps. They also test containment under reduced power.
Otho notices that containment plans don't include vendor-managed systems. He senses third-party exposure.	The leader integrates vendor systems into containment modeling. They also require vendor participation in drills.
Mirelle discovers that containment plans treat all failures as equal severity. She recognizes prioritization gaps.	The leader adds severity tiers and response matrices. They also train teams on tier-based containment.

Process Safety requires clarity during crises, and Emergency Protocol Files define the exact actions teams must take when dangerous conditions arise. They provide structured guidance for evacuation, shutdown, communication, and escalation. This matters because emergencies demand immediate, coordinated action. Emergency Protocol Files prevent chaos by ensuring every employee knows their role. They also reinforce fairness by ensuring all teams receive the same level of preparedness. When ignored, responses become improvised and inconsistent. Teams hesitate or overreact, increasing risk. Over time, Process Safety collapses as emergency readiness erodes.

When executed well, Emergency Protocol Files strengthen Process Safety by creating predictable, disciplined responses to danger. Employees trust that the organization prepares them for high risk scenarios. This improves confidence, reduces harm, and enhances operational stability. Protocols reinforce the principle that preparedness is a moral obligation. They remind leaders that emergencies require clarity, not improvisation. Ultimately, they build an environment where emergency actions are executed with precision.

MORAL COMPASS SCENARIO	LEADERSHIP ACTION
Rina discovers that emergency protocols are stored only on the internal network, inaccessible during outages. She realizes they fail when needed most.	The leader distributes offline copies and posts physical binders at critical stations. They also require redundancy checks.
Jarek notices that evacuation protocols don't account for employees with mobility impairments. He senses exclusion in emergency planning.	The leader adds assisted-evacuation procedures and trains designated helpers. They also test inclusive evacuation drills.
Talia finds that emergency protocols reference equipment removed years ago. She recognizes outdated instructions.	The leader updates all protocol files and enforces annual reviews. They also require engineering sign-off for accuracy.
Corin sees that emergency contacts include former employees. He realizes escalation chains are broken.	The leader refreshes contact lists and automates updates via HR systems. They also require quarterly verification.
Nessa notices that chemical spill protocols lack neutralization steps for new materials. She senses hazard-specific gaps.	The leader adds material-specific procedures and trains teams. They also require chemical-specific protocol updates.
Halden discovers that emergency protocols are written in technical jargon. He worries frontline workers won't understand them under stress.	The leader rewrites protocols in plain language and tests comprehension. They also require readability scoring.

5. SAFETY MONITORING DASHBOARDS

Process Safety requires real time visibility, and Safety Monitoring Dashboards track operational conditions, hazard indicators, and safety critical metrics. They reveal anomalies before they escalate into incidents. This matters because safety risks often emerge gradually. Safety Monitoring Dashboards prevent harm by enabling early detection and intervention. They also reinforce fairness by ensuring all teams have access to the same safety intelligence. When ignored, risks remain invisible until they cause damage. Teams operate without awareness of emerging threats. Over time, Process Safety deteriorates as visibility disappears.

When executed well, Safety Monitoring Dashboards strengthen Process Safety by enabling proactive risk management. Employees trust that the organization monitors safety with discipline. This improves responsiveness, reduces incidents, and enhances operational integrity. Dashboards reinforce the principle that safety requires continuous observation. They remind leaders that real time data is essential to prevention. Ultimately, they build an environment where safety is actively protected through constant monitoring.

MORAL COMPASS SCENARIO	LEADERSHIP ACTION
Kaela sees that dashboards show sensor outages for days without alerts. She realizes monitoring blind spots exist.	The leader adds sensor-health alerts and redundancy. They also require daily dashboard verification.
Orion notices that dashboards aggregate all safety events into a single metric. He senses loss of actionable detail.	The leader breaks events into categories—chemical, mechanical, electrical, human. They also require category-level trend reviews.
Sera finds that dashboards lag by several hours due to slow data ingestion. She recognizes delayed risk detection.	The leader upgrades ingestion pipelines and adds real-time streaming. They also track ingestion latency.
Bram sees that dashboards don't include contractor incidents. He realizes the safety picture is incomplete.	The leader integrates contractor data feeds. They also require contractor-specific safety KPIs.
Elin notices that dashboards show spikes during equipment warm-up cycles. She senses predictable hazard windows.	The leader adds warm-up safety protocols and monitors those windows separately. They also require warm-up risk scoring.
Juno discovers that dashboards are only visible to supervisors. She recognizes that frontline workers lack situational awareness.	The leader deploys floor-level displays and mobile alerts. They also train teams on interpreting dashboard signals.

6. HAZARD CONTROL SHEETS

Process Safety requires clear documentation of risks, and Hazard Control Sheets outline the dangers associated with specific tasks, materials, and environments. They specify exposure risks, control measures, and required protective equipment. This matters because hazards vary widely across operations. Hazard Control Sheets prevent harm by ensuring employees understand the risks before engaging in work. They also reinforce fairness by ensuring all employees receive the same hazard information. When ignored, teams operate without awareness of critical dangers. Mistakes become more likely and more severe. Over time, Process Safety collapses as hazard knowledge disappears.

When executed well, Hazard Control Sheets strengthen Process Safety by making risk information accessible and actionable. Employees trust that the organization communicates hazards transparently. This improves compliance, reduces incidents, and enhances ethical responsibility. Sheets reinforce the principle that safety requires informed action. They remind leaders that hazard clarity protects lives. Ultimately, they build an environment where risks are understood and controlled with discipline.

MORAL COMPASS SCENARIO	LEADERSHIP ACTION
Rafe sees that hazard sheets for a new solvent list outdated exposure limits. He realizes workers are under-protected.	The leader updates exposure thresholds and retrains teams. They also require annual chemical-hazard reviews.
Mireya notices that hazard sheets don't include incompatibility warnings for certain materials stored nearby. She senses a fire risk.	The leader adds incompatibility tables and reorganizes storage. They also require compatibility checks for new materials.
Tovin finds that hazard sheets are text-only and lack diagrams. He recognizes that visual learners may miss critical steps.	The leader adds diagrams and color-coded hazard levels. They also test comprehension with diverse teams.
Selka sees that hazard sheets are only in English despite a multilingual workforce. She realizes comprehension gaps.	The leader provides translated versions and verifies accuracy. They also require multilingual hazard training.
Joren notices that hazard sheets don't specify emergency PPE for rare spill scenarios. He senses incomplete guidance.	The leader adds scenario-specific PPE requirements. They also require spill-scenario drills.
Aveline discovers that hazard sheets are not updated after supplier formulation changes. She recognizes silent risk drift.	The leader integrates supplier change notifications into hazard updates. They also require formulation-change audits.

Ai6x6™ Transformation Blueprint

Resource Equity

"If AI is to shape society, society must shape AI with conscience."

Jacinda Ardern, former Prime Minister of New Zealand known for empathetic leadership.

Resource Equity requires balanced distribution, and Fair Resource Allocation ensures teams receive the tools, time, and support needed to perform effectively. It establishes expectations that prevent favoritism, scarcity manipulation, or inequitable distribution. This matters because resource access directly shapes performance and morale. Fair Resource Allocation prevents operational imbalance by ensuring resources match actual needs. It also reinforces trust by demonstrating that allocation decisions are transparent and principled. When ignored, disparities widen and productivity declines. Teams experience frustration as access becomes inconsistent. Over time, Resource Equity deteriorates as inequities compound. Balanced Access Distribution ensures resources are shared in a way that supports organizational stability.

Self-healing requires a fair and transparent allocation of organizational energy. Capacity Fairness Reviews and Resource Allocation Dashboards allow a leader to see where the organization is most fragile and rebalance for "Bliss." This topic focuses on "Distribution Balance Logs" to ensure that capital and talent are not wasted in silos. By studying "Operational Inclusion Logs" in the blueprint, you identify where the engine is being throttled by internal competition for resources. Equity in operations is a "Mastery" strategy. It ensures that every part of the system is fueled to move as fast as your thoughts. Within the print series, you can map "Resource Transparency Reports" against your "90-Day Roadmap," seeing the whole map of organizational energy. This clinical focus allows you to move capital to the most critical "Survival" and "Growth" areas, ensuring that your "Operations Transformation" is as equitable as it is efficient, making your engine unassailable by competitors who hunt for silos in resource allocation.

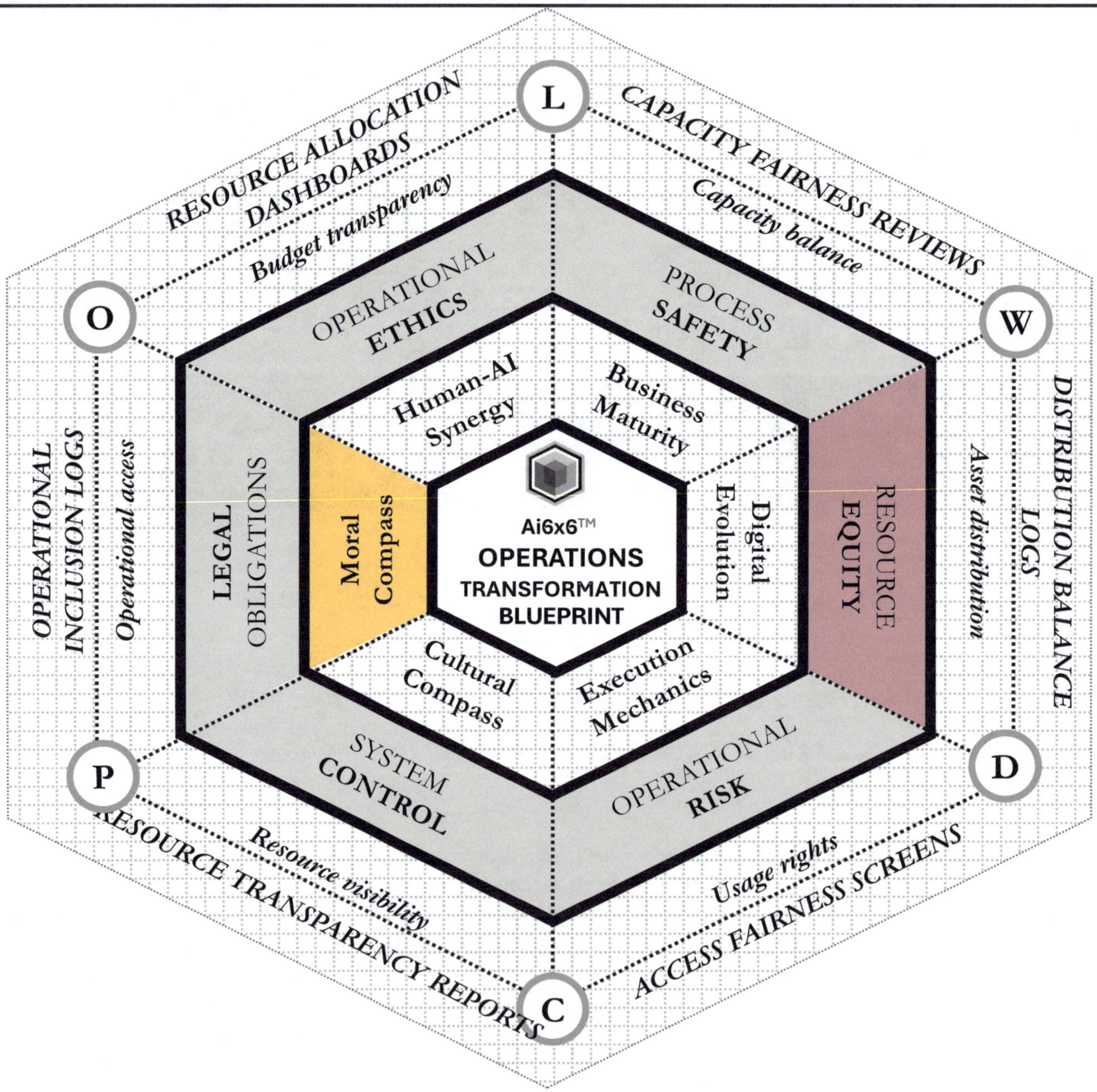

Ai6x6 RESOURCE EQUITY FOR MORAL COMPASS | OPERATIONS TRANSFORMATION

1. RESOURCE ALLOCATION DASHBOARDS

Resource Equity requires transparent distribution, and Resource Allocation Dashboards visualize how time, budget, tools, and staffing are assigned across teams and functions. They reveal imbalances that may otherwise remain hidden inside operational routines. This matters because inequitable allocation creates structural disadvantage. Resource Allocation Dashboards prevent bias by making distribution patterns visible and measurable. They also reinforce fairness by ensuring all teams receive resources aligned with their responsibilities. When ignored, resource decisions become political, opaque, and inconsistent. Teams experience shortages that undermine performance. Over time, Resource Equity deteriorates as inequity becomes normalized.

When executed well, Resource Allocation Dashboards strengthen Resource Equity by enabling data driven fairness. Employees trust that resources are distributed with integrity. This improves morale, reduces conflict, and enhances operational stability. Dashboards reinforce the principle that equity requires visibility. They remind leaders that fairness must be monitored continuously. Ultimately, they build an environment where, resources support performance without favoritism.

MORAL COMPASS SCENARIO	LEADERSHIP ACTION
Nalin sees the dashboard showing that high-visibility teams receive triple the compute resources of back-office teams doing mission-critical processing. He realizes allocation is driven by politics, not need.	The leader reallocates compute based on workload intensity and risk exposure. They also require justification for any resource request above baseline.
Soraya notices that field offices in remote regions consistently appear at the bottom of the allocation list. She senses geographic inequity.	The leader adjusts allocation formulas to weight operational difficulty and service coverage. They also add region-equity scoring to the dashboard.
Idris finds that dashboards hide historical allocation trends, making it impossible to see long-term bias. He recognizes structural imbalance.	The leader adds multi-year trend views and flags persistent disparities. They also require quarterly equity reviews.
Maclin sees that emergency response teams receive resources only after incidents spike. She realizes the system is reactive instead of preventive.	The leader pre-allocates surge capacity to high-risk teams. They also add predictive-risk allocation triggers.
Corvin notices that dashboards show unused resources sitting idle in one department while another is over capacity. He senses silo-driven inefficiency.	The leader implements cross-department resource pooling. They also track idle-resource ratios.
Tova discovers that dashboards don't include accessibility-related resource needs. She realizes teams supporting disabled workers are under-resourced.	The leader adds accessibility-specific resource categories. They also require inclusive allocation

Resource Equity requires balanced workloads, and Capacity Fairness Reviews evaluate whether teams have the staffing, time, and capability needed to meet operational demands. They examine mismatches between expectations and available capacity. This matters because overloaded teams experience burnout and declining quality. Capacity Fairness Reviews prevent inequity by ensuring workloads are distributed responsibly. They also reinforce fairness by protecting teams from chronic overextension. When ignored, capacity gaps widen, and performance becomes unstable. Employees feel unsupported and overwhelmed. Over time, Resource Equity collapses as workload fairness disappears.

When executed well, Capacity Fairness Reviews strengthen Resource Equity by aligning expectations with reality. Employees trust that leadership recognizes and manages capacity constraints. This improves performance, reduces turnover, and enhances operational resilience. Reviews reinforce the principle that fairness must govern workload distribution. They remind leaders that capacity is a moral and operational variable. Ultimately, they build an environment where teams operate within sustainable limits.

MORAL COMPASS SCENARIO	LEADERSHIP ACTION
Jalen discovers that one team is consistently assigned workloads exceeding 120% capacity while others operate at 60%. He realizes burnout is being normalized.	The leader rebalances workloads and enforces capacity ceilings. They also track capacity fairness monthly.
Mirelle notices that capacity models assume all workers have identical skill levels. She senses unrealistic planning.	The leader adjusts models to reflect skill variance and task complexity. They also require skill-weighted capacity reviews.
Otho finds that seasonal spikes are ignored in capacity planning, causing predictable overloads. He recognizes planning blindness.	The leader integrates seasonal forecasting into capacity models. They also add spike-readiness checks.
Selin sees that capacity reviews exclude contractors, masking true workload distribution. She realizes hidden labor is propping up operations.	The leader includes contractors in all capacity calculations. They also require contractor-equity monitoring.
Brann notices that capacity reviews treat remote teams as infinitely flexible. He senses exploitation of time-zone differences.	The leader sets boundaries on off-hour expectations. They also track remote-team load separately.
Fenya discovers that capacity reviews ignore mandatory training hours, inflating available time. She recognizes planning distortion.	The leader incorporates training time into capacity models. They also require protected training blocks.

3. DISTRIBUTION BALANCE LOGS

Resource Equity requires continuous monitoring, and Distribution Balance Logs track how resources shift over time across projects, teams, and priorities. They capture adjustments, reallocations, and emerging imbalances. This matters because resource drift often occurs gradually. Distribution Balance Logs prevent inequity by revealing where resources accumulate or disappear without justification. They also reinforce fairness by ensuring all teams benefit from consistent oversight. When ignored, imbalances become entrenched and difficult to correct. Teams experience chronic shortages or surpluses. Over time, Resource Equity deteriorates as distribution loses discipline.

When executed well, Distribution Balance Logs strengthen Resource Equity by creating a historical record of allocation decisions. Employees trust that resource shifts are monitored and justified. This improves transparency, reduces conflict, and enhances operational alignment. Logs reinforce the principle that equity requires continuous attention. They remind leaders that fairness is dynamic, not static. Ultimately, they build an environment where resource distribution remains balanced and accountable.

MORAL COMPASS SCENARIO	LEADERSHIP ACTION
Kael sees that distribution logs show critical supplies consistently routed to flagship sites first. He realizes smaller sites are perpetually shorted.	The leader implements priority tiers based on operational need, not prestige. They also track distribution fairness.
Orla notices that distribution delays cluster around rural delivery routes. She senses logistical inequity.	The leader redesigns routing algorithms to account for distance and terrain. They also add rural-delivery performance metrics.
Halden finds that distribution logs show repeated shortages caused by inaccurate demand forecasts. He recognizes forecasting bias.	The leader recalibrates demand models using real consumption data. They also require forecast-accuracy reviews.
Suri sees that distribution logs don't track spoilage or waste. She realizes inefficiency is invisible.	The leader adds waste metrics and redesigns replenishment cycles. They also require spoilage-reduction plans.
Bram notices that distribution logs show high-value items being shipped through high-risk routes. He senses unnecessary exposure.	The leader reroutes sensitive shipments and updates risk-based routing rules. They also track route-risk alignment.
Elin discovers that distribution logs are updated manually, causing inconsistent entries. She recognizes data-quality risk.	The leader automates logging and adds validation checks. They also train teams on exception handling.

4. ACCESS FAIRNESS SCREENS

Resource Equity requires equal opportunity, and Access Fairness Screens evaluate whether employees can obtain the tools, systems, and support they need to perform effectively. They examine barriers created by process complexity, permissions, or organizational structure. This matters because unequal access creates performance gaps unrelated to skill. Access Fairness Screens prevent inequity by identifying where access is restricted unfairly. They also reinforce fairness by ensuring all employees operate on a level playing field. When ignored, access disparities widen and certain groups face systemic disadvantage. Teams struggle with avoidable friction. Over time, Resource Equity collapses as access becomes uneven.

When executed well, Access Fairness Screens strengthen Resource Equity by ensuring access is governed by need, not hierarchy. Employees trust that the organization removes unnecessary barriers. This improves efficiency, reduces frustration, and enhances ethical alignment. Screens reinforce the principle that access is a foundational equity variable. They remind leaders that fairness requires structural openness. Ultimately, they build an environment where access supports performance for everyone.

MORAL COMPASS SCENARIO	LEADERSHIP ACTION
Rafe discovers that only senior staff can request specialized equipment, even when junior staff need it for safety. He realizes access is rank-based, not need-based.	The leader opens access requests to all roles and adds need-based approval criteria. They also track access disparities.
Mireya notices that remote workers cannot access certain internal tools due to outdated VPN rules. She senses digital inequity.	The leader updates access policies and modernizes authentication. They also require remote-access fairness checks.
Tovin finds that access screens deny part-time workers entry to training resources. He recognizes structural exclusion.	The leader grants equal access to training and updates eligibility rules. They also track training-access equity.
Selka sees that access screens require high-bandwidth connections, excluding low-connectivity regions. She realizes infrastructure bias.	The leader adds low-bandwidth modes and tests accessibility across connection tiers. They also require connectivity-equity reviews.
Joren notices that access screens rely on outdated role definitions, blocking cross-functional collaboration. He senses bureaucratic drag.	The leader updates role mappings and adds dynamic access rules. They also require quarterly role-access audits.
Aveline discovers that access screens don't support screen readers. She recognizes accessibility failure.	The leader rebuilds screens with WCAG-aligned markup. They also require accessibility testing for all access flows.

5. RESOURCE TRANSPARENCY REPORTS

Resource Equity requires openness, and Resource Transparency Reports disclose how resources are allocated, why decisions were made, and how fairness was evaluated. They provide visibility into the logic behind distribution. This matters because secrecy breeds suspicion and inequity. Resource Transparency Reports prevent misunderstanding by explaining allocation decisions clearly. They also reinforce fairness by ensuring all employees understand how resources are assigned. When ignored, rumors replace facts and trust erodes. Teams perceive favoritism even when none exists. Over time, Resource Equity deteriorates as transparency disappears.

When executed well, Resource Transparency Reports strengthen Resource Equity by building trust in allocation decisions. Employees trust that leadership acts with fairness and clarity. This improves morale, reduces conflict, and enhances operational cohesion. Reports reinforce the principle that transparency is essential to equity. They remind leaders that fairness must be communicated, not assumed. Ultimately, they build an environment where resource decisions are understood and respected.

MORAL COMPASS SCENARIO	LEADERSHIP ACTION
Darin sees that transparency reports summarize resource usage at a high level, hiding which teams consume disproportionate amounts. He realizes accountability is diluted.	The leader adds team-level breakdowns and publishes consumption ratios. They also require justification for outliers.
Liora notices that transparency reports exclude contractor usage entirely. She senses incomplete visibility.	The leader integrates contractor data into reporting. They also require contractor-resource audits.
Fen finds that transparency reports are released annually, long after decisions are made. He recognizes lagged accountability.	The leader moves to monthly reporting cycles. They also add real-time dashboards for critical resources.
Corvin sees that transparency reports omit environmental impact metrics tied to resource use. He realizes sustainability is invisible.	The leader adds carbon, water, and waste indicators. They also require sustainability-aligned resource planning.
Amara notices that transparency reports use vague categories like “miscellaneous.” She senses intentional opacity.	The leader eliminates vague categories and enforces granular classification. They also audit category accuracy.
Jax discovers that transparency reports are not shared with frontline teams. He realizes those most affected lack visibility.	The leader distributes reports widely and trains teams on interpreting them. They also gather feedback for improvements.

Resource Equity requires representation, and Operational Inclusion Logs document whether all teams and groups are included in planning, decision making, and resource discussions. They track participation patterns and identify who is consistently involved or excluded. This matters because exclusion leads to inequitable outcomes. Operational Inclusion Logs prevent bias by ensuring diverse voices shape operational decisions. They also reinforce fairness by ensuring all groups have influence over the resources that affect them. When ignored, certain teams become invisible in planning processes. Decisions skew toward dominant groups. Over time, Resource Equity collapses as inclusion disappears.

When executed well, Operational Inclusion Logs strengthen Resource Equity by embedding representation into operational governance. Employees trust that their perspectives matter. This improves decision quality, reduces blind spots, and enhances ethical alignment. Logs reinforce the principle that inclusion is a structural requirement. They remind leaders that equitable operations require equitable participation. Ultimately, they build an environment where all teams have a voice in resource decisions.

MORAL COMPASS SCENARIO	LEADERSHIP ACTION
Nerin sees that inclusion logs show certain teams never participate in cross-functional pilots. He realizes opportunity is unevenly distributed.	The leader rotates pilot participation and tracks inclusion frequency. They also require equitable pilot selection.
Yara notices that inclusion logs show minimal involvement from night-shift workers in process redesign. She senses time-based exclusion.	The leader schedules redesign sessions across shifts. They also compensate night-shift participation.
Brann finds that inclusion logs show no representation from employees with disabilities in workflow testing. He recognizes accessibility blind spots.	The leader recruits diverse testers and updates testing protocols. They also track accessibility inclusion metrics.
Selin sees that inclusion logs show only senior staff attending strategy workshops. She realizes decision-making is top-heavy.	The leader opens workshops to broader roles and adds structured selection criteria. They also track participation diversity.
Otho notices that inclusion logs show regional offices rarely consulted on global process changes. He senses geographic exclusion.	The leader adds regional consultation checkpoints. They also require region-specific feedback summaries.
Mirelle discovers that inclusion logs show women underrepresented in high-impact operational committees. She recognizes systemic imbalance.	The leader reconstitutes committees with balanced representation. They also track committee-level inclusion metrics.

CHAPTER G4

BLUEPRINT SERIES

Ai6x6™ Transformation Blueprint

Operational Risk

"Sustainability is the art of giving the future a fighting chance."

Wangari Maathai, Nobel Peace Prize laureate and environmental leader.

Operational Risk requires awareness of systemic weaknesses, and Continuity Threat Factors identify where disruptions could jeopardize ongoing operations. They establish visibility into the conditions that threaten stability. This matters because operational continuity underpins organizational reliability. Continuity Threat Factors prevent breakdowns by revealing where reinforcement or redesign is needed. They also reinforce fairness by ensuring all teams benefit from stable, predictable operations. When ignored, disruptions become more frequent and damaging. Teams experience confusion as reliability declines. Over time, Operational Risk increases as vulnerabilities accumulate. System Vulnerability Indicators highlight early signs of potential operational failure.

Identifying the "Kill Zone" of operational fatigue before it leads to extinction. Vulnerability Assessments and Supply Exposure Logs help leaders see the fragility of the organization before the market uses a sledgehammer against it. This topic involves using "Operational Risk Registers" to run a clinical diagnostic on the supply chain and internal systems. By mapping "Escalation Thresholds," you create a protocol for survival. Operational risk management is the "Blueprint" for "Antifragility." It allows the leader to move capital away from legacy "Safe Zones" toward high-growth frontiers, ensuring the organization's long-term enterprise valuation. The print format is essential for this diagnostic; it lets you see the whole "Risk Map" at once, identifying where your operational armor is thin. This clinical focus replaces the "million-dollar guess" of business continuity with a roadmap for resilience, ensuring that your "Operations Transformation" remains robust and your engine moves as cleanly and consistently as your strategic thoughts, protected from the "Kill Zone" of unpredictable systemic failure.

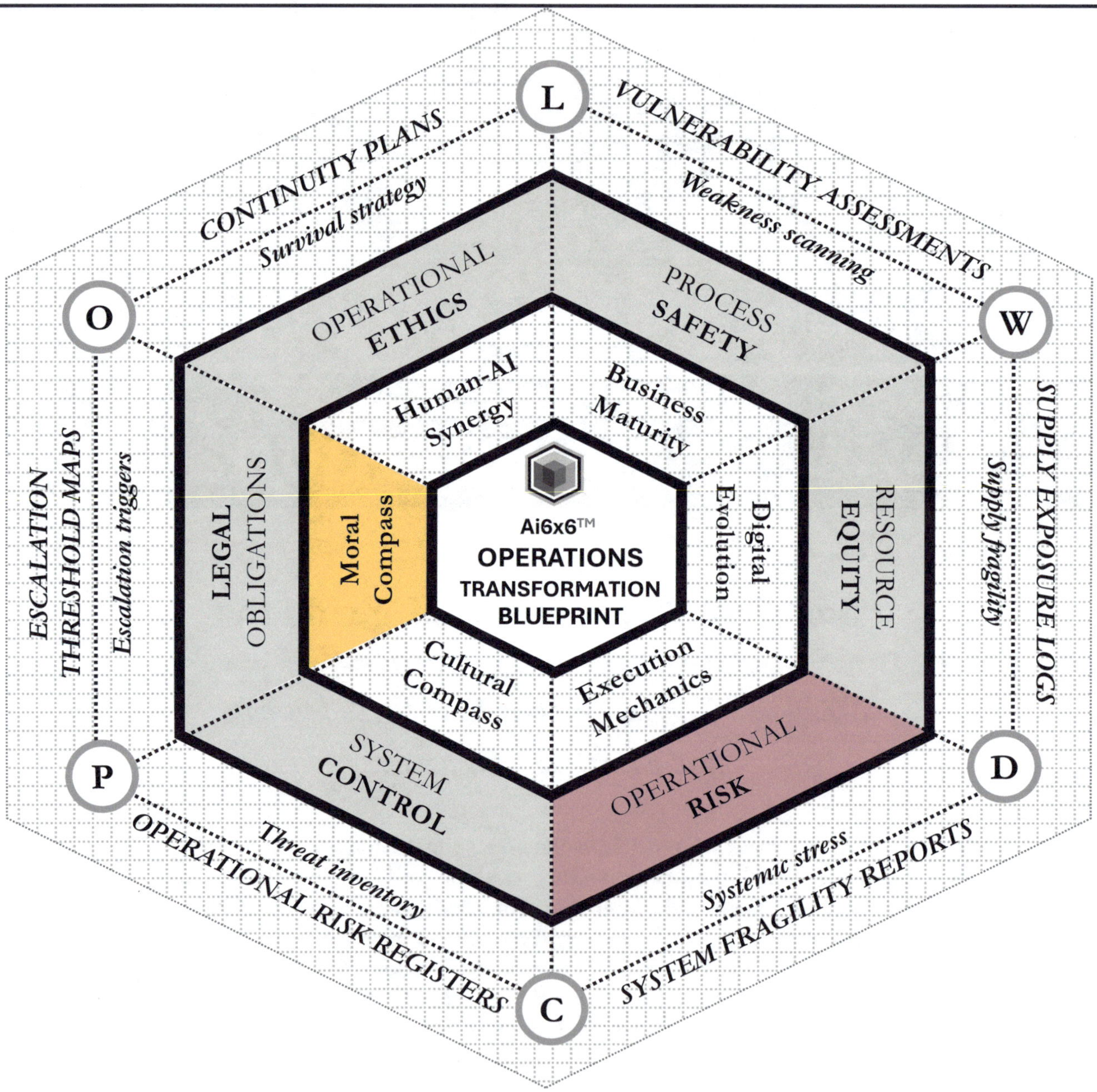

Ai6x6 OPERATIONAL RISK FOR MORAL COMPASS | OPERATIONS TRANSFORMATION

1. CONTINUITY PLANS

Operational Risk requires resilience, and Continuity Plans define how the organization maintains critical operations during disruptions. They outline backup processes, alternative workflows, and recovery priorities. This matters because operational interruptions—whether technical, environmental, or human—can cause cascading failures. Continuity Plans prevent collapse by ensuring essential functions remain available under stress. They also reinforce fairness by protecting all teams and customers from disproportionate impact. When ignored, disruptions become chaotic and unpredictable. Teams scramble without structure, and recovery becomes slow and inconsistent. Over time, Operational Risk escalates as resilience disappears.

When executed well, Continuity Plans strengthen Operational Risk management by creating predictable pathways through uncertainty. Employees trust that the organization can withstand shocks. This improves stability, reduces downtime, and enhances stakeholder confidence. Plans reinforce the principle that resilience must be engineered, not improvised. They remind leaders that continuity is a strategic obligation. Ultimately, they build an environment where operations remain steady even when conditions are unstable.

MORAL COMPASS SCENARIO	LEADERSHIP ACTION
Arlen discovers that the continuity plan assumes a single data-center outage, but the new architecture spans three interdependent regions. He realizes the plan is obsolete.	The leader rewrites the plan to include multi-region failover and tests cross-region resilience. They also require annual architecture-aligned updates.
Mireya sees that continuity plans rely on a vendor whose contract expires next quarter. She senses a looming dependency gap.	The leader secures interim coverage and updates continuity assumptions. They also require vendor-dependency reviews.
Jovan finds that continuity plans assume all staff can work remotely, but several roles require on-site equipment. He recognizes unrealistic expectations.	The leader creates role-specific continuity paths and provisions backup workstations. They also test on-site fallback readiness.
Selka notices that continuity plans don't include communication protocols for multilingual teams. She senses coordination risk.	The leader adds multilingual templates and assigns translation owners. They also test communication clarity during drills.
Torin discovers that continuity plans ignore supply-chain disruptions. He realizes operational continuity is too internally focused.	The leader integrates supply-chain scenarios and adds supplier-failure triggers. They also require cross-functional continuity reviews.
Aveline sees that continuity plans are stored only in a shared drive requiring VPN access. She realizes they'll be inaccessible during outages.	The leader distributes offline copies and posts physical binders. They also require redundancy checks.

Operational Risk requires awareness of weaknesses, and Vulnerability Assessments evaluate where systems, processes, and structures are most exposed to failure. They examine technical gaps, procedural flaws, and environmental threats. This matters because unrecognized vulnerabilities become the root cause of major incidents. Vulnerability Assessments prevent harm by identifying weaknesses before they are exploited or triggered. They also reinforce fairness by ensuring all operational areas receive equal scrutiny. When ignored, vulnerabilities accumulate silently. Teams operate with false confidence. Over time, Operational Risk deteriorates as weaknesses remain unaddressed.

When executed well, Vulnerability Assessments strengthen Operational Risk management by making risk visible and actionable. Employees trust that the organization understands its exposure. This improves preparedness, reduces incidents, and enhances ethical responsibility. Assessments reinforce the principle that risk must be confronted, not avoided. They remind leaders that vulnerability is a signal, not a flaw. Ultimately, they build an environment where weaknesses are identified early and resolved decisively.

MORAL COMPASS SCENARIO	LEADERSHIP ACTION
Nadir discovers that vulnerability assessments focus only on software, ignoring physical access points. He realizes half the attack surface is unexamined.	The leader expands assessments to include physical, environmental, and human vulnerabilities. They also require cross-domain scoring.
Soren notices that assessments rely on outdated threat models from three years ago. He senses blind spots to modern attack patterns.	The leader updates threat models and retrains assessors. They also require annual threat-model refresh cycles.
Lira finds that assessments assume perfect patch compliance, but logs show patching delays of weeks. She recognizes unrealistic baselines.	The leader aligns assessments with real patch latency and accelerates patch workflows. They also track patch-timeliness risk.
Bram sees that assessments ignore insider-risk scenarios. He realizes internal misuse is unmodeled.	The leader adds insider-risk vectors and tests privilege-misuse scenarios. They also require behavioral-risk monitoring.
Orin notices that vulnerability assessments treat all systems equally, ignoring criticality. He senses misaligned prioritization.	The leader weights assessments by business impact and system criticality. They also require risk-tiered remediation.
Fenya discovers that assessments rely on a single scanning tool. She recognizes tool-specific blind spots.	The leader introduces multi-tool scanning and cross-validation. They also require periodic tool-effectiveness reviews.

3. SUPPLY EXPOSURE LOGS

Operational Risk requires visibility into dependencies, and Supply Exposure Logs document where operations rely on specific vendors, materials, or geographic regions. They reveal concentration risks that could disrupt production or service delivery. This matters because supply chains are increasingly fragile. Supply Exposure Logs prevent disruption by highlighting where alternatives or redundancies are needed. They also reinforce fairness by ensuring all operational units receive equal protection from supply volatility. When ignored, single points of failure remain hidden. Teams face sudden shortages without preparation. Over time, Operational Risk escalates as supply fragility becomes systemic.

When executed well, Supply Exposure Logs strengthen Operational Risk management by enabling proactive diversification. Employees trust that the organization monitors supply vulnerabilities with discipline. This improves resilience, reduces delays, and enhances operational stability. Logs reinforce the principle that dependency must be managed, not assumed. They remind leaders that supply chains require continuous oversight. Ultimately, they build an environment where supply risks are understood and mitigated.

MORAL COMPASS SCENARIO	LEADERSHIP ACTION
Kael sees that exposure logs show a single supplier providing 80% of a critical component. He realizes concentration risk is extreme.	The leader diversifies sourcing and adds dual-supplier requirements. They also track concentration thresholds.
Orla notices that exposure logs don't include geopolitical risk scores. She senses blind spots tied to unstable regions.	The leader integrates geopolitical scoring and updates sourcing strategies. They also require region-risk reviews.
Halden finds that exposure logs show long lead times for essential parts, but no contingency stock. He recognizes fragility.	The leader builds buffer inventory and renegotiates lead-time terms. They also track lead-time volatility.
Suri sees that exposure logs ignore transportation bottlenecks like port congestion. She realizes logistics risk is unmodeled.	The leader adds logistics-risk indicators and tests alternative routes. They also require transport-risk monitoring.
Bram notices that exposure logs show multiple suppliers relying on the same upstream mine. He senses hidden single-point failure.	The leader maps upstream dependencies and diversifies at the source. They also require upstream-tier transparency.
Elin discovers that exposure logs are updated manually, causing stale data. She recognizes accuracy risk.	The leader automates updates and adds validation checks. They also require weekly exposure reviews.

4. SYSTEM FRAGILITY REPORTS

Operational Risk requires understanding structural weakness, and System Fragility Reports evaluate how easily processes, systems, or workflows break under stress. They analyze failure points, load thresholds, and interdependencies. This matters because fragile systems fail unpredictably. System Fragility Reports prevent collapse by revealing where reinforcement or redesign is needed. They also reinforce fairness by ensuring all teams operate within stable systems. When ignored, fragility becomes embedded and invisible. Teams experience recurring failures without understanding why. Over time, Operational Risk deteriorates as instability becomes normalized.

When executed well, System Fragility Reports strengthen Operational Risk management by exposing hidden structural risks. Employees trust that the organization invests in system robustness. This improves reliability, reduces downtime, and enhances ethical alignment. Reports reinforce the principle that stability must be engineered. They remind leaders that fragility is a design flaw, not an inevitability. Ultimately, they build an environment where systems withstand stress with confidence.

MORAL COMPASS SCENARIO	LEADERSHIP ACTION
Rafe sees that fragility reports show cascading failures when a single microservice slows down. He realizes the architecture lacks isolation.	The leader adds circuit breakers and redesigns service boundaries. They also require fragility testing for all new services.
Mireya notices that fragility spikes during peak traffic windows. She senses load-related brittleness.	The leader improves load balancing and adds auto-scaling. They also track fragility under stress.
Tovin finds that fragility reports show failures only when legacy systems interact with modern APIs. He recognizes compatibility risk.	The leader builds compatibility layers and tests hybrid interactions. They also require legacy-integration reviews.
Selka sees that fragility increases when teams deploy without coordination. She realizes release timing is destabilizing the system.	The leader enforces coordinated release windows and adds deployment-collision checks. They also track release-related fragility.
Joren notices that fragility reports ignore environmental factors like temperature in edge devices. He senses unmodeled physical stress.	The leader adds environmental stress testing. They also require environmental-risk scoring.
Aveline discovers that fragility reports are not shared with architecture teams. She realizes insights aren't shaping design.	The leader integrates fragility data into architectural planning. They also require fragility-driven design adjustments.

5. OPERATIONAL RISK REGISTERS

Operational Risk requires structured governance, and Operational Risk Registers consolidate all identified risks, their likelihood, impact, and mitigation plans. They create a single source of truth for risk management. This matters because fragmented risk information leads to blind spots. Operational Risk Registers prevent oversight failures by ensuring risks are tracked, prioritized, and addressed systematically. They also reinforce fairness by ensuring all risks—not just high visibility ones—receive attention. When ignored, risks accumulate without coordination. Teams respond reactively instead of strategically. Over time, Operational Risk collapses as governance loses coherence.

When executed well, Operational Risk Registers strengthen Operational Risk management by creating disciplined, transparent oversight. Employees trust that risks are managed intentionally. This improves decision quality, reduces exposure, and enhances organizational resilience. Registers reinforce the principle that risk must be cataloged to be controlled. They remind leaders that governance is essential to stability. Ultimately, they build an environment where risk is managed proactively and comprehensively.

MORAL COMPASS SCENARIO	LEADERSHIP ACTION
Darin sees that the risk register lists dozens of risks with no prioritization. He realizes teams can't distinguish critical threats.	The leader adds impact-likelihood scoring and ranks risks. They also require quarterly reprioritization.
Liora notices that the register hasn't been updated since a major system overhaul. She senses outdated assumptions.	The leader refreshes all entries and aligns them with the new architecture. They also require update triggers tied to major changes.
Fen finds that the register lists risks but no mitigation owners. He recognizes accountability gaps.	The leader assigns owners and deadlines for each risk. They also track mitigation progress.
Corvin sees that the register excludes regulatory risks. He realizes compliance exposure is invisible.	The leader integrates regulatory categories and adds compliance scoring. They also require legal review.
Amara notices that the register treats low-frequency catastrophic risks as low priority. She senses misaligned weighting.	The leader adds severity multipliers for catastrophic events. They also require rare-event modeling.
Jax discovers that the register is stored in a static spreadsheet. He recognizes version-control risk.	The leader migrates it to a controlled system with audit trails. They also enforce change-tracking.

Operational Risk requires clarity in response triggers, and Escalation Threshold Maps define the exact conditions under which issues must be elevated to higher authority. They specify thresholds for severity, impact, and urgency. This matters because delayed escalation magnifies harm. Escalation Threshold Maps prevent confusion by ensuring teams know when to act and whom to notify. They also reinforce fairness by ensuring all incidents receive consistent escalation treatment. When ignored, issues linger unresolved. Teams hesitate or escalate inconsistently. Over time, Operational Risk deteriorates as escalation discipline disappears.

When executed well, Escalation Threshold Maps strengthen Operational Risk management by creating predictable, accountable response pathways. Employees trust that escalation decisions are clear and justified. This improves response speed, reduces damage, and enhances operational integrity. Maps reinforce the principle that escalation must be governed, not improvised. They remind leaders that clarity protects both people and systems. Ultimately, they build an environment where escalation is timely, consistent, and effective.

MORAL COMPASS SCENARIO	LEADERSHIP ACTION
Nerin sees that escalation thresholds are based on outdated service-level expectations. He realizes teams escalate too late.	The leader recalibrates thresholds using real performance data. They also require threshold reviews quarterly.
Yara notices that thresholds differ across regions for identical systems. She senses inconsistent risk tolerance.	The leader standardizes thresholds and documents justified exceptions. They also train regional leads on unified escalation logic.
Brann finds that thresholds rely on manual monitoring, causing delayed escalation. He recognizes automation gaps.	The leader adds automated triggers and alerting. They also test escalation latency.
Selin sees that thresholds don't include customer-impact indicators. She realizes internal metrics miss external harm.	The leader adds customer-impact scoring to escalation logic. They also require customer-impact reviews.
Otho notices that thresholds escalate to individuals instead of teams, creating bottlenecks. He senses single-point dependency.	The leader routes escalations to functional groups. They also add backup escalation paths.
Mirelle discovers that thresholds don't include compound-risk scenarios. She recognizes multi-factor escalation blind spots.	The leader adds compound-risk triggers. They also test multi-signal escalation drills.

Ai6x6™ Transformation Blueprint

System Control

"Morality in technology is not a luxury — it is survival."
Noam Chomsky, linguist and public intellectual.

System Control requires disciplined governance, and Process Oversight Mechanisms define how operational activities are monitored, guided, and evaluated. They establish the controls that prevent drift, inconsistency, or unauthorized changes. This matters because unmanaged processes create instability and ethical risk. Process Oversight Mechanisms prevent deviation by ensuring operations follow defined standards. They also reinforce fairness by ensuring all teams operate under the same oversight structure. When ignored, processes become unpredictable and difficult to manage. Teams experience misalignment as operational behavior diverges. Over time, System Control deteriorates as oversight weakens. Control Enforcement Rules ensure deviations are corrected and standards are upheld.

Total agency over the operational engine. Operational Guardrail Rules and Workflow Control Maps ensure that "Execution Drift" is eliminated through both automated and human-led oversight. This topic uses "System Integrity Audits" and "Permission Logs" to hard-code accountability into every process. By reviewing "Control Enforcement Records" on a printed blueprint, the leader can physically see the boundaries of the system. System control is the definition of "Autonomous Mastery." It provides the "Armor" that protects the organization's legacy, ensuring that operations move with the velocity and precision required for the AI era. In the print series, you can flip between "Monitoring Dashboards" and "System Permission Logs," seeing the whole map of operational intent. This clinical focus ensures that the "How" and the "Why" of every system process are linked, preventing the "Kill Zone" of unmonitored automation and ensuring that your "Operations Transformation" moves as fast and as cleanly as your strategic thoughts, protected by the self-governing engine of the Ai6x6™ blueprint.

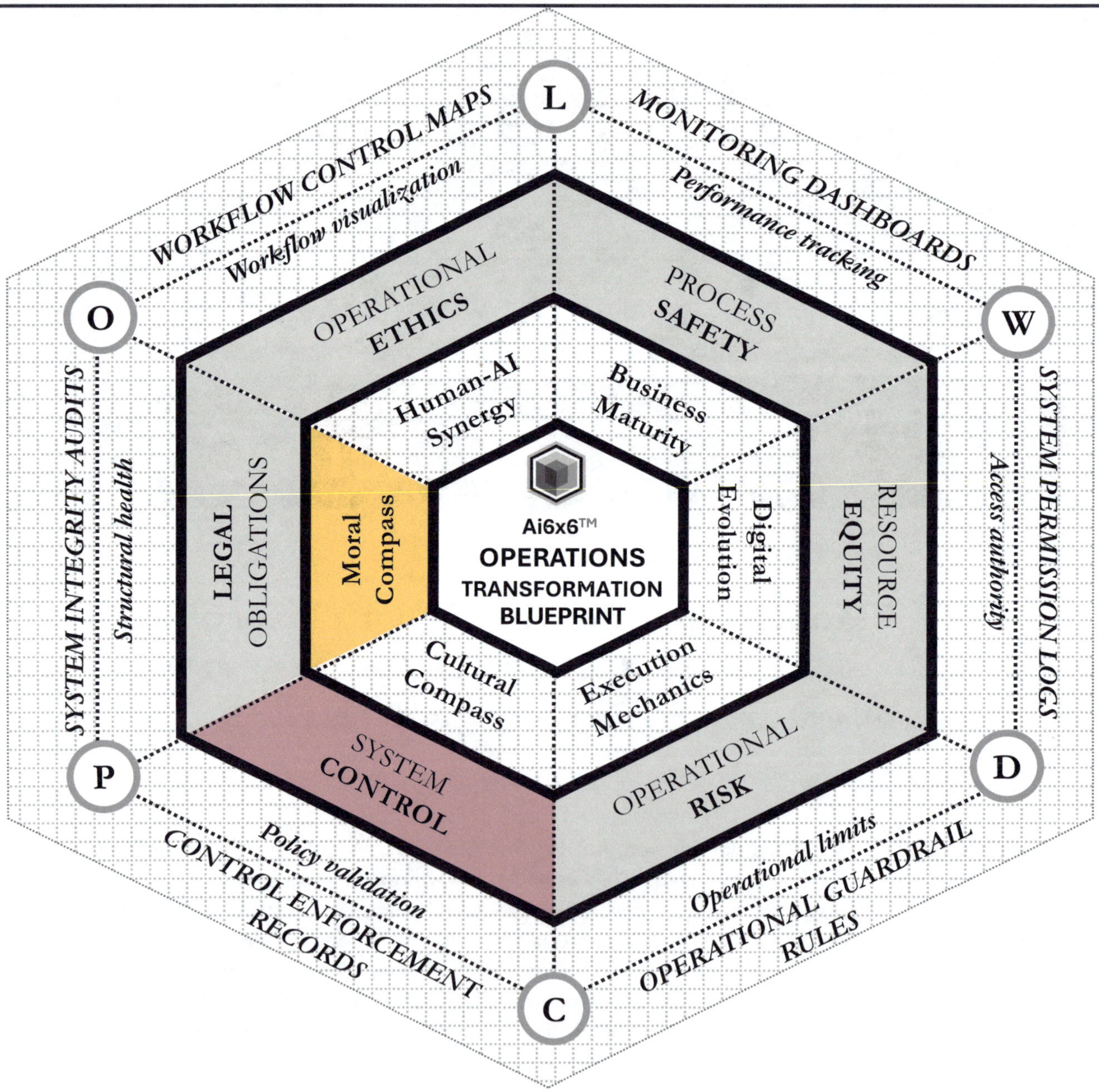

Ai6x6 SYSTEM CONTROL FOR MORAL COMPASS | OPERATIONS TRANSFORMATION

1. WORKFLOW CONTROL MAPS

System Control requires operational discipline, and Workflow Control Maps define how tasks, decisions, and handoffs move through the organization. They visualize dependencies, authority boundaries, and the precise sequence of operational actions. This matters because uncontrolled workflows create inconsistency, delay, and risk. Workflow Control Maps prevent drift by ensuring every step follows a governed path. They also reinforce fairness by ensuring all teams operate under the same structural expectations. When ignored, workflows become fragmented and unpredictable. Teams rely on improvisation instead of design. Over time, System Control deteriorates as processes lose coherence.

When executed well, Workflow Control Maps strengthen System Control by embedding clarity into daily operations. Employees trust that workflows are structured, intentional, and stable. This improves efficiency, reduces errors, and enhances operational reliability. Maps reinforce the principle that control requires visibility. They remind leaders that workflow governance is foundational to consistency. Ultimately, they build an environment where operations flow with precision and accountability.

MORAL COMPASS SCENARIO	LEADERSHIP ACTION
Aris discovers that the workflow map shows parallel steps that actually depend on each other, causing hidden sequencing failures. He realizes the map is misleading.	The leader rebuilds the map with explicit dependency lines. They also require dependency validation before publishing any workflow map.
Liora notices that the workflow map omits exception paths for failed integrations. She senses blind spots in operational routing.	The leader adds exception branches and tests them with simulated failures. They also require exception-path mapping for all workflows.
Bren finds that the workflow map still references a deprecated approval system. He recognizes outdated control logic.	The leader updates all control nodes and enforces version-controlled maps. They also require map reviews after system changes.
Selin sees that the workflow map assumes unlimited queue capacity. She realizes bottlenecks are invisible.	The leader adds queue-capacity indicators and stress-test overlays. They also require capacity modeling in workflow design.
Oren notices that the workflow map shows only automated steps, ignoring manual interventions. He senses incomplete operational visibility.	The leader integrates manual steps and assigns responsible roles. They also require human-in-loop mapping.
Mirelle discovers that workflow maps differ between teams for the same process. She realizes alignment has fractured.	The leader consolidates maps into a single authoritative version. They also enforce cross-team map harmonization.

2. MONITORING DASHBOARDS

System Control requires real time oversight, and Monitoring Dashboards track performance, compliance, and operational health across systems and processes. They consolidate critical indicators into a single, actionable view. This matters because control failures often emerge gradually. Monitoring Dashboards prevent escalation by revealing anomalies early. They also reinforce fairness by ensuring all teams receive the same operational intelligence. When ignored, issues remain hidden until they cause significant disruption. Teams operate without awareness of emerging risks. Over time, System Control collapses as visibility disappears.

When executed well, Monitoring Dashboards strengthen System Control by enabling proactive intervention. Employees trust that the organization monitors operations with discipline. This improves responsiveness, reduces downtime, and enhances ethical alignment. Dashboards reinforce the principle that control requires continuous observation. They remind leaders that real time data is essential to operational stability. Ultimately, they build an environment where systems are governed through constant, informed oversight.

MORAL COMPASS SCENARIO	LEADERSHIP ACTION
Kael sees that dashboards refresh every 15 minutes, making real-time anomalies invisible. He realizes detection is too slow.	The leader upgrades to streaming telemetry and adds real-time alerting. They also track dashboard latency.
Orla notices that dashboards show green status even when underlying logs contain critical errors. She senses misaligned indicators.	The leader recalibrates health thresholds and aligns dashboards with raw log severity. They also require indicator-accuracy audits.
Halden finds that dashboards lack cross-system correlation, forcing teams to manually piece together incidents. He recognizes fragmented visibility.	The leader adds correlation layers and unified incident views. They also require cross-system monitoring integration.
Suri sees that dashboards don't include historical baselines, making it hard to detect slow degradation. She realizes trend blindness.	The leader adds baseline comparisons and anomaly-trend charts. They also require baseline updates quarterly.
Bram notices that dashboards rely on a single data source. He senses single-point monitoring failure.	The leader adds redundant data feeds and cross-checks. They also require feed-health monitoring.
Elin discovers that dashboards are optimized for desktop and unusable on mobile devices. She recognizes accessibility gaps for on-call staff.	The leader deploys mobile-friendly dashboards. They also require mobile-readiness checks.

3. SYSTEM PERMISSION LOGS

System Control requires regulated access, and System Permission Logs document who has access to which systems, data, and operational capabilities. They track changes, approvals, and anomalies in permission assignments. This matters because unauthorized access creates immediate operational and security risk. System Permission Logs prevent misuse by ensuring access aligns with role, responsibility, and need. They also reinforce fairness by ensuring permissions are granted consistently. When ignored, access becomes arbitrary and dangerous. Teams inherit privileges they should not have. Over time, System Control deteriorates as access governance collapses.

When executed well, System Permission Logs strengthen System Control by ensuring access is transparent and justified. Employees trust that permissions are managed responsibly. This improves security, reduces errors, and enhances operational integrity. Logs reinforce the principle that control requires boundaries. They remind leaders that access is a critical control variable. Ultimately, they build an environment where permissions support safety, accountability, and order.

MORAL COMPASS SCENARIO	LEADERSHIP ACTION
Rafe sees that permission logs show dozens of users with admin rights they no longer need. He realizes privilege creep is rampant.	The leader revokes unnecessary privileges and enforces least-privilege rules. They also require quarterly access reviews.
Mireya notices that permission logs don't record failed access attempts. She senses missing security signals.	The leader adds failed-attempt logging and alerting. They also require anomaly detection on access patterns.
Tovin finds that permission logs show shared accounts used across shifts. He recognizes accountability gaps.	The leader bans shared accounts and enforces individual credentials. They also require identity verification for all privileged actions.
Selka sees that permission logs are stored without timestamps. She realizes auditability is compromised.	The leader adds precise timestamping and synchronizes clocks across systems. They also require time-integrity checks.
Joren notices that permission logs don't include API-level access. He senses blind spots in machine-to-machine permissions.	The leader integrates API access logging. They also require API-permission reviews.
Aveline discovers that permission logs are retained for only 30 days. She recognizes insufficient forensic history.	The leader extends retention and secures long-term archives. They also require retention-policy audits.

System Control requires protective constraints, and Operational Guardrail Rules define the limits within which teams must operate. They establish boundaries for decisions, actions, and system behavior. This matters because guardrails prevent small deviations from becoming systemic failures. Operational Guardrail Rules prevent harm by ensuring operations remain within safe, predictable parameters. They also reinforce fairness by ensuring all teams follow the same constraints. When ignored, operations drift into unsafe or inefficient territory. Teams make decisions without structural guidance. Over time, System Control collapses as boundaries dissolve.

When executed well, Operational Guardrail Rules strengthen System Control by embedding discipline into operational decision making. Employees trust that guardrails protect both performance and safety. This improves consistency, reduces risk, and enhances organizational stability. Rules reinforce the principle that freedom must be structured. They remind leaders that constraints enable reliability. Ultimately, they build an environment where operations remain controlled, predictable, and aligned with organizational intent.

MORAL COMPASS SCENARIO	LEADERSHIP ACTION
Darin sees that guardrail rules allow deployments during peak traffic hours. He realizes the system is exposed to unnecessary risk.	The leader restricts deployments to controlled windows. They also require guardrail enforcement in CI/CD pipelines.
Liora notices that guardrail rules don't limit batch sizes for high-risk data operations. She senses potential mass-impact failures.	The leader adds batch-size caps and progressive rollout rules. They also require rollback checkpoints.
Fen finds that guardrail rules allow manual overrides without justification. He recognizes governance gaps.	The leader requires override reasons and multi-party approval. They also track override frequency.
Corvin sees that guardrail rules don't include rate limits for external integrations. He realizes partner systems can be overloaded.	The leader adds rate-limit enforcement and partner-impact modeling. They also require integration-risk reviews.
Amara notices that guardrail rules treat all environments equally, allowing risky actions in production. She senses environment-specific blind spots.	The leader adds environment-tiered guardrails. They also require production-specific safety checks.
Jax discovers that guardrail rules are documented but not enforced. He recognizes compliance theater.	The leader automates guardrail enforcement and blocks non-compliant actions. They also require guardrail-violation reporting.

5. CONTROL ENFORCEMENT RECORDS

System Control requires accountability, and Control Enforcement Records document whether operational controls are followed, violated, or bypassed. They track compliance patterns, exceptions, and corrective actions. This matters because controls lose power when enforcement is inconsistent. Control Enforcement Records prevent erosion by ensuring deviations are visible and addressed. They also reinforce fairness by ensuring all teams are held to the same standards. When ignored, violations multiply and controls become symbolic. Teams operate without consequence for unsafe or noncompliant behavior. Over time, System Control deteriorates as enforcement disappears.

When executed well, Control Enforcement Records strengthen System Control by ensuring controls remain authoritative. Employees trust that governance is applied consistently. This improves compliance, reduces risk, and enhances ethical alignment. Records reinforce the principle that control requires consequence. They remind leaders that enforcement is essential to credibility. Ultimately, they build an environment where controls are respected and upheld.

MORAL COMPASS SCENARIO	LEADERSHIP ACTION
Nerin sees that enforcement records show repeated violations by the same team with no consequences. He realizes enforcement is symbolic.	The leader escalates violations and assigns corrective actions. They also track repeat-offender metrics.
Yara notices that enforcement records don't include severity ratings. She senses misaligned responses.	The leader adds severity tiers and response matrices. They also require severity-aligned enforcement.
Brann finds that enforcement records are updated manually, causing delays and errors. He recognizes data-quality risk.	The leader automates enforcement logging. They also add validation checks.
Selin sees that enforcement records exclude third-party operators. She realizes external teams operate without oversight.	The leader integrates third-party enforcement tracking. They also require vendor compliance reviews.
Otho notices that enforcement records don't include remediation deadlines. He senses open-ended risk.	The leader adds deadlines and assigns owners. They also track remediation completion rates.
Mirelle discovers that enforcement records are siloed by department. She recognizes cross-team blind spots.	The leader centralizes enforcement data. They also require cross-department enforcement reviews.

System Control requires structural assurance, and System Integrity Audits evaluate whether systems, processes, and controls function as designed. They examine alignment between intended behavior and actual performance. This matters because systems degrade over time. System Integrity Audits prevent silent failure by identifying where controls weaken or break. They also reinforce fairness by ensuring all operational areas receive equal scrutiny. When ignored, integrity gaps widen unnoticed. Teams rely on systems that no longer behave reliably. Over time, System Control collapses as structural reliability erodes.

When executed well, System Integrity Audits strengthen System Control by validating the health of the operational ecosystem. Employees trust that systems are maintained with rigor. This improves stability, reduces incidents, and enhances organizational confidence. Audits reinforce the principle that control requires verification. They remind leaders that integrity must be tested continuously. Ultimately, they build an environment where systems remain trustworthy and resilient.

MORAL COMPASS SCENARIO	LEADERSHIP ACTION
Arlen sees that integrity audits reveal configuration drift across identical servers. He realizes the environment is no longer deterministic.	The leader enforces configuration baselines and adds drift-detection alerts. They also require periodic configuration reconciliation.
Nessa notices that integrity audits ignore firmware versions on edge devices. She senses unmonitored hardware risk.	The leader adds firmware-level auditing. They also require hardware-integrity checks.
Jovan finds that integrity audits show mismatched encryption settings across services. He recognizes inconsistent security posture.	The leader standardizes encryption policies and validates compliance. They also require encryption-integrity reviews.
Selka sees that integrity audits don't include dependency libraries. She realizes supply-chain vulnerabilities are invisible.	The leader adds library-integrity scanning. They also require dependency-risk assessments.
Torin notices that integrity audits are performed annually despite rapid system changes. He senses outdated assurance.	The leader moves to quarterly audits and adds automated interim checks. They also require audit-trigger rules tied to major changes.
Aveline discovers that integrity audits exclude access-control misconfigurations. She recognizes a major oversight.	The leader integrates access-control checks into integrity audits. They also require privilege-alignment reviews.

Ai6x6™ Transformation Blueprint

Legal Obligations

"AI will not replace humanity, but it may reveal what humanity truly values."
Esther Dyson, technology investor and futurist.

Legal Obligations require strict adherence to regulatory requirements, and Mandatory Regulatory Duties define the legal responsibilities operations must meet to remain compliant. They establish the minimum standards for lawful process execution. This matters because noncompliance exposes organizations to penalties, shutdowns, and reputational harm. Mandatory Regulatory Duties prevent violations by ensuring operational activities align with legal expectations. They also reinforce integrity by demonstrating that compliance is a core operational priority. When ignored, legal exposure increases rapidly. Teams experience uncertainty as decisions drift outside regulatory boundaries. Over time, Legal Obligations deteriorate as adherence becomes inconsistent. Required Compliance Filings ensure documentation and reporting meet regulatory mandates.

The final layer of the blueprint and the ultimate anchor of operational sovereignty. Mastering AML, Sarbanes-Oxley Section 404, and IRS Reporting ensures the organization is not only antifragile but self-disrupting and legally unassailable. This topic treats "Legal Obligations" as the "Clinical Proof" of operational health. By reviewing "Federal Acquisition Regulations" and "Bank Secrecy Act" compliance, the leader secures the firm's standing in the global market. Legal compliance is the final "Shift" toward "Bliss," providing the "Self-Disruption" insurance policy that protects the enterprise from the threat of market extinction. Using the print series, you can sit with these complex obligations away from distractions, seeing how they form the "Moral Compass" of your operational strategy. This clinical mastery ensures that your "Operations Transformation" is not just scalable, but legally sound and morally unshakeable, providing the ultimate insurance policy against the market extinction that follows the failure to meet the sovereign reporting and monitoring standards of the global industrial landscape.

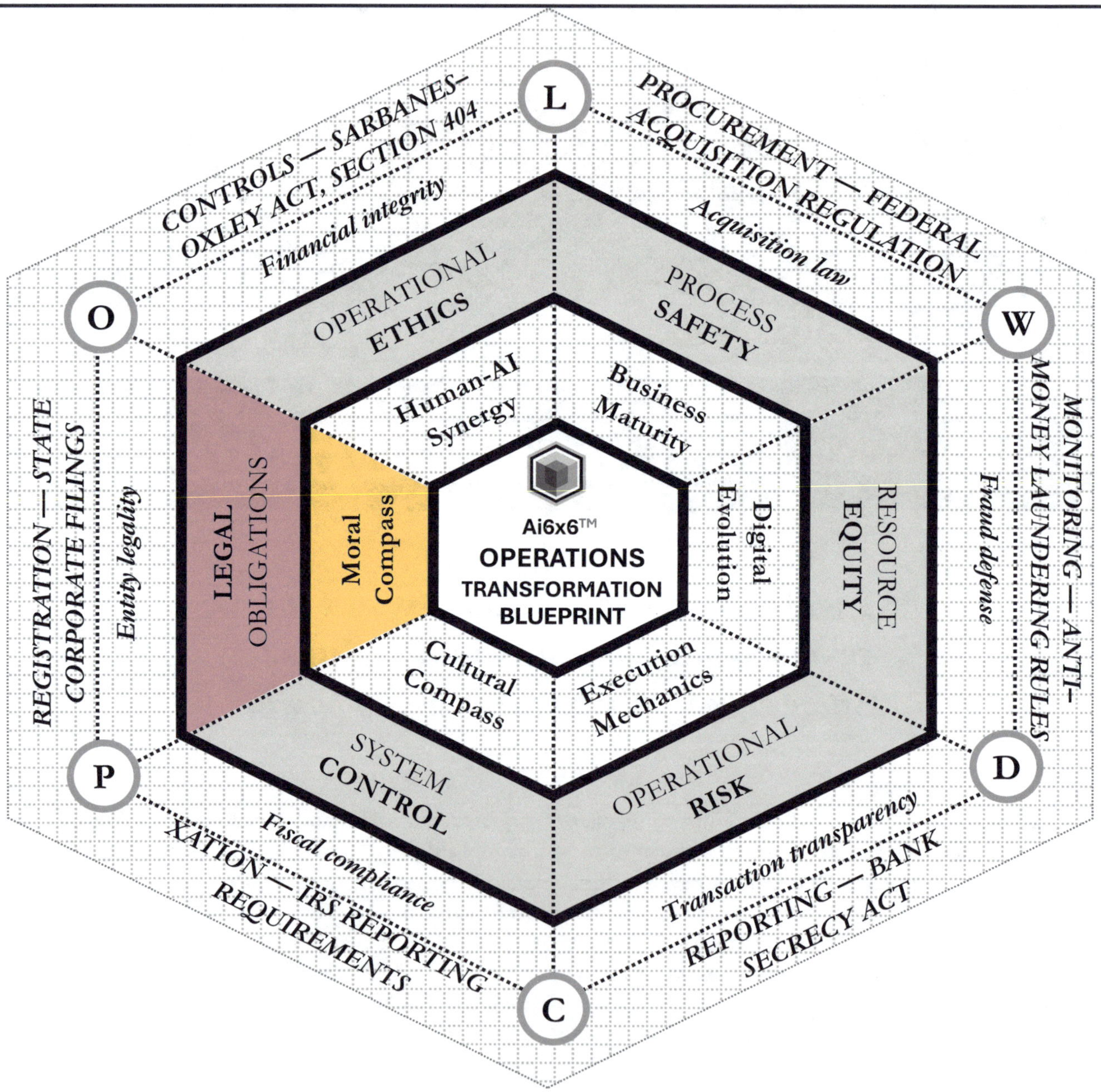

Ai6x6 LEGAL OBLIGATIONS FOR MORAL COMPASS | OPERATIONS TRANSFORMATION

1. CONTROLS — SARBANES–OXLEY ACT, SECTION 404

Legal Obligations require financial integrity, and Section 404 of the Sarbanes–Oxley Act mandates rigorous internal controls over financial reporting. It requires organizations to document, test, and certify the effectiveness of these controls. This matters because weak financial controls create opportunities for fraud, misstatement, and operational manipulation. Section 404 compliance prevents financial misconduct by enforcing disciplined oversight. It also reinforces fairness by ensuring all stakeholders receive accurate, trustworthy financial information. When ignored, financial systems become vulnerable to error and abuse. Investors lose confidence in the organization's governance. Over time, Legal Obligations deteriorate as financial transparency collapses.

When executed well, Section 404 compliance strengthens Legal Obligations by embedding accountability into financial operations. Stakeholders trust that financial data reflects reality. This improves credibility, reduces regulatory exposure, and enhances ethical alignment. Controls reinforce the principle that financial truth is non negotiable. They remind leaders that governance requires structure, documentation, and verification. Ultimately, they build an environment where financial reporting is reliable, auditable, and protected from manipulation.

MORAL COMPASS SCENARIO	LEADERSHIP ACTION
Ardan discovers that financial-system access logs show unreviewed admin overrides during quarter-close. He realizes internal controls are being bypassed.	The leader enforces dual-approval for overrides and adds automated alerts. They also require override reviews in every 404 audit cycle.
Mirelle notices that reconciliation steps are performed manually despite an approved automated workflow. She senses elevated control-failure risk.	The leader reinstates automation and validates reconciliation accuracy. They also require manual-override justification logs.
Jovan finds that evidence for key controls is stored in personal folders. He recognizes audit-trail fragility.	The leader centralizes evidence storage with immutable logs. They also enforce retention policies aligned with 404 requirements.
Selka sees that control owners are unaware of updated risk classifications. She realizes controls no longer match current risk levels.	The leader retrains owners and updates control matrices. They also require risk-aligned control reviews.
Torin notices that quarterly control testing is skipped when deadlines collide with product launches. He senses compliance erosion.	The leader enforces non-negotiable testing windows. They also add escalation paths for missed tests.
Aveline discovers that segregation-of-duties conflicts are flagged but never remediated. She recognizes latent fraud exposure.	The leader resolves conflicts and adds automated SoD monitoring. They also require monthly SoD attestation.

Legal Obligations require disciplined procurement, and the Federal Acquisition Regulation (FAR) establishes the rules governing how organizations engage in government related purchasing. It defines standards for fairness, competition, transparency, and ethical conduct. This matters because procurement misconduct creates legal, financial, and reputational risk. FAR compliance prevents abuse by ensuring procurement decisions follow strict, accountable procedures. It also reinforces fairness by ensuring all vendors receive equal opportunity. When ignored, procurement becomes vulnerable to favoritism, fraud, and conflict of interest. Regulators intervene and trust erodes. Over time, Legal Obligations collapse as procurement governance weakens.

When executed well, FAR compliance strengthens Legal Obligations by ensuring procurement operates with integrity. Stakeholders trust that purchasing decisions are ethical and transparent. This improves operational stability, reduces legal exposure, and enhances public credibility. Regulations reinforce the principle that procurement must be governed, not improvised. They remind leaders that public facing procurement carries heightened responsibility. Ultimately, they build an environment where procurement reflects fairness, discipline, and accountability.

MORAL COMPASS SCENARIO	LEADERSHIP ACTION
Nadir sees that a team awarded a contract without competitive bidding “to save time.” He realizes FAR competition rules were ignored.	The leader reopens the procurement and retrains staff on competition requirements. They also require justification memos for exceptions.
Soren notices that vendor files lack conflict-of-interest disclosures. He senses FAR compliance gaps.	The leader collects disclosures and updates procurement templates. They also require annual COI re-attestation.
Lira finds that subcontractors are added without approval. She recognizes unauthorized supply-chain expansion.	The leader enforces subcontractor approval workflows. They also require subcontractor visibility in procurement records.
Bram sees that FAR-required clauses are missing from several contracts. He realizes legal exposure is growing.	The leader amends contracts and updates clause libraries. They also require contract-compliance checks before signature.
Orin notices that procurement teams rely on expired vendor certifications. He senses due-diligence failures.	The leader adds certification-expiry alerts. They also require certification verification before award.
Fenya discovers that procurement decisions are documented only in email threads. She recognizes auditability risk.	The leader centralizes procurement decisions in a controlled system. They also enforce structured decision logs.

3. MONITORING — ANTI–MONEY LAUNDERING RULES

Legal Obligations require vigilance, and Anti–Money Laundering (AML) rules mandate monitoring systems that detect suspicious financial activity. They require organizations to identify patterns that may indicate fraud, terrorism financing, or illicit transactions. This matters because financial systems are targets for criminal exploitation. AML compliance prevents harm by ensuring suspicious activity is flagged and investigated. It also reinforces fairness by protecting legitimate customers from the consequences of financial crime. When ignored, illicit activity infiltrates operational systems. Regulators impose severe penalties and trust collapses. Over time, Legal Obligations deteriorate as monitoring discipline disappears.

When executed well, AML compliance strengthens Legal Obligations by embedding continuous surveillance into financial operations. Stakeholders trust that the organization protects the integrity of its financial ecosystem. This improves security, reduces legal risk, and enhances ethical responsibility. Monitoring reinforces the principle that vigilance is essential to financial stewardship. It reminds leaders that prevention requires constant attention. Ultimately, it builds an environment where financial crime is detected early and addressed decisively.

MORAL COMPASS SCENARIO	LEADERSHIP ACTION
Kael sees that AML alerts are dismissed with generic notes like “low risk.” He realizes analysts aren’t documenting rationale.	The leader enforces detailed disposition notes and adds quality checks. They also require escalation for vague dismissals.
Orla notices that transaction-monitoring thresholds haven’t been updated in years. She senses outdated risk calibration.	The leader recalibrates thresholds using recent typologies. They also require annual threshold reviews.
Halden finds that AML monitoring excludes cross-border transfers under a certain amount. He recognizes structuring risk.	The leader adds cumulative-transaction monitoring. They also require cross-border risk scoring.
Suri sees that AML alerts spike during weekends but staffing is minimal. She realizes detection is delayed.	The leader adds weekend coverage and automated triage. They also track alert-response latency.
Bram notices that AML rules don’t include new digital-wallet channels. He senses blind spots.	The leader integrates digital-wallet monitoring. They also require channel-specific risk models.
Elin discovers that AML logs are retained for only one year. She recognizes regulatory non-compliance.	The leader extends retention to mandated durations. They also enforce retention policy audits.

Legal Obligations require transparency, and the Bank Secrecy Act (BSA) mandates reporting of financial activities that may indicate criminal behavior. It requires organizations to file Suspicious Activity Reports and Currency Transaction Reports. This matters because reporting gaps enable illicit activity to flourish. BSA compliance prevents criminal exploitation by ensuring regulators receive timely, accurate information. It also reinforces fairness by protecting the financial system from abuse. When ignored, organizations become conduits for illegal transactions. Regulators respond with penalties and increased scrutiny. Over time, Legal Obligations collapse as reporting integrity erodes.

When executed well, BSA compliance strengthens Legal Obligations by ensuring transparency in financial operations. Stakeholders trust that the organization fulfills its reporting responsibilities. This improves regulatory relationships, reduces risk, and enhances ethical alignment. Reporting reinforces the principle that transparency protects the system. It reminds leaders that disclosure is a legal and moral duty. Ultimately, it builds an environment where financial reporting supports public safety and institutional integrity.

MORAL COMPASS SCENARIO	LEADERSHIP ACTION
Rafe sees that suspicious-activity reports (SARs) are filed late because investigators wait for "more evidence." He realizes timeliness rules are being violated.	The leader enforces statutory filing windows and adds escalation triggers. They also train investigators on timely reporting.
Mireya notices that currency-transaction reports (CTRs) are missing for several large cash deposits. She senses operational oversight.	The leader audits cash-handling workflows and adds automated CTR triggers. They also require teller-level compliance training.
Tovin finds that BSA reports contain inconsistent customer identifiers. He recognizes data-quality risk.	The leader standardizes identifier fields and validates historical records. They also require data-quality checks before submission.
Selka sees that BSA reporting relies on a single analyst. She realizes continuity risk.	The leader cross-trains additional analysts. They also add workload balancing.
Joren notices that BSA reports exclude foreign-branch activity. He senses incomplete regulatory coverage.	The leader integrates foreign-branch data feeds. They also require global BSA alignment.
Aveline discovers that BSA reports are stored without encryption. She recognizes confidentiality risk.	The leader encrypts all reports and restricts access. They also require encryption-integrity checks.

Legal Obligations require fiscal accuracy, and IRS Reporting Requirements mandate truthful, complete, and timely submission of tax related information. They govern income reporting, deductions, payroll taxes, and compliance documentation. This matters because tax misreporting creates legal and financial exposure. IRS compliance prevents penalties by ensuring financial activities are documented correctly. It also reinforces fairness by ensuring organizations contribute their lawful share. When ignored, tax liabilities accumulate and credibility erodes. Regulators intervene with audits and sanctions. Over time, Legal Obligations deteriorate as fiscal discipline collapses.

When executed well, IRS compliance strengthens Legal Obligations by embedding accuracy into financial operations. Stakeholders trust that the organization manages its tax responsibilities responsibly. This improves financial stability, reduces legal exposure, and enhances ethical credibility. Requirements reinforce the principle that taxation is a shared civic obligation. They remind leaders that fiscal integrity is foundational to public trust. Ultimately, they build an environment where tax reporting is precise, transparent, and compliant.

MORAL COMPASS SCENARIO	LEADERSHIP ACTION
Darin sees that 1099 forms are generated with outdated vendor addresses. He realizes tax filings may be misdelivered.	The leader updates vendor records and adds address-verification steps. They also require annual vendor-data audits.
Liora notices that payroll tax withholdings differ between systems. She senses reconciliation failures.	The leader aligns payroll systems and adds automated cross-checks. They also require monthly withholding audits.
Fen finds that R&D tax-credit documentation is incomplete. He recognizes risk of disallowed claims.	The leader rebuilds documentation and assigns owners for evidence collection. They also require quarterly documentation reviews.
Corvin sees that sales-tax filings omit certain digital products due to misclassification. He realizes compliance gaps.	The leader corrects tax codes and updates product-tax mappings. They also require classification audits.
Amara notices that IRS deadlines are tracked manually on spreadsheets. She senses deadline-miss risk.	The leader automates deadline tracking and adds alerts. They also require compliance-calendar reviews.
Jax discovers that tax-related data is stored without audit trails. He recognizes traceability issues.	The leader adds immutable logging and restricts edits. They also require audit-trail reviews.

Legal Obligations require organizational legitimacy, and State Corporate Filings ensure that entities maintain their legal standing through required registrations, renewals, and disclosures. They document ownership, governance, and operational status. This matters because failure to maintain filings jeopardizes the organization's legal existence. State Corporate Filings prevent administrative dissolution by ensuring compliance with state requirements. They also reinforce fairness by ensuring all organizations operate under the same legal framework. When ignored, entities lose good standing and face operational restrictions. Customers and partners question the organization's legitimacy. Over time, Legal Obligations collapse as legal identity erodes.

When executed well, State Corporate Filings strengthen Legal Obligations by maintaining the organization's legal foundation. Stakeholders trust that the entity operates within the law. This improves continuity, reduces risk, and enhances institutional credibility. Filings reinforce the principle that legal identity requires upkeep. They remind leaders that compliance extends beyond operations into corporate structure. Ultimately, they build an environment where legal standing is preserved through disciplined governance.

MORAL COMPASS SCENARIO	LEADERSHIP ACTION
Nerin sees that several state registrations have lapsed due to missed renewal dates. He realizes the company is operating out of compliance.	The leader files reinstatements and automates renewal reminders. They also require annual compliance checks.
Yara notices that registered agent information is outdated in multiple states. She senses service-of-process risk.	The leader updates agent records and verifies accuracy with each state. They also require agent-information audits.
Brann finds that new subsidiaries were formed without updating state filings. He recognizes legal-entity misalignment.	The leader updates all filings and enforces entity-change notifications. They also require legal-entity governance reviews.
Selin sees that filings list former executives as officers. She realizes governance records are stale.	The leader updates officer information and aligns HR and legal systems. They also require officer-record synchronization.
Otho notices that foreign-qualification filings are missing for states where the company now operates. He senses exposure to penalties.	The leader completes foreign-qualification filings and adds expansion-trigger rules. They also require geographic-footprint reviews.
Mirelle discovers that filing documents are stored in personal drives. She recognizes auditability and continuity risk.	The leader centralizes filings in a controlled repository. They also enforce document-governance standards.

BONUS CHAPTER

Ai6x6™ Transformation Blueprint

The Resilience Paradox

"We cannot have infinite growth on a finite planet." -- **Greta Thunberg**, the environmental activist who sparked the global "Fridays for Future" movement, delivers the blunt reality check for the **Resilience Paradox**. Her message forces operations leaders to confront the physical limits of the Earth, shifting the focus from the fragile "efficiency" of infinite throughput to the robust "resilience" of a localized, regenerative supply chain that respects planetary boundaries.

For thirty years, the "Lean" philosophy was the gold standard of operations, removing every ounce of "waste" to maximize short-term efficiency. However, the climate crisis has exposed the fatal flaw of this traditional thinking: a hyper-efficient system is a hyper-brittle one. To survive in a volatile, post-AI world, leaders must embrace the **Resilience Paradox**. This requires a multi-dimensional shift from a "command and control" model to a "Glo-Cal" network that balances global standards with local infrastructure reality. It means re-introducing "Just-in-Case" redundancy and mastering the logistics of taking products back. Traction is found when operations are no longer just about moving goods at the lowest cost, but about maintaining moral and physical integrity through constant, climate-driven flux.

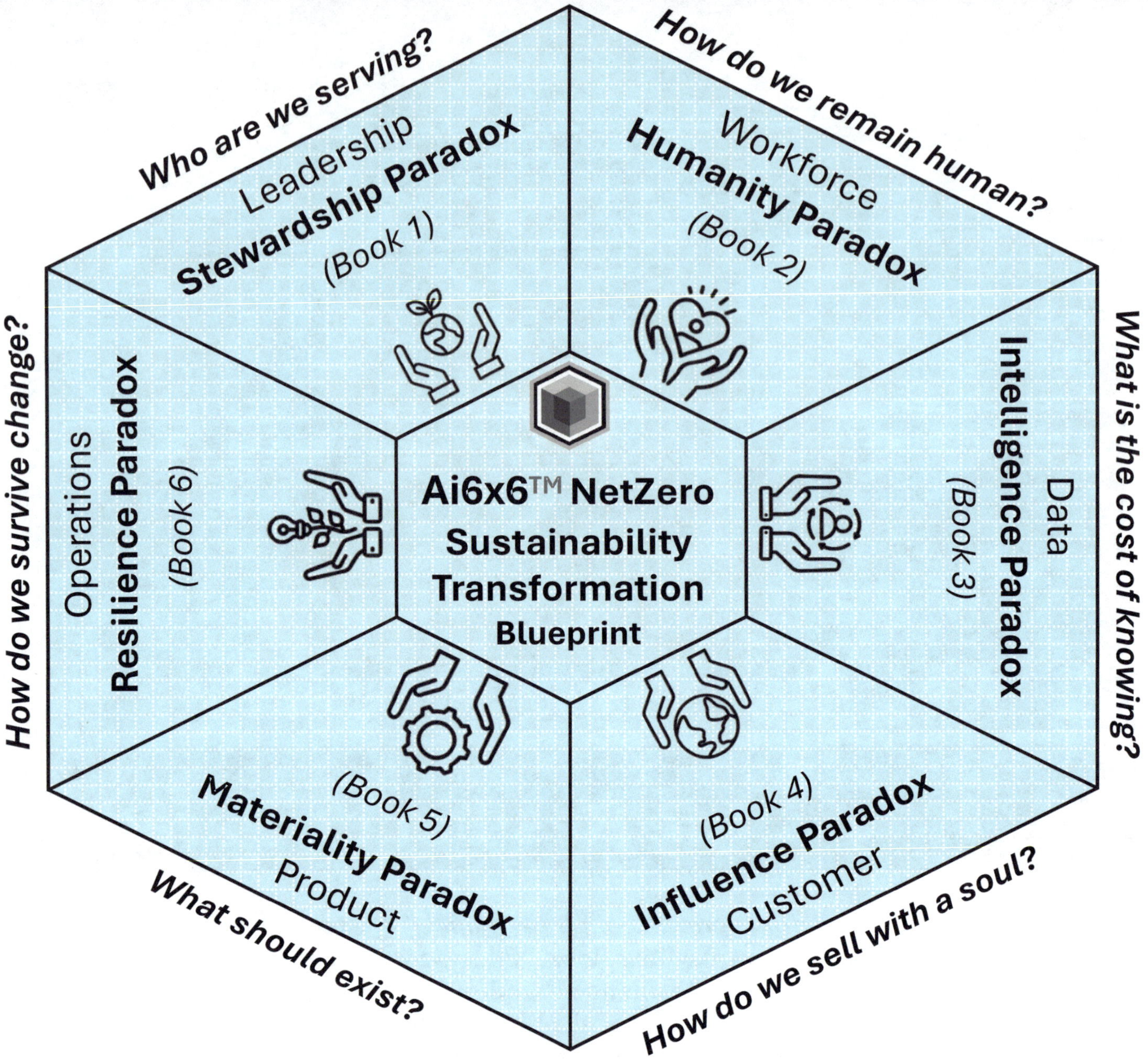

The **Ai6x6 NetZero Sustainability Transformation Blueprint** redefines the path to Net Zero by replacing the traditional "check-the-box" compliance mindset with a rigorous framework of **paradoxical leadership**. In a world accelerated by AI, sustainability is rarely a choice between a clear right and wrong; instead, it is a constant navigation of thirty-six competing tensions across six organizational lenses. By framing the journey through these paradoxes—such as the conflict between **Digital Intelligence and Energy Consumption** or **Global Targets and Local Infrastructure**—the framework prepares leaders for the messy, non-linear realities of systemic change.

Rather than viewing these tensions as obstacles to be avoided, the **Ai6x6 methodology** utilizes them as engines for innovation. It guides executives to move past the paralysis of "either/or" thinking, teaching them to embrace a "both/and" strategy that maintains fiduciary health while securing long-term planetary survival. This approach ensures that a company's moral compass remains calibrated even as technology and climate volatility shift the ground beneath them. Ultimately, the series provides a diagnostic roadmap that turns systemic friction into strategic advantage, empowering leaders to achieve radical Net Zero goals without sacrificing the operational resilience or human dignity required to thrive in the post-AI era.

THE RESILIENCE PARADOX is the ultimate reality check for any Net Zero strategy: the tension between the global reach of a business and the local physical limits of the planet. While strategy happens in the boardroom, transformation happens on the factory floor, in the warehouse, and on the road. This creates a constant struggle between the drive for a unified, worldwide standard of "Green" and the messy, fragmented reality of local energy grids and infrastructure. To transform, an organization must move from a "command and control" model to a resilient network that can maintain its moral integrity even when the physical world is in a state of constant, climate-driven flux.

1. THE GLOBAL-LOCAL PARADOX (Global Goals vs. Local Reality)

Multinational companies often set a single, aggressive Net Zero target for all their offices and factories worldwide. The paradox is that while the goal is unified, the local reality is not; a factory in a region with a coal-heavy energy grid cannot move as fast as one powered by wind. Forcing a global standard can lead to "carbon leakage" or the closing of essential local hubs. Success requires a "Glocal" approach—maintaining a strict global moral compass while allowing for local transition paths that respect the specific energy and economic limitations of each region.

2. THE ZERO-WASTE SUPPLY PARADOX (Saving Waste vs. Saving Stock)

Using AI to create a "Lean" zero-waste supply chain is the best way to eliminate daily carbon and excess inventory. However, the paradox is that a perfectly lean system has no "buffer" to survive the unexpected. In a world of climate-driven storms and port closures, "zero-waste" can quickly lead to "zero-stock," leaving communities without essential food or medicine. Leaders must balance the environmental win of lean operations against the moral requirement of "Just-in-Case" redundancy, ensuring the supply chain is robust enough to protect the people who depend on it.

3. THE ROUTE-OPTIMIZATION PARADOX (Green Routes vs. Fast Shipping)

The market demands that products arrive "yesterday," but the greenest route is rarely the fastest one. The paradox is that optimizing for carbon often requires slower speeds, consolidated loads, and non-direct paths that clash with the "Need-it-Now" customer culture. Operations leaders are caught between the logistics team's target for speed and the sustainability team's target for emissions. Resolving this requires a fundamental shift in the definition of "performance"—where a "successful" delivery is measured by its carbon footprint rather than just the minutes it took to arrive.

4. THE VALUE-CHAIN AUDIT PARADOX (Checking Suppliers vs. Trusting Partners)

To hit Scope 3 targets, companies must ensure their suppliers are meeting strict environmental standards. The paradox is that replacing long-term, trust-based relationships with constant, AI-driven satellite and data monitoring can erode the very collaboration needed for innovation. If suppliers feel like they are being "policed" rather than "partnered with," they are less likely to share the risks of new green technology. A moral operations strategy must balance the need for hard data with the human need for trust, building a "Value-Chain Community" rather than just a list of audited vendors.

5. THE CIRCULAR-LOGISTICS PARADOX (Taking Things Back vs. More Trucks)

Committing to a "Cradle-to-Cradle" model means a company must take back every product at the end of its life. However, the paradox is that the logistics of collecting and transporting fragmented waste back to a central facility can sometimes generate more carbon than the original delivery. We risk polluting the air to save the landfill. Operations transformation requires a radical rethinking of "Reverse Logistics"—using local micro-hubs and partner networks to process materials where they are, rather than shipping trash across the globe in the name of circularity.

6. THE PREDICTIVE-MAINTENANCE PARADOX (Keeping Old Tools vs. Getting New Ones)

Using AI to predict when a machine will fail allows companies to keep old industrial equipment running for decades, saving the "embodied carbon" of building a new machine. The paradox is that these legacy machines are often significantly less energy-efficient than modern, green alternatives. Leaders must decide: Is it better to keep the old "low-manufacturing-cost" tool or buy the new "low-operating-cost" tool? This requires a sophisticated lifecycle analysis that looks at the total carbon impact of the asset, from the factory that built it to the power plant that feeds it.

1. THE GLOBAL-LOCAL PARADOX (World Standards vs. Local Reality)

- **Food & Beverage:** A global soda brand localizes 90% of its ingredient sourcing to within a 200-mile radius of each regional bottling plant to bypass global shipping volatility.
- **Construction:** A developer uses 3D-printing technology to build with "Local Soil" and recycled rubble, adhering to global safety standards while eliminating the need for imported cement.
- **Apparel:** A fashion giant utilizes "Micro-factories" in major hubs like New York and London to produce only what is sold that day, cutting out the carbon of transcontinental logistics.
- **Pharmaceuticals:** A vaccine manufacturer moves to "Modular Lab" units that can be deployed to produce life-saving medicine locally, ensuring supply chain integrity during border closures.
- **Solar Energy:** A renewable firm sources glass and silicon from the same continent as the final "Solar Farm" installation to reduce the transit carbon of heavy components.
- **Furniture:** A retailer produces flat-pack items in regional hubs rather than one central mega-factory, drastically reducing the carbon footprint of shipping "air" in large boxes.

2. THE ZERO-WASTE SUPPLY PARADOX (Lean Inventory vs. Circular Stock)

- **Heavy Manufacturing:** An industrial firm maintains "Safety Stocks" of recycled steel and aluminum, insulating itself from the price shocks of volatile global virgin-ore markets.
- **Hospitality:** A hotel chain implements a "Closed-Loop Kitchen" where 100% of organic waste is transformed into compost for the hotel's on-site herb and vegetable gardens.
- **Heavy Industry:** A steel mill captures its "Waste Heat" and pipes it to a neighboring industrial greenhouse, turning an operational byproduct into a revenue-generating heat source.
- **Retail:** A department store implements "Zero-Waste Returns," where returned items are inspected, refurbished, and resold at the same local branch rather than being shipped back to a central hub.
- **Mining:** A copper miner re-processes its old "Tailings" (waste piles) using new AI-sorting tech to extract rare earth minerals needed for EV motors without digging new pits.
- **Chemicals:** A plant installs "Closed-loop Water" systems that recycle and purify 98% of operational water on-site, making the facility independent of local municipal water stress.

3. THE ROUTE-OPTIMIZATION PARADOX (Fast Transit vs. Low-Carbon Flow)

- **Ocean Freight:** A logistics firm equips its cargo fleet with high-tech "Wind-Sails," accepting slightly longer transit times in exchange for a 30% reduction in heavy fuel oil consumption.
- **Postal Services:** A national mail carrier replaces its inner-city van fleet with "Cargo Bikes" for the last mile, bypassing gridlock and reducing carbon in dense urban centers.
- **Aviation:** An airline uses AI-driven "Contrail-avoidance" software to adjust flight altitudes in real-time, reducing the non-CO2 warming impact of its operations.
- **Rail Logistics:** A freight operator synchronizes its train schedules to run only during windows when renewable energy generation on the grid is at a surplus.
- **Waste Management:** A city uses sensor-enabled smart bins that signal trucks to follow a dynamic route only when bins are full, reducing unnecessary miles driven by 40%.
- **Cold Chain:** A food distributor switches to "Passive Cooling" containers that maintain temperature for 72 hours without needing fuel-powered refrigeration units on trucks.

4. THE VALUE-CHAIN AUDIT PARADOX (Deep Visibility vs. Supplier Privacy)

- **Apparel Manufacturing:** A brand provides low-interest "Solar Loans" to its Tier 3 weaving mills, requiring energy transparency in exchange for capital to eliminate coal-powered steam.
- **Electronics:** A hardware firm uses AI-powered satellite imagery to verify that its suppliers are not encroaching on protected forest lands, without requiring invasive on-site physical audits.
- **Automotive:** A manufacturer uses blockchain to create a "Digital Thread" for every component, allowing them to verify carbon data while protecting the supplier's proprietary manufacturing "recipes."
- **Cosmetics:** A brand implements a "Supplier Scorecard" where contract extensions are tied to the supplier's year-over-year reduction in water intensity and hazardous waste.
- **Agriculture:** A food giant provides "Digital Agronomy" tools to smallholder farmers, gathering soil-health data to verify carbon sequestration while helping the farmers increase their yields.
- **Furniture:** A retailer mandates that all Tier 2 wood-processing plants must have ISO 14001 certification, using third-party verification to maintain global standards across fragmented local suppliers.

5. THE CIRCULAR-LOGISTICS PARADOX (Outbound Sales vs. Inbound Returns)

- **Consumer Tech:** A national electronics retailer uses its existing delivery van network to collect "E-Waste" from residential porches on the return leg of its daily delivery routes.
- **Beverage Distribution:** A brewery implements a "Bottle-Return" logistics loop, where delivery trucks pick up empty crates and glass from bars to be sterilized and refilled at the source.
- **E-Commerce:** A retailer partners with local "Pickup Points" to serve as return hubs, consolidating thousands of individual returns into a single, efficient weekly bulk shipment.
- **Industrial Gas:** A supplier uses "Smart Cylinders" that signal when they are 90% empty, allowing the logistics team to optimize the pickup of empties before the customer even calls.
- **Apparel:** A brand launches a "Mending-as-a-Service" loop, where the existing outbound shipping infrastructure is used to send garments back to a central repair hub and return them to the user.
- **Logistics:** A pallet provider shifts to a "Pallet-Pooling" model, where they manage a massive shared network of high-quality, repairable pallets that never leave the circular loop.

6. THE PREDICTIVE-MAINTENANCE PARADOX (Fixing Early vs. Using Fully)

- **Renewable Energy:** A wind farm operator uses drone-based AI to detect and repair microscopic turbine blade cracks *in situ*, extending their operational life by a decade and preventing premature scrap.
- **Public Transit:** A subway system uses vibration sensors on tracks to predict wheel-wear, allowing for targeted maintenance that prevents service-disrupting breakdowns and energy-heavy emergency repairs.
- **Manufacturing:** A factory uses "Acoustic AI" to listen to motor bearings, ensuring machines are serviced only when they actually show signs of wear, rather than following a wasteful, calendar-based schedule.
- **Real Estate:** A building manager uses "Digital Twin" simulations to balance elevator loads, reducing mechanical wear-and-tear and extending the replacement cycle of heavy cable systems by years.
- **Water Utilities:** A utility uses AI to predict pipe bursts before they happen, allowing for "Sleeve-lining" repairs that avoid the carbon-intensive process of digging up streets and replacing entire sections.
- **Shipping:** A maritime firm uses "Hull-Cleaning Bots" to remove barnacles in real-time, maintaining optimal hydrodynamics and reducing fuel drag by 15% throughout the entire voyage.

Ai6x6 COMPREHENSIVE TRANSFORMATOION BOOK SERIES

Most comprehensive business transformation book series worldwide

Ai6x6™ Transformation Engine

The AI6x6 Transformation Engine™

Your Operating System for the AI Era

In an age of autonomous systems, fragmented strategy is a liability. The **AI6x6 Transformation Engine™** is a precision-engineered, 18-volume masterwork designed to move your enterprise from organizational inertia to scalable momentum. Built on a proprietary 6×6 geometry, this ecosystem unifies architecture, execution, and calibration into one high-performance unit.

The journey begins with the **Transformation Blueprint Series**, your **Intelligence Layer**. It codifies a "Navigation Deck" of Six Pillars and Six Lenses to anchor your strategic North Star. Next, the **Consulting Playbook Series** acts as the **Transmission Layer**, utilizing the IGNITE™ methodology to translate executive intent into high-torque, repeatable action. Finally, the **Maturity Diagnostic Series** serves as the **Sensor Layer**, providing a data-driven intake to quantify readiness and calibrate performance.

This isn't just a book collection; it is a rigid **Chassis** for institutional excellence. Secure the definitive architecture for sustained, high-stakes digital evolution. Drive the future with the Engine.

Transformation Series

To ensure this book series benefits your company holistically, we strongly recommend you buy the series and then distribute it to the appropriate transformation sponsors in each of these departments. Don't forget to browse through consulting offers at the end of the book.

WE NEED YOU!

WELCOMING OUR NEXT SET OF THOUGHT LEADERS

The remarkable success of the Ai6x6 Organization Transformation Series is built upon the **hard-won wisdom** of executive leaders who have successfully navigated the AI transformation within their industries, turning theory into a deeply practical, actionable guide. As we expand the series, we are extending a highly selective invitation to a new cohort of pioneers—individuals whose professional journey is at the **leading edge of AI strategy and execution**. We are looking for executives who possess a proven track record of rewriting business models, mastering unforeseen ethical and competitive challenges, and offering a **clear blueprint** for the next 3-5 years of the AI-Era.

Contributing to the Ai6x6 Series is a unique opportunity to **cement your professional legacy**, joining an exclusive, curated network of global authorities whose insights are distilled into executive-level knowledge cards, directly shaping the governance and innovation decisions of the world's most influential companies. Our selection process is based strictly on **pure expertise and demonstrated leadership** in the AI-Era, reinforcing our commitment to editorial integrity and authentic thought leadership—there is no monetary exchange involved. Select contributors will also receive an invitation to blog on our expert network microsite, further amplifying their platform. To confidentially inquire about contributing your expertise, please contact us via our dedicated portal at **www.Ai6x6.com/contact** or by scanning the adjacent QR code.

Contact

SUBSCRIBE TO OUR NEWSLETTERS: To stay ahead of the curve and receive continuous, executive-level intelligence on the rapid evolution of the AI-Era, we invite you to register for our **Ai6x6 Weekly Newsletter**. Each edition delivers curated insights directly from our thought leaders and highlights critical shifts in AI strategy, governance, and provides relevant excepts from our website and books. Keep your finger on the pulse of the future and ensure your company remains competitive by signing up on the same page

PLEASE REVIEW US ON AMAZON!

Operations Blueprint

We know your time is valuable, but if the Ai6x6 Organization Transformation Series delivered the real-world, actionable strategies you needed, please consider leaving a brief review on Amazon. If we missed the mark, we would also like to know, and that is the only we we can improve. Your honest feedback is crucial; it helps fellow executives discover this unique framework and benefit from the insights of our thought leaders. Thank you for supporting our mission!

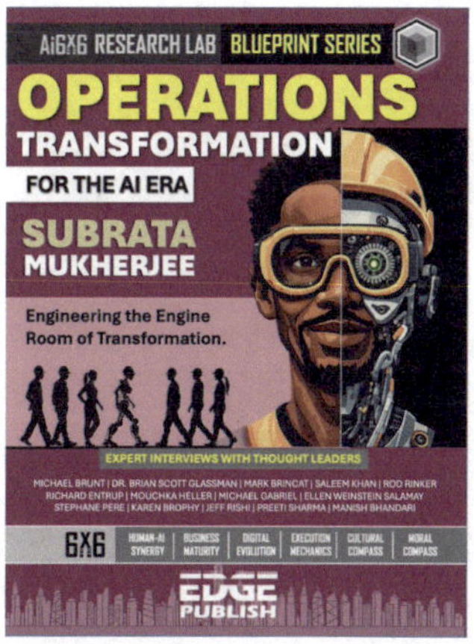

THE Ai6x6™ CERTIFIED TRANSFORMATION STRATEGIST

Certified for Thinking, Not Reciting

The **Ai6x6™ CTS (Certified Transformation Strategist)** is an elite designation for MBAs and strategic leaders with a minimum of **five years of professional experience**. This is not a test of your ability to regurgitate theory; it is a clinical evaluation of your ability to orchestrate high-stakes business transformation in real-time. It signals to the market that you are a strategist who can think on your feet, navigate ambiguity, and utilize a world-class clinical framework to deliver results.

The Non-Traditional Certification

We don't believe in "black box" testing. We believe in the "Open Book" reality of the modern boardroom. To earn your CTS, you must demonstrate mastery through a **60-minute Board-Level Defense** before a member of the Ai6x6™ Global Leadership Team.

$1000 per defense

The Preparation

Mastery of the Series: Candidates must familiarize themselves with all six volumes in the **Ai6x6™ Transformation Series**. You choose **two Retreat Exercises** from any two **Consulting Playbooks** within the series that best align with your expertise. Best practices for the presentation are provided on our website.

The 60-Minute Defense

1. **The Presentation (30 Minutes):** You have 15 minutes per exercise to present your strategic approach and expected outcomes. 60% of your score is for this exercise.
2. **The Situational Audit (30 Minutes):** Our leadership team will pose **six situational transformation scenarios**. You must demonstrate how to apply the **6 Lens Sequence™** to solve these complex frictions.
3. **The Open-Resource Advantage:** You are **encouraged** to refer to any of the six Ai6x6™ books during the exam. We test your ability to navigate and apply the framework under pressure, not your memory.

The Induction: Beyond the Credential

Passing the Board Defense is only the beginning. Successful candidates are immediately inducted into the **Ai6x6™ Executive Mentor Network**, unlocking:

- **Thought Leadership:** Opportunities to contribute to the Ai6x6™ books, knowledge articles and gain visibility among global industry leaders.
- **Engagement Pipeline:** Direct access to a network of future consulting engagements and strategic partnerships.
- **The Fast Track:** Exceptional participants may be invited to join our elite **Practitioner Program**, bypassing the standard entry requirements to lead their own white-label boutique practice.

SCHEDULE YOUR BOARD DEFENSE AT:

Ai6x6.com/certification

Consulting Certification

THE Ai6x6™ PRACTITIONER PROGRAM

Your Authority. Our Infrastructure. A Shared Legacy.

The **Ai6x6™ Practitioner Program** is designed for the high-level strategist—with proven industry leadership experience of over thirty years—who seeks a modern, standardized engine to deploy their operational wisdom. We provide the institutional infrastructure of a global consulting firm, delivered as a white-label solution that honors your professional identity and independence.

From $2000 per month membership

The "Un-Franchise" Model

We have re-engineered the consulting partnership for the modern era. Think of this as a **franchise without the friction or the high fees to get started**:

- **No Long-Term Commitments:** Maintain your agility with a model built for the independent architect.
- **No Loyalty Fees or Royalties:** You keep 100% of your engagement fees. We don't tax your success.
- **A Partnership of Mutual Trust:** Our relationship is powered by a transparent monthly membership, ensuring our incentives are perfectly aligned with your practice's growth.

The Foundations of Practice

All Practitioners gain immediate access to the **Ai6x6™ "Firm-in-a-Box"**:

- **White-Label Ecosystem:** Seamlessly integrate your own brand into our proprietary presentations, client pitches, and diagnostic audit templates.
- **The Clinical Engine:** Augment your strategic intuition with fact-based, quantitative scoring systems that replace subjective brainstorming with measurable transformation.

The White Glove Partnership

For those who wish to move from independent advisor to institutional authority, our **White Glove** tier offers a deep-tissue integration into the Ai6x6™ ecosystem:

- **Monthly 1-on-1 Strategic Check-ins:** Direct access to the **Ai6x6™ leadership team** to stress-test your pipeline, refine high-stakes offerings, or audit active client delivery.
- **Institutional Visibility:** Gain a global platform as a **Co-Author or Featured Expert** in upcoming volumes. We showcase your success stories on **Ai6x6.com**, cementing your legacy as a master of the 6 Lens Sequence.
- **The Bulk Purchase Program:** Exclusive access to deep discounts on the **Ai6x6™ Book Series**, enabling you to provide every client stakeholder with the "source code" for orchestration.

APPLY TO OUR PRACTITIONER PROGRAM AT:

Ai6x6.com/practitioner

Consulting Practitioner

THE Ai6x6™ TRANSFORMATION RETREAT

One Playbook. Your Leadership. Total Orchestration.

The **Ai6x6™ Business Transformation Series** is designed to eliminate the need for traditional, slow-moving consulting teams. Using our "Clinical Blueprints," your organization can self-organize, self-diagnose, and self-disrupt. Everything you need to facilitate a high-stakes, 2.5-day executive offsite is contained within the books.

The Self-Guided Mastery (100% Internal & Free)

You have the "Source Code." Your internal Offsite Coordinator or Chief of Staff can lead the entire 2.5-day agenda end-to-end using the **Consultant Playbook (Part 2)** and our **Supplemental Workbooks**.

- **No External Fees:** Run the entire diagnostic without "billable hours."
- **Full Control:** Use the 6-book series to "mix and match" exercises that fit your specific departmental frictions.
- **Tactile Execution:** Move from whiteboards to prioritized MoSCoW cards and RACI matrices using the physical templates provided in the series. Our Advisory section has all the inspiration your team needs.

Packages from $4999

The Orchestrated Upgrade (White-Glove Support)

For leadership teams seeking institutional weight and an objective "Cold Eye" review without the overhead of travel, we offer the **Orchestrated Retreat** package. This provides three critical layers of external authority:

- **Phase 1: The 30-Minute Architect's Briefing (Pre-Retreat):** A high-velocity strategic sprint with your internal Facilitator. we help you select the exact modules from across the 6-book series to target your unique friction points and set the tactical "Rules of Engagement."
- **Phase 2: The 30-Minute Executive Keynote (Live Remote):** A virtual address delivered by the **Ai6x6™ Leadership Team** to kick off Day 1. This keynote provides the external authority required to break through internal political silos and align the room's energy.
- **Phase 3: The 14-Day Clinical Roadmap Audit (Post-Retreat) :** Submit your digital outputs—prioritized lists, RACI matrices, and visualized Roadmaps. Our team performs a rigorous review and delivers a **Comprehensive Strategic Report** within **two weeks**, stress-testing your plan for diagnostic integrity and self-disruption readiness.

UPGRADE TO AN ORCHESTRATED RETREAT:

Ai6x6.com/retreat

Consulting Retreat

"Peace is its own reward."

Mahatma Gandhi

Ai6x6 Research Lab
Publishing Team

NEW YORK OFFICE
(Editorial Hub)

Editorial Lead: Marisa Thompson
Research Lead: Talia Rosen
Line Editor: Jordan Alvarez

LONDON OFFICE
(Design Hub)

Design Lead: Keiko Tanaka
Layout Specialist: Oliver Whitmore
Illustrator: Aisha Rahman
Design Team from Sumu Arts

KOLKATA OFFICE
(Support Hub)

Project Manager: Sohini Chatterjee
Production : Arindam Ghosh
Quality Lead: Priya Deshmukh
Fact Checker: Ravi Patel
Research Assistant: Ananya Sengupta
Research Assistant: Rohan Mukherjee
Project Assistant: Madhumita Roy
Production Assistant: Sagnik Dutta
Quality Assistant: Ishita Banerjee
Series Lead (Blueprint): Amit Verma
Series Lead (Playbook): Neha Reddy
Series Lead (Diagnostic): Karan Mehta

The End

Made in the USA
Monee, IL
05 November 2026

67814285R00210